MODERN DAIRY TECHNOLOGY

Volume 2

Advances in Milk Products

MODERN DAIRY TECHNOLOGY

Volume 2

Advances in Milk Products

Edited by

R. K. ROBINSON

M.A., D.Phil.

Department of Food Science, University of Reading, UK

ELSEVIER APPLIED SCIENCE PUBLISHERS
LONDON and NEW YORK

ELSEVIER APPLIED SCIENCE PUBLISHERS LTD
Crown House, Linton Road, Barking, Essex IG11 8JU, England

Sole Distributor in the USA and Canada
ELSEVIER SCIENCE PUBLISHING CO., INC.
52 Vanderbilt Avenue, New York, NY 10017, USA

WITH 149 ILLUSTRATIONS AND 51 TABLES

© ELSEVIER APPLIED SCIENCE PUBLISHERS LTD 1986

British Library Cataloguing in Publication Data

Modern dairy technology.
Vol. 2: Advances in milk products.
1. Dairying — Great Britain — Technological
innovations
I. Robinson, R. K.
338.1'77'0941   S494.5.I5

Library of Congress Cataloging-in-Publication Data

Modern dairy technology.

Includes bibliographies and index.
Contents: v. 1. Advances in milk processing — v. 2.
Advances in milk products.
1. Dairy processing.   I. Robinson, R. K. (Richard
Kenneth)   II. Title: Dairy technology.
SF250.5.M63   1986   637   85-20414

ISBN 0-85334-394-2

Phototypesetting by Tech-Set, Gateshead, Tyne & Wear.
Printed in Great Britain by Page Bros (Norwich) Ltd.

Preface

Retail sales of most dairy products are still on the increase world-wide, and this expansion is, at least in part, a reflection of the fact that prices have tended to remain at a competitive level. This relative stability has been achieved either through the introduction of major changes in technology, as in the case of the territorial cheeses, or by a massive scale-up of a traditional process like yoghurt making, but whatever the chosen route, the enhanced productivity has been to the benefit of the consumer. Improved methods of product control have also been instrumental in raising the efficiency of the various manufacturing procedures, and the intention of this second volume is to record the current 'state of the art' in respect of the major dairy products. Obviously processes will continue to become sophisticated, but if this text can provide a background to future developments, then the endeavours of the contributors will have been worthwhile.

R. K. ROBINSON

Contents

List of Contributors

Mr T. ANDERSEN
A/S N. Foss Electric, 69 Slangerupgade, DK 3400 Hillerød, Denmark.

Mr N. BREMS
A/S N. Foss Electric, 69 Slangerupgade, DK 3400 Hillerød, Denmark.

Mr M. M. BØRGLUM
A/S N. Foss Electric, 69 Slangerupgade, DK 3400 Hillerød, Denmark.

Mr S. KOLD-CHRISTENSEN
A/S N. Foss Electric, 69 Slangerupgade, DK 3400 Hillerød, Denmark.

Mr E. HANSEN
A/S N. Foss Electric, 69 Slangerupgade, DK 3400 Hillerød, Denmark.

Mr J. T. HOMEWOOD
c/o Nestec Ltd, Avenue Nestlé 55, 1800 Vevey, Switzerland.

Mr J. H. JØRGENSEN
A/S N. Foss Electric, 69 Slangerupgade, DK 3400 Hillerød, Denmark.

Dr M. J. LEWIS
Department of Food Science, Food Studies Building, University of Reading, Whiteknights, PO Box 226, Reading RG6 2AP.

Mr H. L. MITTEN
Crepaco International Inc., 8303 West Higgins Road, Chicago, Illinois 60631, USA. Present address: 336 Kenilworth Avenue, Glen Ellyn, Illinois 60137, USA.

Mr J. NEIRINCKX
Crepaco International Inc., Avenue de Tervuren 36, Bte 9, 1040 Brussels, Belgium.

Mr L. NYGAARD
A/S N. Foss Electric, 69 Slangerupgade, DK 3400 Hillerød, Denmark.

Dr G. L. PETTIPHER
Cadbury Schweppes plc, The Lord Zuckerman Research Centre, The University, Whiteknights, Reading RG6 2LA.

Dr R. K. ROBINSON
Department of Food Science, Food Studies Building, University of Reading, Whiteknights, PO Box 226, Reading RG6 2AP.

Mr M. B. SHAW
Dairy Crest Foods, Research and Development Division, MMB Crudgington, Telford, Shropshire TF6 6HY.

Dr A. Y. TAMIME
Department of Dairy Technology, West of Scotland Agricultural College, Auchincruive, Ayr, Scotland.

Chapter 1

Recent Developments in Yoghurt Manufacture

R. K. Robinson
Department of Food Science, University of Reading, UK

and
A. Y. Tamime
West of Scotland Agricultural College, Ayr, Scotland, UK

Although yoghurt is a product that can be manufactured in small volumes, and with a modest level of technology, the growing demand within the industrialised countries has meant that the scale of many operations is now extensive. Outputs of thousands of litres per day are now commonplace, and as volumes have increased, so the plants have become increasingly more sophisticated.

Nevertheless, the basic stages of production are common to all the various systems, and these stages can be summarised as follows:

(i)   Preparation of a basic process milk with around 12–14 per cent milk solids-not-fat (MSNF).

(ii)   Heating this milk to 85–95°C, preferably in a unit that allows the temperature to be held for some 10–20 min.

(iii)   Inoculating the milk with a culture in which *Lactobacillus bulgaricus* and *Streptococcus thermophilus* are the principal organisms present.

(iv)   Incubating the inoculated milk at 42°C (or 30°C overnight) until a smooth coagulum has formed, together with the desired level of acidity and flavour.

(v)   Cooling the finished product and, unless the milk has been incubated in retail cartons (set yoghurt), mixing with fruit or other ingredients.

(vi)   Packaging the stirred yoghurt into containers prior to dispatch under chilled conditions.

1

Most plants for making yoghurt are built, quite simply, to accommodate these six stages.

It will be entirely appropriate, therefore, to employ these same divisions as a basis for discussing recent changes in manufacturing practice, for while the technology has improved, the basic intention of each stage has remained unaltered.

PREPARATION OF THE BASIC MIX

Most factories employ liquid milk as their 'main ingredient', and this material will be delivered, in bulk, employing road tankers. The reception tests applied to such milk, and the standards of hygiene expected in the unloading bays, are similar to those associated with any milk processing plant (Luck, 1981), and assuming that the normal criteria are met, the milk will be transferred to storage silos at < 5 °C.

Beyond this stage, however, the processing of the milk for yoghurt does adopt a pattern of its own, for few manufacturers include more than a trace of butterfat in their finished yoghurt (see Table I). The next essential step involves, therefore, the passage of the milk through a centrifugal separator to give a stream of skim-milk (0·5–0·7 per cent fat) and cream. If the manufacturer wishes to produce a medium fat yoghurt, then some of the separated cream will be fed back into the skim-milk, and automatic monitoring will ensure that the final process stream is of constant composition.

Once the fat content has been standardised, the next stage is to raise the level of MSNF, for without this additional protein, the gel produced during fermentation will be unacceptable to the consumer, i.e. thin and 'watery', and prone to syneresis. The excessive use of stabilisers could, of course, mask this deficiency, but as most countries ban (Robinson and

TABLE I
Some typical values of the major components of yoghurt

Component (g/100 g)	Yoghurt (natural)	
	Full fat	Low fat
Protein	3·9	5·0
Fat	3·4	1·0
Carbohydrate	4·9	6·5

Adapted from Deeth and Tamime (1981).

Tamime, 1976), or at least frown upon, the use of stabilisers in natural yoghurt, the tendency is to minimise their use across the entire range of retail products. The nutritional value of yoghurt is also enhanced by raising the non-fat solids, and hence most manufacturers resort to adjustment by one of two alternative approaches.

Addition of Milk Powder

The powder most widely used in this context is antibiotic-free, skim-milk powder, because apart from being an adequate source of the types of protein required, it is readily available, and at a modest price. The description — antibiotic-free — is usually rephrased into the form 'CYP-powder suitable for cheese and yoghurt', because total freedom from antibiotics is, given normal farming practice, an almost impossible state to achieve. However, a specification of not more than 0.02 IU g^{-1} of inhibitory substances can be met, and on a routine basis, and it is milk of this type that is commonly used for yoghurt. This level is selected, in part, because it is around the lowest level of detection for many of the simple laboratory tests for 'antibiotics' in milk, but it also represents the tolerance limit for a number of strains of yoghurt bacteria. Thus, residues above 0.02 IU ml^{-1} of reconstituted milk can seriously inhibit the activity of the starter cultures, and because the incoming milk may itself contain antibiotics, adherence to firm limits in any materials employed for fortification can be a worthwhile precaution.

It is also important that the powder shall be:

(a) easily dispersed into the aqueous phase;
(b) fully soluble, and leave no gritty particles; and
(c) free from scorch particles, for black or brown specks will show clearly in natural yoghurt.

In order to meet these various aims, the selection of most manufacturers would be a 'semi-instantised' powder, particularly as such powders are often derived from certain types of spray-drier without recourse to additional processing.

The method of incorporation of the powder varies with the scale of the operation but, in principle, the aim is always to achieve rapid mixing of the powder with the minimum of opportunity for lump formation, or the excessive introduction of air. The most efficient system for the smaller plant is probably the 'pump and hopper' unit illustrated in Fig. 1, but for larger factories, a special mixing tank may be employed to

Fig. 1. A typical 'pump and hopper' unit for the incorporation of milk powder into process milk. (Courtesy of APV International Ltd, Crawley, UK.)

standardise all the milk required to meet the entire capacity of the plant. The Crepaco 'Blender' shown in Fig. 2 is an example of the type of mixer that has been developed specifically to handle an operation of this nature, and it is of note that:

(i) the base of the tank is conical to allow rapid unloading;

(ii) the high-speed agitator has a dual action design incorporating a swirling motion, with the creation of a deep vortex, that quickly disperses the milk powder into the process mix with a minimum of foaming.

A number of variants of this system are available, and the quality of the milk powder may be one factor that determines, in part, the ultimate choice. Thus, a powder which contains numerous heat-damaged or insoluble particles will reconstitute poorly, and the resultant milk will be unacceptable in that:

(i) the particles may damage the orifices of an homogeniser, or lead to soil formation in a heat exchanger; and

(ii) the finished yoghurt may be spoilt by the presence of scorched particles, or a gritty mouthfeel.

Hence removal of the insoluble material can become a primary

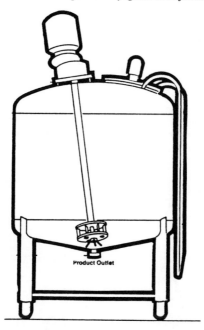

Product Outlet

Fig. 2. This Crepaco 'Blender' is designed to mix milk powder or other ingredients into a liquid base. (Courtesy of APV International Ltd, Crawley, UK.)

consideration. Centrifugal clarifiers can be employed, but the less expensive option is to use stainless-steel/nylon filters installed in the lines carrying the reconstituted milk (see Fig. 3). If large volumes of milk are involved, then a system of interchangeable filters will be needed so that in the event of a filter becoming clogged, an alternative can brought into operation without delay.

Although skim-milk powder is the obvious choice for most manufacturers, there is an extensive range of milk-based powders that could be considered as alternatives. Some of these powders are shown in Table II, and it is noticeable that the levels of protein, either casein or whey protein, can be adjusted to meet almost any requirement. The cost of the products *vis-à-vis* skim-milk powder or whey powder is a reflection of the protein content, but as the functional properties of the powder are also, in large measure, a reflection of the protein content, some manufacturers prefer the modified materials. However, the need to incorporate the powder into the liquid base still remains, and the plant requirements will be broadly similar.

TABLE II

The approximate chemical composition of some available powders based on either casein or whey protein

Product	Protein	Fat	Ash	Lactose
Whey powder (standard)	13·0	0·8	11·0	71·0
Whey powder (high protein)	76·0	4·0	4·3	12·0
Caseinate (sodium)	89·3	0·9	4·5	0·2

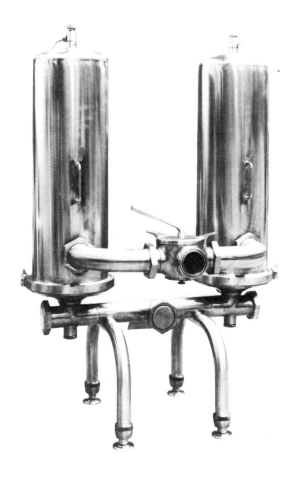

Fig. 3. A filter unit capable of removing particulate material from large volumes of fresh or reconstituted milk. (Courtesy of APV International Ltd, Crawley, UK.)

Concentration of the Milk

Raising the total solids of the standardised milk can be achieved by use of an evaporator, and in order to avoid damage to the milk constituents, the process is normally handled under vacuum. Single-effect evaporators are the most widely used, and the milk is usually preheated in a conventional plate heat exchanger. In some systems, a circulation cycle is established such that a fixed percentage of water is removed on each cycle until the desired total solids has been achieved, and plants of this type can handle up to 8000 litres h^{-1}.

The advantage of this system is that the quality of the end-product is reported to be excellent, but the exact validity of this claim is difficult to substantiate. The same problem exists for yoghurt made from milk concentrated by ultrafiltration (UF) or reverse osmosis (RO), but nonetheless, milk can be modified by membranes to give a process mix of the desired total solids. It is also possible, depending on the type of membrane selected, to alter the relative proportions of the components in the milk, so that while the RO membrane brings about an overall concentration, UF processes give an increased level of protein at the expense of lactose and minerals. The influence of these changes on the quality of yoghurt has been discussed elsewhere (Glover, 1971; Tamime and Robinson, 1985), and an example of the type of plant required is shown in Fig. 4. It is relevant, however, that membrane processing is not a cheap alternative, and it may be that economic considerations alone will limit the extent to which yoghurt milk is modified by this system in commercial practice.

Homogenisation

If a full-fat, set yoghurt is being made, i.e. above 3·0 per cent fat, then homogenisation of the mix is essential to prevent separation of a cream line during incubation, but even for low-fat varieties, homogenisation is reported to offer certain advantages, namely:

(i) it reduces the average diameter of the fat globules to $<2\,\mu$m;
(ii) the viscosity of the yoghurt is improved by an increased adsorption of fat globules onto the casein micelles;
(iii) syneresis is reduced;
(iv) the yoghurt milk becomes 'whiter' due to the increased number of fat globules, and enhanced light scattering; and
(v) the process ensures uniform mixing of any dry ingredients added to the milk.

Fig. 4. An ultrafiltration unit of the type that could be used to modify the composition of milk. (Courtesy of Pasilac Silkeborg Ltd, Preston, UK.)

The process itself is accomplished by forcing the milk, under high pressure, through a fine orifice, and the shearing effect of the passage is sufficient to reduce large fat globules to an acceptable size. Further shattering of the globules occurs when the milk impacts on the homogeniser rings (Anon., 1980), and the final milk is both homogeneous and stable. A typical unit is shown in Fig. 5, and although the capital cost is high, many producers believe the expenditure to be justified.

HEAT-TREATMENT OF THE YOGHURT MILK

Some of the aims of this stage of the process can be summarised as follows:

(i) destruction of any micro-organisms present in the vegetative state, so avoiding the risk of competition between the starter culture and miscellaneous bacteria; potential spoilage organisms like yeasts are also eradicated — the surviving spore-formers are not likely to cause problems;

(ii) expulsion of oxygen from the mix, so providing the micro-aerophilic conditions needed by the starter organisms;

(iii) redistribution of minerals, especially calcium, between the soluble and colloidal forms, so leading to a decrease in coagulation time;

but the most important change involves the proteins. In general terms, the effect of heating the yoghurt milk is to:

(i) denature the whey proteins, so enabling both the β-lactoglobulin and α-lactalbumin to interact with the caseins (Davies *et al.*, 1978); and

(ii) encourage a physical expansion/uncoiling of the casein micelles, so giving rise to a 'softer' coagulum than would result from the acid precipitation of unheated milk.

Fig. 5. An APV-Gaulin homogeniser (MC 45) capable of handling in excess of 20 000 litres of milk per hour. (Courtesy of APV Co. Ltd, Crawley, UK.)

Fig. 6. These multi-purpose tanks can be employed for heat-treating the process mix, incubation and, finally, cooling of the finished product. (Courtesy of Goavec SA, Alencon Cedex, France.)

These two changes combine to give a firm, homogeneous coagulum that effectively immobilises the aqueous phase within the protein matrix, so reducing the risk of whey separation. This benefit is best achieved when the milk is heated to 85 °C and held at this temperature for 30 min, but modern practice is often unable to cope with this demand. Thus, the sheer volume of milk that has to be processed in any medium/large dairy means that high temperature–short time treatments are much more popular. Nevertheless, some manufacturers still prefer to retain the extended holding time, and hence the types of equipment used for the heating of yoghurt milk can be grouped as:

(i) for batch purposes, e.g. pasteurising or multi-purpose tanks (see Fig. 6); and

(ii) for continuous processes, e.g. plate or tubular heat exchangers.

The Batch Process

The tanks employed at this stage are usually water-jacketed, and range in capacity from a few hundred litres up to several thousand for a typical model from Goavec SA (see Fig. 6). In practice, a series of these tanks could be used at regular intervals for the production of yoghurt on a semi-continuous basis (Tamime and Robinson, 1985), and a typical usage cycle would be:

(i) fill the tank with process milk (i.e. standardised, fortified, pre-heated and homogenised);
(ii) heat the milk to 80–85 °C for 15–30 min;
(iii) cool the milk to incubation temperature;
(iv) inoculate, and allow the fermentation to proceed to the desired acidity; and
(v) cool the yoghurt to below 20 °C prior to further processing.

It is also important that a tank of this type should be fitted with an effective, slow-speed agitator, and total sweep blades that scrape the walls of the tank can be an advantage. A cone-shaped base to the tank provides a convenient aid for discharging the coagulum, and can also assist in improving the heat transfer characteristics of the tank. This latter aspect is sometimes carried a stage further by fitting the tank with internal cooling coils and/or a hollow agitator that can be filled with an appropriate coolant. As a result of efficient design, the heating and cooling stages of the operation can take as little as 30–40 min, and the volume throughput can easily cope with the demands of a small dairy.

Continuous Process

The unit that is most widely used for this process is the standard plate heat exchanger, in which the heating medium (hot water) flows along a channel separated from the yoghurt milk by a thin partition of stainless steel. Details of this system are included elsewhere (Lewis, 1985), but the advantages to the producer can be summarised as follows:

(i) lower energy costs; and
(ii) a considerable saving in the time required to process a given volume of milk; for although the milk should still be held in a bulk tank for 30 min at 80–85 °C (see Fig. 7), the lengthy procedures of heating the milk from ambient to 80 °C and cooling again to 42 °C become operations that can be handled with ease.

An alternative type of unit is the tubular heat exchanger, in which a tube (or tubes) containing the yoghurt milk are suspended in an outer tube containing the heating medium. It has been suggested that this type of

Fig. 7. A pasteurisation plant with associated holding tanks. (Courtesy of APV Co. Ltd, Crawley, UK.)

plant may be less damaging to sensitive products than the plate heat exchanger, but in the present context, the choice will be governed by economic and/or engineering considerations alone. Although an extended holding-time at 80–85°C is the most desirable procedure, many manufacturers opt for a holding time of around 10 min at a temperature of 90–95°C. This approach allows the plate or tubular heat exchanger to be used more efficiently, as the incorporation of holding tubes of the type shown in Fig. 8 allows for a continuous flow of process milk through the unit.

If the intention is to heat-treat the finished yoghurt as well, then the preferred equipment may be the scraped/swept surface heat exchanger, which consists of a jacketed cylinder fitted with a scraper blade. The action of this blade is to remove continuously the product from the heated surface, a motion that both prevents heat damage to the ingredients, and allows the diameter of the cylinder to be increased considerably in contrast with the conventional tubular system. This latter aspect allows more viscous products to be handled e.g. yoghurt, as well as the fluid process mix.

However, the overall aim must be to meet the main objectives of the heat-treatment outlined earlier, and to ensure that the plant does not contaminate the milk. To this latter end, hot water is usually circulated

Fig. 8. A heating unit for yoghurt milk including a balance tank (centre), a plate heat exchanger, and a 'zig-zag' holding tube (foreground) that gives a residence time of 8 min. (Courtesy of the Northern Ireland Milk Marketing Board.)

through the plant at the start of the heat treatment process to sanitise the pipework, and incidently, warm the plant to the desired processing temperature.

INOCULATION AND INCUBATION

At this stage, the process milk is cooled to incubation temperature, which would be in the region of 40–45°C for a short fermentation (3–3½ h), or around 30°C for incubation overnight. Once this temperature has been achieved, the milk is ready for inoculation with the desired culture.

Starter Cultures

The complex microbiology of the yoghurt fermentation has been described in detail elsewhere (Tamime and Robinson, 1985), but for practical purposes, it is sufficient to record merely that starter cultures for yoghurt consist, almost universally, of two species of bacterium — *Streptococcus thermophilus* and *Lactobacillus bulgaricus.* Thus, the unique synergism between these two organisms ensures not only that the required level of lactic acid is reached within the allotted time, but also that the end-product has the flavour and consistency associated with yoghurt. Obviously these attributes vary with the precise strains of bacteria that have been selected and combined to give the working culture, but once the combination has been established, it should behave, within a given plant, in a thoroughly predictable manner.

The next critical point is establishing that the cultures required for daily use are produced in a manner that provides inocula that are uniformly viable and active, and this requirement implies an additional aim that the constituent species shall be present in a prescribed ratio. The usual balance is one chain of *Streptococcus* to one chain of *Lactobacillus,* and although other ratios can be employed to alter the characteristics of the final yoghurt, a balanced culture is the easiest to sustain.

Once the culture has been selected in terms of the organoleptic features of the retail product, the purely practical question of handling the cultures in the factory attains priority. In general, two options merit active consideration, namely the production of a bulk starter in a volume equivalent to around 2 per cent of anticipated factory output, or the use of a direct-to-vat system purchased from an outside supplier. Further details of these alternatives are given in Table III, and the systems available for the production of bulk starters have been recently reviewed by Tamime (1981). The ultimate choice of any given system will depend on the availability of plant and the preference of the individual manufacturer, but whatever the route, the process milk must be furnished with an abundant and viable microflora. Once this goal has been achieved, the fermentation stage can begin. The plant required at this point in the process is designed to provide and maintain the temperature conditions required by the bacteria, and the equipment will reflect the type of yoghurt being produced, i.e. set yoghurt or stirred yoghurt.

Production of Set Yoghurt

In this case, coagulation of the milk takes place in the retail containers, and hence the process involves:

— cooling the yoghurt milk to incubation temperature;

— adding the starter culture, and if appropriate, flavouring and colouring ingredients;

— filling into the retail containers — for 'Sundae' style yoghurts, the fruit is placed in the carton before the milk;

— incubating the milk to achieve the desired acidity.

This latter phase can be carried out in a number of ways.

TABLE III

The alternative forms of starter culture that are available for the production of yoghurt, and some observations on the method of utilisation

Type of culture	Comments
Liquid cultures	Expensive in terms of laboratory facilities Requirement for bulk starter and intermediate culture vessels Demanding on creamery personnel High risk of infection during transfers, e.g. bacteria, yeasts Culture characteristics liable to change with frequent sub-culturing
Frozen cultures (a) Short-term −20 to −40°C	Requirement for central laboratory within company, but also available from independent suppliers Requirement for rapid and reliable transport service Requirement for bulk starter (and perhaps intermediate) culture vessels Low risk of infection if bulk starter facilities are well maintained Less demanding on creamery personnel
(b) Long-term at −196°C	Expensive culture storage facilities required Requirement for bulk starter vessels Easy to use Low risk of infection
Dried cultures (a) Freeze-dried (standard)	Eliminates need for maintenance of stock cultures Standard culture characteristics Otherwise handled as liquid cultures (see above)
(b) Freeze-dried (concentrates)	Requirement for domestic deep-freeze (−20°C) for storage Requirement for bulk starter vessels Easy to use Can be employed as direct-to-vat cultures, and have proved especially useful for small-scale production of yoghurt

Water baths

In this system, the filled containers, which are usually made of glass, are placed in trays and immersed in shallow tanks of warm water. The water level is held just below the tops of the containers, and once the desired acidity has been reached, circulating cold/chilled water is used to rapidly cool the finished product (Crawford, 1962). The trays of cool yoghurt are then transferred to a refrigerated store for final chilling.

However, although water is an extremely efficient medium for heat transfer, the need for expensive packaging has severely restricted interest in the system, and most manufacturers prefer air as the heating/cooling agent.

Incubation cabinets

These cabinets are, as the name implies, small, insulated chambers with capacities ranging up to 750 litres. Warm air is circulated during the fermentation stage, and at the desired acidity, chilled air is employed to attain rapid cooling of the retail product. A typical cabinet is shown in Fig. 9.

The limitation of this approach is mainly in relation to size, in that the cabinets must be small in order to:

(i) ensure that the air circulates throughout the chamber, so avoiding the risk of high or low temperature pockets of stagnant air;

(ii) enable the trays or pallets of yoghurt to be stacked quickly at the start of incubation, so avoiding too long a time lag between the first and last cartons entering the chamber.

If large volumes of yoghurt are required, then a battery of incubators offers one solution, but a more satisfactory process can be instigated by use of a tunnel system. The tunnel consists of two sections, one with circulating warm air and the other with chilled air, and the pallets of retail cartons are placed on a conveyor that runs throughout the length of the unit. The speed of the conveyor is regulated to reflect the rate of acid production in the yoghurt milk, so that entry to the cooling section only occurs at the required pH of 4·4–4·5. After passage through the chilled section, the pallets are transferred to a cold store for the final temperature reduction to 4–5°C. The main disadvantage with this approach is that the cartons are in motion during coagulation, and especial care is required to avoid jarring the delicate structure of the new

Fig. 9. An electrically operated cabinet used for the incubation of set or stirred yoghurt. (Notice the internal glass doors providing good insulation and minimising heat losses. These units can also have a refrigeration system incorporated so that the cabinet can then be used as an incubator/cooler. The floor area is used for the production of yoghurt in cans, and the upper shelves are used for set yoghurt.)

yoghurt. Nevertheless, Cottenie (1978) concludes that the tunnel system is enjoying increased popularity among the major manufacturers.

Production of Stirred Yoghurt

The coagulum of stirred yoghurt is produced in bulk, and the gel structure is then broken to provide a smooth homogeneous base for fruit or other ingredients. A standard, water-jacketed tank of the type illustrated in Fig. 6 is often used for this purpose, so that after heat-treatment of the milk, warm water can maintain the temperature of the mix during incubation, and circulating, chilled water cool the finished yoghurt. This approach is adequate for small batches of 400–500 litres per day, but beyond that, attention must be turned to the separation of the heating stage from the fermentation phase.

This separation is usually achieved by employing special incubation tanks, in which the outer jacket is packed with polystyrene or some similar material. This barrier is sufficient to maintain the desired temperature throughout the period of incubation. The advantage of this system is that, in comparison with the multipurpose tanks, insulated vessels are less expensive. It is quite feasible, therefore, to increase the throughput of a factory by servicing a number of fermentation tanks from just one heat-treatment unit, and at a level of capital investment that is not prohibitive.

Control of the process can, with all these systems, be based on the level of acidity in the yoghurt, a parameter that can be monitored continuously by the inclusion of pH electrodes into the basic design of the tank.

Cooling the Yoghurt

At the required acidity, i.e. around 0·8–1·0 per cent lactic acid, cooling of the coagulum commences, and the intention is to reduce the temperature of the coagulum to below 20 °C within an acceptable time-span. Thus, below 20 °C, the metabolic activity of the starter organisms is sufficiently reduced to prevent the yoghurt becoming unpalatable through excess acidity, and hence initiation of cooling depends on:

(a) the level of lactic acid required in the end-product — usually between 1·2 and 1·4 per cent lactic acid;

(b) the rate of cooling that can be achieved with the available equipment, and in a manner that does not damage the texture of the yoghurt (Robinson, 1981).

In most factories, two approaches to cooling are employed, namely air cooling, and cooling before packaging.

Air cooling

The use of chilled air in cabinets or tunnels employed for the manufacture of set yoghurt has already been mentioned, but the importance of efficient circulation in cold stores or transport vehicles must not be overlooked (Boast, 1985). In general, yoghurt should always be stored at $< 7\,^{\circ}\text{C}$ throughout the distribution chain, for otherwise the keeping quality of the product may be severely reduced.

Cooling before packaging

In-tank cooling necessitates the circulation of chilled water through the outer jacket of the tank, and the rate of cooling depends on:

— the surface area in contact with the yoghurt;
— the degree of agitation that can be achieved;
— the difference in temperature between the coolant and the product, and the flow rate of the cooling agent through the outer jacket.

The majority of manufacturers rely on the sides and base of their tanks to provide sufficient contact surfaces, and it is estimated that, in a vessel of this type, 5000 litres of yoghurt could be cooled from $42\,^{\circ}\text{C}$ to $5\,^{\circ}\text{C}$ in 4 h (Jay, 1979). The critical fall to $< 20\,^{\circ}\text{C}$ can, of course, be achieved much more rapidly, and for this reason, cooling to storage temperature is normally accomplished in a chill room or cold store.

Apart from the saving in time, it is advantageous that stirring be kept to a minimum to avoid undue structural damage to the coagulum. The design of the agitator can go some way towards reducing this danger (Tamime and Greig, 1979), and the overall aim of the available tanks is to achieve 'effective' mixing with as little shearing as possible. However, although in-tank cooling eliminates the need for additional plant, more rapid processing of the yoghurt can be accomplished by the incorporation of plate or tubular heat exchangers into the system. According to Piersma and Steenbergen (1973), minimal structural damage to the coagulum will occur in a tubular cooler, but conventional plate heat exchangers can be successfully modified to attain the same end. The key feature is the provision of large gaps between the plates, but the avoidance of high back-pressures is a valuable corollary. This reduction involves either restricting the flow-rate of the yoghurt, or progressively increasing the gaps between the plates across the unit, and the use of a number of small units in parallel rather than one large item is a further precaution.

The cooled yoghurt is then delivered to storage tanks prior to further processing. These vessels may be merely insulated if the yoghurt is to be packaged immediately, or they may allow further chilling of the product for storage overnight. This latter facility does allow the manufacturer to operate a more flexible schedule, and as shown in Fig. 10, the time-scale for production argues in favour of retaining some room for manoeuvre.

Packaging of Yoghurt

The packaging of yoghurt is an important step during production, and the purpose of packaging can be summarised as follows:

— protection of the product against dirt, micro-organisms and the environment, e.g. gases (oxygen) and light;
— provide relevant information to the consumer, e.g. the food labelling guidelines: name and origin of the food, ingredients, instructions for use, expiry date;
— the packaging material must be non-toxic, and no chemical reaction should take place between the material and the yoghurt (this may not be true in all cases — see Ministry of Agriculture, Fisheries and Food (MAFF), 1983).

A detailed review of the general background to the packaging of yoghurt, i.e. packaging materials and equipment, has been published

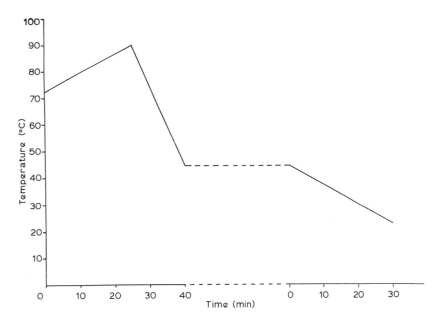

Fig. 10. A summary of the times and temperatures associated with a typical manufacturing procedure for yoghurt using the Goavec tank; the broken line represents period of incubation. (After Tamime and Robinson, 1985.)

recently by Tamime and Robinson (1985), but under commercial practice, the handling of yoghurt (set or stirred type) may vary slightly. The primary steps involved during the handling of stirred yoghurt should include consideration of the following aspects.

Type of packaging materials
The unit container may consist of glass, earthenware, plastics or laminate. The latter two types are widely used for packaging yoghurt, and Fig. 11 illustrates some typical examples.

Fruit mixing machines and related equipment
The processed fruit is mainly received in metal cans, polypropylene drums or buckets, and/or stainless steel tanks. The former two methods of packaging are very popular with small- and medium-scale manufacturers, but in large dairies, the fruit is obtained in bulk or processed on site.

The handling of the processed fruit in the dairy is dependent on the degree of automation and/or the type of container used to package the

Fig. 11. Some containers for the packaging of yoghurt. (1, 5: thermoformed plastic container; 2: glass bottle sealed with aluminium 'pull-ring' type; 3: Pure Pak laminated carton; 6: preformed plastic containers sealed with 'snap-on-lids' (family size); 4, 7: preformed plastic cups heat sealed with aluminium foil.

fruit. For example, if metal cans are used, the normal approach is to open the can (i.e. hand operated, semi-automatic or automatic equipment — see Fig. 12) and then meter the fruit directly to the fruit/yoghurt blending equipment. However, if the fruit ingredients are received in stainless steel tanks, the normal procedure is to meter them directly into the yoghurt line prior to packaging.

Many types of fruit/yoghurt blending machines are being employed,

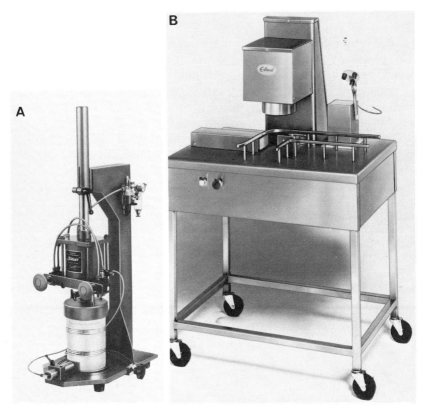

Fig. 12. Different types of metal can openers. (A) Pneumatic opener 'Blitzer PD-10' which is a semi-automatic opener operated by compressed air. (Reproduced by courtesy of Karl Engelhardt, Bremen, West Germany.) (B) Edlund Model 825 which is a heavy duty can opening device and can be automated to handle up to 1500 cans h^{-1} (see Tamime and Robinson, 1985). (Reproduced by courtesy of Peter Holland Food Machinery Ltd, Lincolnshire, UK.)

and the approaches to mixing could be classified as manual, batch or continuous.

Packaging equipment

There is a wide range of yoghurt filling machines available on the market, and the speed of filling may vary from < 1000 to 50 000 cups per hour. The principle of filling involves the use of a positive, reciprocating displacement pump, and according to Tamime and Robinson (1985), the factors which influence the choice of a certain packaging machine are as follows:

— capital cost;
— method of filling (e.g. form/fill/seal or fill/seal);
— method of sealing (e.g. heat sealing, crimping or press-on-lids);
— degree of automation and speed of filling;
— need for a high standard of hygiene;
— desirability of cleaning-in-place (CIP), and the time required to change from one flavour to another;
— the reliability and versatility of the machine;
— power and labour requirements;
— accuracy of filling and safety measures;
— desirability of filling under a controlled atmosphere.

The latter aspect is taken into account in the classification of automatic filling machines (i.e. no-controlled atmosphere, controlled atmosphere, or aseptic), and it is against this background that some examples will be reviewed.

A yoghurt packaging line consists of the following equipment:

— Buffer or intermediate tank (this tank is made of stainless steel, and is normally insulated to hold the cool yoghurt for a short period).
— Fruit mixing and blending equipment.
— Filling machines.

EQUIPMENT FOR FRUIT MIXING AND YOGHURT BLENDING

The method of fruit/yoghurt mixing could be manual or controlled automatically, so that the process can, in practice, become continuous. The *manual* and *batch* blending methods of fruit/yoghurt mixing are similar (see Fig. 13), and consist usually of two tanks which are used in parallel. Where manual blending is practised, the sequence of operations is as follows.

— Empty into each tank, the amount of fruit required for a given volume of yoghurt;
— mix gently with a plunger, and pump the yoghurt/fruit blend to the filling machines;
— if the first tank is emptied as the second one is being prepared the process becomes, in effect, continuous.

In principle, the batch blending system is similar to the method described above, except that the mixing of the fruit and yoghurt is carried out mechanically in larger tanks fitted with specially designed agitators. The fruit and yoghurt are metered into each tank, and the provision of an enclosure system can minimise aerial contamination.

Continuous fruit/yoghurt mixing can be achieved using an in-line static mixer or a special blender. In the former type, a length of stainless

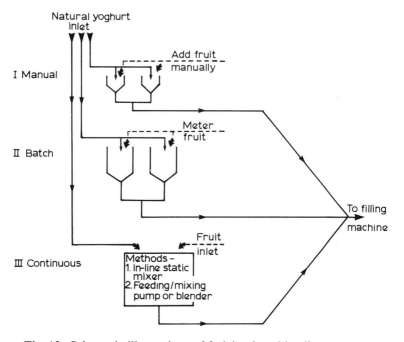

Fig. 13. Schematic illustrations of fruit/yoghurt blending systems.

steel pipe, enclosing a set of welded twisted blades, is mounted vertically with a T-piece at the bottom, and fruit and yoghurt are metered through the side-arms of the 'T' from their respective tanks. However, special blenders are also available.

INDAG Mixing System

This is an automatic fruit/yoghurt mixer, and is manufactured by INDAG GmbH, Heidelberg, West Germany. In principle, the mixer consists of two positive displacement pumps to introduce the respective volumes of fruit and yoghurt. Blending is achieved in a geared mixing chamber, and the fruit/yoghurt mix is fed directly to the hopper of the filling machine. Incidently, the fruit pump has an adjustable speed which can be altered to suit the characteristics of the fruit. Since the fruit pump is connected to a series of fruit tanks, the change over from one 'flavour' to another is simply achieved by pressing a button.

The Burdosa Fruit Blender

This blender is driven by a duplex metering pump which simultaneously moves the fruit base and the natural yoghurt to the mixing chamber. The pump is adjustable, so that the ratio of fruit base to yoghurt can be pre-set and maintained. The maximum rate of fruit introduction is 250 cm^3 litre^{-1} of yoghurt, and the maximum throughput, depending on the model and fruit to yoghurt ratio at blending is 4500 litres h^{-1}. This type of blender is manufactured by Ing. Herwig Burget in West Germany.

The Gasti Feed and Mixing Pump

Gasti-Verpakungsmaschinen GmbH of West Germany manufactures mixing and feeding pumps which are suitable for blending fruit and yoghurt. The DOGAmix 60 unit consists of two feed pumps that draw the yoghurt and fruit from their storage tanks in pre-set proportions, and a mixing chamber fitted with a dynamic agitator of variable drive. The fluctuations in product flow caused by the plunger-type pumps are eliminated in the mixing chamber, which also acts as a pressure equaliser. The mixed product (fruit/yoghurt) is homogeneous, and flows directly to the hopper of the filling machine.

The feed rates and mixing ratios are 75 litres min^{-1} and 1:5 to 1:20 respectively, and the mixing accuracy is $\pm 0.5\%$ of the final volume. This feed and mixing pump can be CIP cleaned without dismantling, and suitable systems are 1·2 per cent soda lye or nitric acid at 80°C, and steam sterilisation at 140°C. Incidentally, the moving parts, i.e. the pump rods and mixer drives, are isolated from the surrounding atmosphere with a sterile air chamber, so minimising microbial contamination during the mixing stages.

Crepaco Ingredient Feeders

These feeders can be obtained in three different models, and can be used for the continuous blending of fruit/flavours with the yoghurt. Model S-410 (see Fig. 14) has a combined base-product and ingredient throughput ranging from 228 to 3800 litres h^{-1}. The ingredient and the yoghurt are accurately metered, and the metering of the former product is via an agitator/auger feed combination which transfers the fruit from the hopper to an enrobing rotor at a controlled speed, e.g. 2, 10 or 50 rpm. The vertical tube situated along the side of the unit (see Fig. 14) is equipped with integrally mounted and hydraulically-driven stainless steel blenders. It is claimed by the manufacturer that the design of the dashers assures gentle mixing of the fruit/yoghurt mixture to give a final product which is uniform and free from mottling and ingredient damage.

EQUIPMENT FOR PACKAGING YOGHURT

A multitude of high-speed filling machines are available on the market, and in general, it is recommended that the packaging capacity is equivalent to the hourly throughput of the yoghurt processing plant. In large dairies, more than one filling machine is normally installed, and of different filling rates, so that while the larger machines are used for packaging yoghurt with fruit of 'regular' demand, the smaller machines are used for fruited yoghurts of more limited appeal.

Some examples of filling machines for yoghurt are as follows.

Filling Machines — No-controlled Atmosphere

Most manufacturers of filling machines offer a system of packaging in which the yoghurt is exposed to the atmosphere during the filling stage. Some typical examples are: Trepco of Denmark — Hansen (1984) reported

Fig. 14. The Crepaco Ingredient Blenders can continuously combine yoghurt with fruit at throughputs of up to 3800 litres h^{-1}. (1: fruit feeder hopper; 2: rotor chamber funnel; 3: auger/agitator on/off switch; 4: auger/agitator feed control; 5: enrobing rotor speed control; 6: yoghurt inlet; 7: blender unit; 8: yoghurt/fruit mixture outlet.) (Courtesy of Crepaco International Inc., Brussels, Belgium.)

that the latest four track model can be adjusted for four different cup sizes and has a capacity of 10 000 cups per hour; Colunio of UK, who produce 2000, 6000 and 8000 models, which range from one to four lanes, and have a capacity per hour equivalent to their model number.

Filling Machines — Controlled Atmosphere

An example of such a filling machine (see Fig. 15) is manufactured by E. P. Remy & Co. in France. The cup filling and closure area is enclosed within a cabinet with a sterile, laminar air flow. This particular filler is equipped for cups of 500 g capacity fitted with snap-on plastic lids. However, there is also an aseptic version in which the cups are sterilised with hydrogen peroxide (H_2O_2), and then dried with hot, sterile air; the aluminium-foil (diaphragm) dispensing unit is sterilised by UV-C lamps. The output of such machines, depending on the number of filling lanes, ranges from 3000 to 24 000 containers per hour. This machine can be cleaned by CIP and sterilised with steam.

Fig. 15. An E. P. Remy yoghurt filling machine, Type 54. (1: cup dispenser; 2: laminar flow cabinet; 3: filling head; 4: plastic lids dispenser; 5: cup loader and carton-track stacker. (Reproduced by courtesy of E. P. Remy & Co. Ltd, Reading, UK.)

Aseptic Filling Machines

The Gasti DOGAseptic 42 (see Fig. 16) is a typical example of an aseptic, yoghurt filling machine. The plastic containers, e.g. poly-propylene cartons and aluminium foil lids, are sterilised by H_2O_2 and hot air. The sterile air in the drying, filling and heat-sealing stations of the machine is maintained above atmospheric pressure, and the care-fully pressurised/exhaust air system prevents any air containing H_2O_2 from escaping through the cup-feed inlet or outlet stations. The filling machine can be cleaned by CIP, has a four-lane filling line, and an output capacity between 4000 and 9000 cups per hour; for certain dessert products, the DOGAseptic 42 can be equipped to 'multilayer' the product.

An alternative type of aseptic filling machine is the Hamba BK 10010/10, which has an output capacity of 36 000 cups per hour, and the plastic cups and aluminium foil lids are sterilised by UV-C lamps (see Möller, 1982).

Fig. 16. The DOGAseptic yoghurt filling machine. (1: cup dispenser; 2: cup sterilising, filling and sealing section; 3: cup loader and carton-track loader.) (Reproduced by courtesy of Gasti Verpackungsmaschinen GmbH & Co., Schwabisch, West Germany.)

Miscellaneous Filling Machines

Other systems which could be used for yoghurt filling are:

(1) Carton and paper containers, e.g. Tetra Brik, Tetra Rex and Pure Pak (some of these machines could be of the aseptic type).

(2) Form/fill/seal machines, e.g. the Illig FS 32 which can also handle the multi-flavour family pack, have been recently described by Tamime and Robinson (1985). A different type of form/fill/seal machine is the Tetra King, and in this case, the packaging material consists of expanded polystyrene, coated on both sides with a layer of plastics suitable for packaging yoghurt.

A recent innovation in the packaging of yoghurt is known as the 'piggy-back' (Colangelo, 1980), or in the United Kingdom, as the 'snack-pack' which was launched in 1983–84 (Anon., 1983). In principle, the natural yoghurt is packaged in a pre-formed plastic container and heat sealed by using laminated, aluminium foil. The muesli or similar flavouring material is then packaged separately in a transparent cup, and the yoghurt container and the cup are slotted together (see Fig. 17).

Fig. 17. A typical 'snack-pack' of the type that separates the yoghurt from other ingredients until immediately prior to consumption. (Reproduced by courtesy of Sweetheart International Ltd, Hampshire, UK, and Express Dairy-UK Ltd, Middlesex, UK.)

The filling of these products can take place on the same machine, or on two different units.

The final stage in the handling of the yoghurt containers is the crating and stacking of the pots. In small to medium scale factories, the operation can be carried out manually, but in large factories the final packaging is normally automated. The crates can be made out of metal, plastic or cardboard, but the latter two types are widely used for economic reasons, and to avoid the collection of the empty crates. The sequence of operation can be summarised as follows:

— nest the yoghurt pots in plastic or cardboard trays;
— stack the trays in large cardboard boxes or shrink wrap in polythene;
— stack the trays onto wooden pallets, or wire cages fitted with castors;
— transfer the yoghurt to the cold store, and after 24–48 h, dispatch the product in refrigerated lorries.

MISCELLANEOUS TREATMENTS OF YOGHURT

The shelf-life of yoghurt can be extended for a few months, and in some cases almost indefinitely, by the application of selected techniques. Obviously such treatments may alter the nature of the end-product very considerably, but for certain markets, the changes are readily accepted. For example, the application of heat can be employed, and different time–temperature combinations ranging from flash heating at 60 °C to more rigorous treatments at more than 100 °C have been advocated; applications of the Alfa-Laval Aseptjomatic and ILL Frau systems have been reported by Tamime and Robinson (1985).

In the Middle East, yoghurt is used as a raw material for the manufacture of other traditional dairy products, and Fig. 18 illustrates some typical examples. 'Labneh' or concentrated yoghurt is particularly popular in that part of the world, and traditionally this product is concentrated using the cloth bag method. Recently, however, the process has been mechanised (see Fig. 19), and such installations are to be found in Saudi Arabia and other Arab countries. In brief, the warm low fat natural yoghurt is concentrated to around 18 per cent total solids using a nozzle separator, and cream is added at a later stage so that the 'Labneh' meets the existing legal standards, e.g. 24 per cent total solids and 10 per cent fat (Lebanese Standards, 1965).

R. K. Robinson and A. Y. Tamime

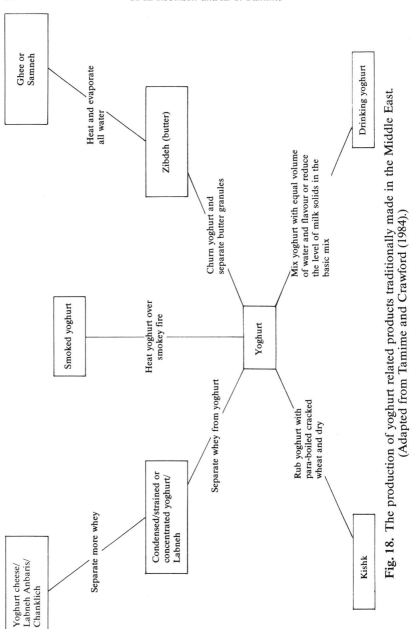

Fig. 18. The production of yoghurt related products traditionally made in the Middle East. (Adapted from Tamime and Crawford (1984).)

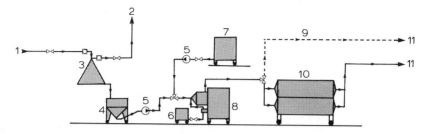

Fig. 19. A mechanised system for the manufacture of 'Labneh'. (1: inlet of yoghurt at 40–45 °C; 2: whey outlet; 3: Westfalia separator; 4: funnel or holding tank for 'Labneh'; 5: positive pump; 6: cream tank; 7: optional tank for fruits or other added ingredients; 8: Westfalia Quark mixer; 9: hot packaging line; 10: tubular cooler to cool 'Labneh' to 6°C; 11: filling machines.) (Reproduced by courtesy of Westfalia Separator Ltd, Milton Keynes, UK.)

REFERENCES

Anon. (1980). *Dairy Handbook,* Alfa-Laval A/B, Lund, Sweden.
Anon. (1983). *Packaging,* **54**, 16.
Boast, M. F. G. (1985). The technology of freezing. In: *Microbiology of Frozen Foods* (Ed. R. K. Robinson), Elsevier Applied Science Publishers, London, pp. 1–39.
Colangelo, M. (1980). *Dairy Field,* **163**, 95.
Cottenie, J. (1978). *Cultured Dairy Products Journal,* **13**(4), 6.
Crawford, R. J. M. (1962). *Journal of Dairy Engineering,* **79**, 4.
Davies, F. L., Shankar, P. A., Brooker, B. E. and Hobbs, D. G. (1978). *Journal of Dairy Research,* **45**, 53.
Deeth, H. C. and Tamime, A. Y. (1981). *Journal of Food Protection,* **44**, 78.
Glover, F. A. (1971). *Journal of Dairy Research,* **38**, 373.
Hansen, R. (1984). *North European Dairy Journal,* **50**, 193.
Jay, J. L. (1979). Personal communication.
Lebanese Standards (1965). 'LS:24 Milk and Milk Products'. Lebanese Standards Institution, Beirut, Lebanon.
MAFF (1983). In: *Survey of Styrene Levels in Food Contact Materials and in Foods,* Food Surveillance Report No. 11, HMSO, London, UK
Möller, E. (1982). *Dairy Science Abstracts,* **44**, 25.
Lewis, M. J. (1985). In: *Modern Dairy Technology, Vol. I* (Ed. R. K. Robinson), Elsevier Applied Science Publishers, London, pp. 1–50.
Luck, H. (1981). In: *Dairy Microbiology, Vol. II* (Ed. R. K. Robinson), Applied Science Publishers, London, pp. 279–324.
Piersma, H. and Steenbergen, A. E. (1973). *Official Orgaan FNZ,* **65**, 94.
Robinson, R. K. (1981). *Dairy Industries International,* **46**(2), 31.

Robinson, R. K. and Tamime, A. Y. (1976). *Journal of the Society of Dairy Technology*, **29**, 148.

Tamime, A. Y. (1981). In: *Dairy Microbiology, Vol. II* (Ed. R. K. Robinson), Applied Science Publishers, London, pp. 113–56.

Tamime, A. Y. and Crawford, R. J. M. (1984). *Egyptian Journal of Dairy Science*, **12**, 299.

Tamime, A. Y. and Greig, R. I. W. (1979). *Dairy Industries International*, **44**(9), 8.

Tamime, A. Y. and Robinson, R. K. (1985). *Yoghurt — Science and Technology*, Pergamon Press, Oxford.

Chapter 2

Modern Cheesemaking: Hard Cheeses

A. Y. Tamime

West of Scotland Agricultural College, Ayr, Scotland, UK

THE HISTORY AND ORIGIN(S) OF CHEESEMAKING

Cheesemaking is one of the oldest methods practised by man for the preservation of a highly perishable and nutritional foodstuff, e.g. milk, into a product which is not likely to deteriorate. The exact origin(s) or method of cheesemaking is difficult to establish, but from definite archaeological evidence, cheese was produced around 6000–7000 BC. According to Pederson (1979), the continued existence of man was primarily influenced, some 10–15 thousand years ago, when he had changed his way of life from being a 'food gatherer' to a 'food producer'. It is possible that such a transition was gradual and occurred at different times in different parts of the world.

Animals such as the cow, goat, sheep or buffalo, had been domesticated at that time, and the milk was utilised as food. However, some of man's recorded civilisations, e.g. the Sumarians and the Babylonians in Mesopotamia, the Egyptians in north-east Africa, and the Indians in Asia, illustrate that they were well advanced in husbandry methods, and in the production of fermented food products including cheese and yoghurt.

The oldest methods of preserving food known to mankind are: concentration, drying, fermentation and salting, and cheese is a fermented dairy product which is particularly concentrated and salted. It is safe to assume that cheese was first produced in the eastern part of the Mediterranean, and a summary of some cheese varieties with the date first noted is shown in Table I.

TABLE I
The archaeological references to cheese and their names with the date
first recorded

Year	Cheese variety
BC 9000	Rock drawings in the Sahara desert illustrate cow worshipping and milking
7000–6000	Bread and cheese have been identified as the staple food of early civilisations in the 'fertile crescent' which is situated along the eastern region of the Mediterranean (Lebanon and Syria) and Iraq (between the Tigris and the Euphrates rivers)
4000	Detailed records of the Egyptian civilisation appear to illustrate great developments in husbandry and dairy processing
3000	Cheese has been found in the tomb of Hories-Aha (Sumarian civilisation)
	The act of killing the cow for meat has been regarded sinful by the Vedic Hymns of India; instead milk and other dairy products have been used as food
2000	Reference to cheese can be found in the Babylonian records
1800	Woven reed baskets have been used in Asia to separate the curds from the whey (at present the same process is used to manufacture Surati Panir and Decca curds); similar baskets have been discovered in Dorset dated to the same period which indicate cheesemaking in England may have taken place prior to the Roman occupation
1500	Evidence of cheese has been referred to in Biblical times
1200–0	The Greeks and the Romans have been reputed to eat cheese and drink wine during their banquets
	Homer (1184 BC) wrote about cheese which was manufactured from the milk of the sheep and the goat (at present Feta and Halloumi cheeses are widely produced in Greece, Cyprus and Bulgaria)
	Herodotus (484–408 BC) referred to Scythian cheese made from mare's milk, and Aristotle (384–322 BC) reported the use of mare's and asses' milks for the manufacture of Phrygian cheese
	Varro (116–27 BC) noted the differences in the nutritional properties of cheeses, e.g. laxative effect, and the nourishing qualities in cheeses made from cow's milk (most), sheep (intermediate) and goat (least)
	Rameses tomb (100 BC) has scenes of goats and milk stored in skin bags

TABLE I—*contd.*

Year	Cheese variety
AD 0-100	Columella (50 AD) in his De Re Rustica reported in detail the process of cheesemaking and pointed out that hygiene in milk production was essential during the production stages
	Pliney (23-79 AD) wrote about sour milk cheeses which may have been the ancestors of present day 'pickled' cheeses produced in the eastern Mediterranean
200-300	The Roman emperor Diocletian enforced maximum prices for cheese, e.g. Lunar which later came to be known as Parmesan
400	The Romans brought the art of cheesemaking to Britain

Year	Cheese variety	
879	Gorgonzola	
1000	Schabzieger	
1070	Roquefort	The movement of trade and tribesmen
1174	Maroilles	from the Middle East through the
1178	Schwangenkäse	Balkan peninsula, the Mediterranean
1200	Grana/Parmesan	basin, or to India in the east helped to
1282	Taleggio	spread the art of cheesemaking and
1288	Gruyère	the development of different varieties
1500	Cheddar	known today (see text). However,
1579	Parmesan	monasteries and other establishments
1622	Emmental	at different dates helped to maintain
1688	Dunlop	proper records of the cheesemaking
1697	Gouda	process where the identity of the
1783	Gloucester	modern varieties have been preserved
1785	Stilton	
1791	Camembert	
1800	Limburg	
1861	St. Paulin	

Data compiled from: Marth (1953), Crawford (1960), Davis (1965), Kosikowski (1977), Pederson (1979) and Scott (1981).

It is possible to suggest that modern cheesemaking could have evolved in two main stages: firstly, the manufacture of sour milk products, e.g. Leben (Laban), Ayran or yoghurt, and secondly, by partial separation of the whey and the addition of salt, e.g. yoghurt cheese, concentrated yoghurt or soft cheese. In sub-tropical climatic conditions, e.g. the Middle East, milk sours very rapidly in a few hours after milking, due to the high ambient temperature and the presence of micro-organisms in the milk. These bacteria may have originated from the animal, the hands of the milker, the surfaces of utensils used to hold the milk or the environment. These organisms can produce two different types of fermentation. Firstly, the non-lactic fermentation is brought about by micro-organisms other than lactic acid bacteria, and the product is normally stale, insipid or of bad taste when consumed. Secondly, a fermentation produced by the so-called lactic acid bacteria gives a more desirable product which is pleasant to eat.

Traditionally the containers used for carrying or storing milk were made from animal skins or stomachs. As the milk is left undisturbed, clotting of the milk may occur due to developed acidity as a result of bacterial activity, and possibly due to the presence of clotting enzymes which originated from the stomachs. A soft coagulum is formed, and some of the liquid phase of milk (whey) is absorbed into the skin, or seeps through and is lost by evaporation. The coagulum is concentrated further by hand-squeezing and sun drying. This dairy product was found to have better keeping quality compared to the original milk due to higher concentration of lactic acid, i.e. limiting or preventing the growth of bacteria producing severe taints, etc. However, longer shelf-life was achieved by preserving the concentrated curd in a salt solution (brine) which also improved its palatability. This fermented dairy product was later known as 'pickled cheese', which is still manufactured in parts of the Middle East, e.g. Feta, Domiati or Halloumi cheeses.

As pickled cheese became popular in the Eastern Mediterranean region, its popularity spread to western countries, e.g. in some parts of Europe via tradesmen from the east. It is possible to suggest that as 'pickled cheese' became an acceptable dairy product in Europe, efforts were made to learn how to manufacture such cheeses locally. With the establishment of dairies in Europe, i.e. in a comparatively colder climate than the Middle East, the preservation of cheese in brine was replaced by partial brining, e.g. Dutch cheese variety, or by using dry salting because the curd produced was much drier. It could be argued that because local manipulations in cheesemaking methods had taken place, the soft cheese has evolved into a new kind of cheese, i.e. the semi-hard cheese varieties. Furthermore, the production of even drier curd cheese resulted in the production of different types, e.g. hard pressed varieties, which can be stored for a long period of time at ambient temperature. Typical examples of such products are Cheddar cheese and other British territorial varieties.

CHEESE NAMES AND NOMENCLATURE

Throughout the world the names applied to cheese could almost reach 2000 (Scott, 1981). In the United States of America (USDA, 1978) more than 800 cheese varieties have been described, but some of these cheeses have different local names and are practically the same; thus, a more accurate list would include no more than 400 varieties. This confirms the view of Davis (1965). More recently, the International Dairy Federation (IDF, 1981) has produced a catalogue in collaboration with its National Committees describing 510 different cheeses, and according to Davis (1965) the exact origin of cheese names could be attributed to the following:

— region and/or towns of cheese manufacture;
— religious institutions;
— type of milk (cow, goat, sheep or buffalo);
— borrowed or made-up names;
— shape, appearance, type of cheese, method of ripening or the addition of additives.

National and international bodies have been involved for the past few decades in preparing or improving the existing specifications of cheese. Different standards exist in each country, and the majority of regulations take into account the following factors: chemical composition, method of manufacture, coagulation of milk, species of starter bacteria and/or method of ripening. These standards are essential on a national level both to the practical cheesemaker and the consumer; however, on the international level, it facilitates marketing of cheese between different countries.

It is clear that the above criteria are used to discuss the definition and classification of cheese, and it is appropriate to consider, therefore, some of the manufacturing procedures available, and to assess the relevance of these different processes in relation to the end product.

DEFINITION OF CHEESE

The word 'cheese' in the English language is presumably derived from different sources. For example, in Urdi (chiz) means the 'perfect thing', and 'cheese' could have come into the English language by Anglo-Indian route, from the Old English (cese), or from Latin (caseus). Cheese in other languages is also derived from Latin, e.g. Irish (cais), German (käse), Dutch (kaas), Spanish (queso) and Portuguese (quijo). In some languages, i.e. French and Italian (fromage and formaggio respectively), the word cheese is derived from the Latin word 'forma' meaning shape or form.

In the United Kingdom, The Food and Drugs Act of 1955 provides a detailed description of the cheese regulations; these regulations may differ slightly, or come into operation at different times in Scotland, England and Wales and Northern Ireland. For example, The Cheese (Scotland) Regulations (1966/98-S.8), (1967/93-S.8), (1979/108-S.4), (1974/1337-S.115) and (1984/847-S.84) provide the definition of cheese, and a summary of these regulations is shown in Table II. A list of 29 varieties of cheese has been provided, and includes detailed chemical analysis, e.g. the permitted level of fat (expressed as minimum per cent of milk fat in dry matter (FDM)), and the moisture content (expressed as maximum per cent of water). Some other analytical qualities in cheese which are of importance, but not covered in these regulations are: the salt percentage (in some instances it is calculated as per cent salt in water) and the percentage of moisture in fat free cheese (MFFC).

Other ingredients, which are sometimes used during the manufacture of cheese and processed cheese, are colour additives, emulsifiers and stabilisers, preservatives and enzymes, and the relevant regulations and/or recommendations are as follows: The Colouring Matter in Food (Scotland) Regulations (1973/1310-S.100), (1975/1595-S.229), (1976/2232-S.184) and (1979/107-S.12); The Emulsifiers and Stabilisers in Food (Scotland) Regulations (1980/1888-S.175), (1982/514-S.65) and (1983/1815-S.171); The Food Labelling (Scotland) Regulations (1981/137-S.20) and (1982/1779-S.192); The Miscellaneous Additives in Food (Scotland) Regulations (1980/1889-S.176) and (1982/515-S.66); The Preservative in Food (Scotland) Regulations (1979/1073-S.96) and the UK Food Additives and Contaminants Report (1982).

The latest amendment to The Cheese (Scotland) Regulations (1984/847-S.84) covers the specifications of a few more cheese varieties including Sage Derby, Mozzarella, Pecorino, Romano and Feta. The latter type has been recently produced in Scotland at the Galloway factory (Scottish Milk Marketing Board), and the estimated production in 1984 was in excess of 5500 tonnes (Hynd, 1984).

TABLE II
A summary of cheese regulations in the United Kingdom

Type of cheese	Fat in dry matter (%)	Milk fat (%)	Maximum water (%)
1. Hard cheese*			
Full fat	≥48	—	48
Medium fat	<48->10	—	48
Skimmed milk	≤10	—	48
2. Soft cheese			
Full fat	—	≥20	60
Medium fat	—	<20->10	70
Low fat	—	<10->2	80
Skimmed milk	—	≤2	80
3. Cream cheese			
Cream	—	≥45	—
Double cream	—	≥65	—
4. Whey cheese			
Full fat	≥33	—	—
Whey	<33->10	—	—
Skimmed whey	≤10	—	—
5. Processed cheese*			
Full fat	≥48	—	48
Medium fat	<48->10	—	48
Skimmed milk	≤10	—	48
6. Cheese spread	—	≥20	60

*After 1st January, 1973 any hard or pressed cheese may have an alternative description to the above regulation, e.g.:

x % fat in dry matter (minimum);
y % moisture (maximum);
z % milk fat (minimum).

However, for exact specification of FDM and moisture content in cheese refer to the cheese variety list in the regulations or Tables VII, VIII and IX.

CLASSIFICATION OF CHEESE

The classification of cheese may vary from one country to another, and different systems have been used which may include one or more of the following technical aspects:

(a) Type of milk.
(b) Shape and weight of cheese.
(c) Type of rind.
(d) Method of coagulation.
(e) Consistency of cheese.
(f) Fat content.
(g) Method of preparation and maturation.

Milk from different species of mammals has also been used, and the world production figures of these types of milk are shown in Table III. However, the data of world production of cheese (Table IV) illustrate that around 87 per cent of cheese is manufactured in north America, Europe and Oceania, which are the major producers of cow's milk. It is possible to suggest that the word 'cheese' could be reserved for the product manufactured from cow's milk, and the cheese produced from milk of other species of mammals should imply the type used, e.g. buffalo's, sheep's or goat's cheeses. A typical example is the labelling of some French soft cheese varieties produced from different milks.

The specifications regarding the shape and weight of cheese could provide some useful information, but it would appear that the data is of limited value to a cheese scientist or technologist, because most of the cheeses could be produced in different shapes or weights. However, the type of rind does provide some technical knowledge regarding the starter culture, particularly in respect of the soft cheese varieties, e.g. surface mould or slime and blue-vein cheeses.

The coagulation of milk could be achieved by one of the following methods: firstly, the use of acids (e.g. mainly lactic acid by the starter organism for the production of Sauermilchkäse, Fromage Frais and Quark), secondly, the addition of a coagulant (e.g. rennet only for the manufacture of Domiati cheese) and thirdly, a combination of acid and a coagulant which is widely employed for the production of most cheeses. It is apparent, therefore, that such an approach to the classification of cheese would be rather limited.

A scheme for the classification of cheese, which is accepted by both cheesemakers and scientists, is based mainly on details of the method of manufacture and chemical analysis of the product. In view of the scheme proposed by the FAO/WHO (1978), some existing cheese regulations, and the IDF (1981), a generalised method for the classification of cheese is illustrated in Table V. It can be observed that such an approach to the classification of cheese is applicable to all varieties, and

TABLE III

Production figures of milk of different species of mammals in different parts of the world (million tonnes)

	Cow					Buffalo					Sheep					Goat				
	1970	1980	1981	1982	1983	1970	1980	1981	1982	1983	1970	1980	1981	1982	1983	1970	1980	1981	1982	1983
Africa	9·8	10·3	10·5	10·4	10·7	1·0	1·2	1·3	1·2	1·3	0·7	0·7	0·7	0·7	0·7	1·5	1·4	1·4	1·4	1·4
America (N and S)	66·8	76·5	78·6	80·3	82·4	—	—	—	—	—	—	—	—	—	—	0·2	0·3	0·3	0·3	0·3
America (S)	18·0	23·0	23·5	24·0	24·3	—	—	—	—	—	0·02	0·03	0·04	0·04	0·04	0·1	0·1	0·1	0·1	0·1
Asia	27·8	37·0	38·1	39·8	41·3	23·1	25·8	26·5	28·3	29·5	3·0	3·4	3·6	3·9	4·1	3·2	3·6	3·7	3·4	3·5
Europe*	230·0	267·9	265·6	270·9	282·7	0·05	0·09	0·09	0·09	0·09	3·0	3·4	3·6	3·6	3·6	2·1	1·9	2·0	1·9	1·9
Oceania	13·5	12·4	11·9	12·0	12·3	—	—	—	—	—	—	—	—	—	—	—	—	—	—	—
World	366·3	427·1	428·2	437·4	453·7	24·15	27·09	27·89	29·59	30·89	6·27	7·73	7·94	8·24	8·44	7·1	7·3	7·5	7·1	7·2

*Data include production figures in USSR.
After FAO (1972, 1982, 1984).

TABLE IV
World production figures of all types of cheeses (thousand tonnes)

	1970	1980	1981	1982	1983
Africa	246	364	370	369	372
America (N & C)	1 557	2 482	2 586	2 720	2 839
America (S)	341	463	457	457	443
Asia	1 917	672	688	789	820
Europe*	4 258	7 116	7 291	7 450	7 605
Oceania	177	254	221	264	274
World	8 496	11 351	11 613	12 049	12 353

*Data include production figures in the USSR.
After FAO (1972, 1982, 1984).

an illustrated example regarding the description of cheese in terms of these categories (I to V) could be as shown in Table VI.

It is impossible to describe in detail all known cheese varieties, and in view of the scheme of classification illustrated in Table V it was decided that some of these varieties (very hard, hard and semi-hard) would be discussed in this chapter, and the semi-soft and soft/fresh cheeses, including Stilton would be dealt with in Chapter 3.

CHEESE SPECIFICATIONS AND STANDARDS

The primary objectives of cheese standards in any one country are to protect the health of the consumer, to produce a quality product, and to describe precisely individual cheeses so as to help ensure fair practices in international trade. The FAO/WHO Codex Alimentarius Commission, which consists of 122 member countries, has been established to provide standards for the major milk products including cheese. The recommended international standards for 25 cheeses and their acceptance by 19 government bodies have been published by FAO/WHO (1972), and the latest report (FAO/WHO, 1984) provides an up-to-date specification for 35 cheeses.

More recently the 29 National Committees of the International Dairy Federation (IDF, 1981) made available information regarding 510 cheese varieties produced in their countries. A selection of cheese

TABLE V
Classification of cheese

	I Consistency		II Fat content		III Moisture		IV Scalding		V
Firmness	MFFC (%)	Designation	FDM (%)		content (%)		temperature (°C)		Method of ripening	
1. Very hard	<51	High	>60	Very high	55–80	High	55	Starter bacteria → No gas holes / With gas holes	
2. Hard	49–56	Full	45–60	High	45–55	Medium	40		
3. Semi-hard	54–63	Medium	25–45	Medium	34–45	Low	35	Mould → Surface (white) / Internal (blue) / Surface (white) + Internal (blue)	
4. Semi-soft	61–69	Low	10–25	Low	<34	No scald	30		
5. Soft/fresh	>67	Skim	<10					Miscellaneous (surface slime) Unripened	

MFFC: moisture in fat free cheese.
FDM: fat in dry matter.
Adapted from Kosikowski (1977), FAO/WHO (1978) and Scott (1981).

TABLE VI

Cheese type	I MFFC	II FDM	III Moisture	IV Scald	V Ripening
Parmesan	Very hard	Medium	Low	High	Starter (no gas holes)
Cheddar	Hard	Full	Medium	Medium	
Cheshire	Hard	Full	Medium	Low	
Gouda	Semi-hard	Full	Medium	Low	
Roquefort	Semi-soft	Full	High	No scald	Mould (internal)
Cottage	Soft/fresh	Low	Very high	High	Unripened
Brick	Semi-hard	Full	Medium	Low	Miscellaneous
Emmental	Hard	Full	Medium	High	Starter (with gas holes)

specifications and/or standards (very hard, hard and semi-hard cheeses) are illustrated in Tables VII, VIII, and IX). In general, the compositional quality of an individual variety of cheese is rather similar around the world, but differences still exist in different countries. For example, Cheddar cheese (Table VIII) the FDM (min. per cent), moisture (max. per cent) and MFFC (mean per cent) ranged between 48 and 50, 37 and 39, and 50 and 56 respectively. The majority of Cheddar cheese is, therefore, manufactured in different parts of the world to comply with these specifications, but such standards are somewhat limited, because the specifications do not provide the cheesemaker with adequate parameters to produce a 'quality' Cheddar. Under commercial practice, for example, for a long holding, mature Cheddar or cheese intended for export, i.e. from New Zealand and Australia, the chemical composition of such cheese ought to consist of 50 per cent FDM (minimum), 37 per cent moisture (maximum) and 53-55 per cent MFFC.

WORLD PRODUCTION AND MARKETING OF CHEESE

In 1983 the world production figure for cheese was 12·35 million tonnes, and Table IV shows the trend of cheese production in various continents. It can be observed that since 1970 cheese production has increased by around 4 million tonnes, and 87 per cent of the cheese has been produced in Europe, north America and Oceania. It is difficult to obtain an exact breakdown of all the cheese varieties produced in the world; however, Eck (1984) reported that, in 1981, hard cheese types (e.g. Cheddar, Emmental and Gruyère) and the semi-hard cheeses, i.e. Gouda and Edam, made up about 60 per cent of total world output. World cheese production over the last four years, has risen by 3·1 per cent (IDF, 1984b), and it is estimated that, in 1984, 40 per cent of the world's milk supply would be utilised for the manufacture of cheese, compared with butter (30 per cent), liquid milk and others (25 per cent) and preserved milk (5 per cent) (IDF, 1982b). However, assuming that nothing dramatic happens to milk production, and taking into consideration the trend of milk utilisation during the period of 1974–80, cheese and butter in 1990 will remain the largest utilisers of milk, e.g. 34 and 33 per cent respectively (IDF, 1982b; Schelhaas, 1982).

In a recent report (IDF, 1982a), the structure of the international

TABLE VII

Specifications of very hard cheese varieties

Cheese variety	Country	Raw material	Description of cheese		Weight (kg)	Chemical analysis (%)		
			Interior	Exterior		FDM (min)	Moisture (max)	MFFC (mean)
Parmesan	Australia		IO	HDR	15–20	32	32	43·7
	Canada		NO	HDR	2·27–22·7	32	32	41·1
	Japan	Cow's milk	NO	SDR	15	35	30	41·1
	New Zealand		IO		10	32	32	43·0
Parmeasao	Brazil		IO		4–20	35	30	46·2
*Parmiggiano Regiano ●○	Italy	Cow's milk/Cn	SRO/NO		30	32	33	54·1
*Grana Padano ●○	Italy	Cow's milk/Cn	SRO/NO		24–40	32	34	52·3
Montasio	Belgium		IO	HDR	7	35	30	46·7
Romano	Australia	Cow's milk	IO		3–6	38	35	48·6
	New Zealand		IO		10	38	34	54·5
	Canada	Cow's/goat's milk	NO		5·4–22·7	37·8	34	45·3
Bra	Canada	Cow's milk	NO		0·9–5·4	40·5	30	48·6

			SRO/NO	SRSS				
*Pecorino Siciliano ● ○	Italy	Sheep's milk/Cn	NO	SDR	4–12	40	33	56·9
*Pecorino Romano ● ○	Italy	Sheep's milk	NO	NR	8–20	36	33	53·2
*Fløtemysost ●	Norway	Cow's milk/Cn/La/Lg	NO	NR	0·25–4	33	20	26·2
*Geitost ●	Norway	Sheep's milk/Cn/La/Lg	NO	NR	0·2–4	33	20	26·2
*Getost ●	Sweden				0·2–0·5	30	19	—
*Gudbrands-dalsost ●	Norway				0·2–4	35	20	25·7

*Statutory standards.
● Cheese originally produced in country indicated.
○ Cheese with international or national protection of origin.
FDM: fat in dry matter.
MFFC: moisture in fat free cheese.
Cn: casein.
La: lactalbumin.
Lg: lactoglobulin.
Data compiled from IDF (1981).

SRO: small round opening.
NO: no opening.
IO: irregular opening.
HDR: hard dry rind.
SDR: soft dry rind.
NR: no rind.
SRSS: soft rind with smeary surface.

TABLE VIII

Specifications of hard cheese varieties

Cheese variety	Country	Description of cheese			Chemical analysis (%)		
		Interior	Exterior	Weight (kg)	FDM (min)	Moisture (max)	MFFC (mean)
*Emmentaler	Austria			60		38	51·9
*Emmenthaler	Denmark	LRO	HDR	—		40	53·0
*Emmental	France			45–130		38	52·7
*Emmentaler/ Emmental ●○	Switzerland			60–130		38	52·5
Gruyère	Canada			18:15	45	38	52·8
*Gruyère	France	MSRO	HRSS	20–45		38	52·7
Gruyère	Poland			30		40	55·7
*Gruyère Greyèrzer Gruviera ●○	Switzerland			20–45		38	52·3
*Cheddar	Australia	IO/NO	HDR/NR	4·5–36	50	38	53·4

	Country						
	Canada		NR	8–18·5	50·8	39	56·5
*	Denmark	NO	NR	—	50	38	50–54
	France		HDR	30–35	50	39	56·0
	Ireland		HDR	—	48	36	52·9
*	New Zealand		HDR/NR	2·25–36	50	37	55·1
•	UK		HDR/NR	4–30	48	39	55·2
Cheshire	Australia	IO	NR	19–20		44	57·1
	Ireland	NO	SDR	—		44	60·3
	New Zealand	IO	SDR	20		44	55·5
*	UK	IO	HDR/NR	4–20	48	44	60·2
• *Derby	UK		HDR	4–15		42	58·2
• *D. Gloucester	UK	NO	HDR	4–20		44	60·2
• *Leicester	UK			4–15		42	58·2
• *Wensleydale	UK		HDR/NR	4–10		46	62·1

For abbreviations see Footnote to Table VII.
LRO: large round opening.
MSRO: medium sized round opening.
HRSS: hard rind with smeary surface.
Cheshire cheese (Ireland and New Zealand) is referred to as semi-hard variety.
All of the above cheeses are produced from full fat cow's milk.
Data compiled from IDF (1981).

TABLE IX
Specifications of some semi-hard cheese varieties

| Cheese variety | Country | Description of cheese | | | Chemical analysis (%) | | |
		Interior	Exterior	Weight (kg)	FDM (min)	Moisture (max)	MFFC (mean)
Caerphilly	Ireland	NO	SDR	—	48	46	62·1
•	UK	NO	SRWM	4·5	40	48	54·5
Edam	Australia	SRO	HDR	3	40·7	46	59·0
	Canada	SRO	SRP/NR	18	40	48	60·5
	France	SRO/NO	HDR/SRP	1·7–2·5	40	45–47	57–59
*	Netherlands†	SRO/NO	HDR/NR	0·8– >6	40	45	56
	New Zealand	SRO	HDR	5	40	45	54·1
Gouda	Australia	MSRO	HDR/NR	5–10	48	45	59·7
	Canada	SRO	SRP	1·4–18·2	49·1	43	61·1
	France	SRO/NO	HDR/SDR	4–5	48	45	57·6–61·1
•	Netherlands†	SRO/SDR/NR	HDR/SDR/NR	0·18– >6	48	41·5–45·5	56–58
	New Zealand	SRO	HDR	10	46	45	58–62
*Fynbo	Denmark†	MSRO	HDR	—	30–45	46–51	58–62
*Tybo •	Denmark†	MSRO	HDR	—	30–45	46–54	65·3
*St. Paulin	Belgium	NO	SDR	1·75	45	52	63
	Canada	NO	SRSS/SRP	0·45–2·26	50	46	66·7
*	France	NO	SRSS	1–3·2	40	56	62·8
	Norway	SRO/NO	SDR	1·5	45	50	

For abbreviations see Footnote to Tables VII and VIII.   †Illustrate different specification within a cheese variety.
SRP: soft rind with paraffin.   Data compiled from IDF (1981).
SRWM: soft rind with white mould.
All the above cheeses are produced from full fat cow's milk.

cheese trade, i.e. imports and exports, has been briefly analysed. Over the past decade, the cheese imports have been dominated by north America, western Europe and Japan. However, this trend is changing, and between 1975 and 1980, cheese imports to these regions have been reduced from 68 per cent to 56 per cent of total world cheese imports, and during the same period, imports to Iran have increased more than five-fold, i.e. from 2 per cent to 11 per cent. According to FAO figures for 1980 (IDF, 1982a), the total world cheese imports were 1·37 million tonnes, and the 10 major importing countries (e.g. Belgium/Luxemburg, France, Germany, Iran, Italy, Japan, Netherlands, Saudi Arabia, UK and USA) accounted for 77 per cent of the total world imports. The most significant varieties of cheese imported are similar to the types reported by Eck (1984) plus Feta (to Iran and Saudi Arabia) and to a lesser degree some semi-soft and soft/fresh cheeses.

In 1980, the 10 largest cheese exporting countries (e.g. Australia, Austria, Denmark, France, Finland, Germany, Ireland, Netherlands and New Zealand) accounted for 86·7 per cent of total exports (IDF, 1982a). The cheeses mentioned above are the most significant types exported, including some processed cheese.

CHEESE CONSUMPTION

The consumption of different varieties of cheese varies from one country to another, and Table X shows *per capita* annual consumption of all cheeses in some European countries. It is evident that there is a steady and uniform increase in the consumption of cheese. Such a trend can be used to classify these countries into two groups: firstly, countries of low *per capita* cheese consumption (e.g. <10 kg/head), and secondly, countries of high *per capita* cheese consumption (e.g. >10 kg/head). The reason(s) for such differences in cheese consumed in these countries is highly complex, and could be due to the following aspects.

Eating Habits and Personal Preference

In some countries (Germany, Iceland, Iran, Israel and Poland), soft/fresh cheeses constitute a large proportion of total cheese consumption, and food habits, once formed, are difficult to break. However, there is a significant correlation between cheese and wine consumption, and no correlation between cheese and milk consumption. For example, in

TABLE X
Per capita annual consumption of all cheeses (kg/head)

Country	Year			
	1970	1980	1981	1982
Australia	3·7	6·6	6·6	7·0
Austria	5·8	8·2	8·7	9·1
Belgium	8·8	13·4	13·6	13·9
Brazil	NA	1·5	NA	NA
Canada	6·3	8·8	9·2	8·9
Chile	NA	1·7	2·1	1·8
Czechoslovakia	5·5	9·0	10·2	10·4
Denmark	9·5	9·6	10·9	10·7
Finland	4·5	7·9	8·6	9·0
France	14·0	18·4	18·9	19·3
Germany	9·9	13·7	14·1	14·4
Iceland	8·8	14·0	14·4	16·1
Ireland	2·5	3·3	3·6	3·4
Israel	10·3	13·2	13·8	14·9
Italy	NA	14·2	14·3	15·2
Japan	0·4	0·8	0·7	0·7
Luxemburg	8·3	9·6	10·2	10·4
Netherlands	8·4	13·1	13·4	13·7
New Zealand	4·3	8·9	8·5	9·3
Norway	8·8	12·4	12·5	12·2
Poland	7·5	11·5	12·3	11·7
South Africa	1·2	1·3	1·3	1·2
Spain	3·0	3·9	3·8	3·5
Sweden	9·1	13·7	13·9	NA
Switzerland	10·0	13·4	13·2	13·3
UK	5·4	5·7	6·4	6·7
USA	7·6	10·1	9·8	10·1
USSR	4·0	4·8	4·6	4·8

NA: not available.
After IDF (1982a, 1983, 1984a).

France, Italy and Germany (cheese consumption is high, but milk consumption is low) the per capita annual wine consumption in 1981 was 92, 87 and 25 litres/head respectively. In Ireland and the United Kingdom (cheese consumption is low, but milk consumption is high), the wine consumption during the same period was 3 and 6 litres/head respectively (Anon., 1982a). Furthermore, in France and Germany, everyday wine was, historically, in plentiful supply and cheap, while milk had a very poor keeping quality, and hence the habit of drinking wine was established. By comparison, in the United Kingdom (until recently), wine was relatively expensive, while milk was cheap and in plentiful supply; thus milk became the preferred drink.

Uses of Cheese

In Scandinavia, Netherlands and France, cheese is consumed at any meal including breakfast, while in the United Kingdom cheese is not normally eaten during the morning meal. Certain dishes, for example fast foods like hamburgers and pizza pies, are major contributors to the increased cheese consumption in the United States of America. It is possible to assume that the low cheese consumption in the UK could be mainly attributed to the fact that cheese is normally served after the dessert (i.e. less cheese is eaten), compared with the custom of most countries where cheese is consumed after the main dish before the dessert (i.e. more cheese could be eaten).

Miscellaneous Factors

Factors such as climatic conditions, availability of different cheese varieties on the market and economic standards can affect the level and type of cheese consumption; health or dietetic reasons may increase fresh and/or low fat cheese consumption; advertisements and other promotional activities could increase consumption; a multitude of other factors, which differ from one country to another, could affect cheese consumption, and a comprehensive study of such factors in some IDF member countries has been recently published (IDF, 1982a).

NUTRIENTS IN CHEESE

The nutritional properties of cheese are excellent, and the major constituents of some cheeses discussed in this chapter are shown in Table XI. The nutrients present in cheese are protein, fat, carbohydrate, vitamins, minerals and salts. The level of these constituents in cheese may vary, due mainly to the quality or type of milk used and the variety of cheese produced. In general, very hard, hard and semi-hard cheeses contain high levels of protein (mainly casein) which is a rich source of essential amino acids required by man. The energy is provided by the fat and carbohydrate fractions; the latter component is low in matured cheese, because most of the lactose is lost in the whey during processing, or has been utilised by the starter culture bacteria for the production of lactic acid. The high figures for lactose reported in Table XI could be attributed to the type of cheese, or to the fact that the cheese was analysed at a relatively 'young' age, i.e. a few days after manufacture. Indeed, some patients, who show symptoms of lactose intolerance when they consume liquid milk, can eat cheese without having any allergic response.

The energy (calorific value) in cheese is mainly derived from the fat, but medical opinion may still regard the fat in cheese as a potential source of coronary heart disease, and physicians may advise their patients not to eat cheese without taking into account the trend of cheese consumption in their country. For example, in the United States of America (cheese consumption is low, and it is equivalent to 25 g day^{-1}), the average recommended daily intake of cholesterol is 500 mg, and cheese only contributes 26 mg of cholesterol/day, i.e. not significant (Speckmann, 1979). The cholesterol content of some cheeses (see Table XI) may be considered rather low, and cheese can still be eaten within certain therapeutic guidelines especially in countries of low cheese consumption.

Cheese is considered to be a good source of certain vitamins, but deficient in ascorbic acid (Vitamin C) which is lost during the manufacturing stages. A varied diet can supplement such a deficiency if cheese is consumed with vegetables, e.g. lettuce.

Cheese contains appreciable quantities of certain elements or minerals (e.g. calcium and phosphorus) which are essential for teeth and bone formation. The salt level, i.e. sodium chloride, in cheese varies with the type of cheese produced, method of salting employed (dry or brining) and/or the amount used. Salt is used as a preservative and flavour enhancer, but recently cheese has been considered by the medical profession as a contributing factor to high blood pressure due to its 'high' salt content. Recently, Lindsay *et al.* (1982) have carried out some trials to produce Cheddar cheese low in salt, or with a mixture of sodium/potassium chloride (1:1 molar basis). The results showed that the mean consumer preference was towards 'salty' cheese, and some bitterness was observed in cheese containing 1·5 per cent salt mixture.

CHEESEMAKING PROCESS

Although there are many varieties of cheese, the basic concepts of

TABLE XI

Proximate composition, energy values, inorganic constituents and vitamins in some cheese varieties (figures are quantities per 100 g of cheese)

Cheese type	Water (g)	Protein (g)	Fat Total (g)	Polyunsaturated	Cholesterol	Carbohydrates (g)	Energy kcal	Energy MJ	a + β carotene (IU)	Thiamine	Riboflavin	B₆	Niacin	Pantothenic acid	Ascorbic acid	Tocopherol	Folic acid (free)	Biotin	B₁₂ (U/g)	Sodium*	Potassium	Calcium	Magnesium	Iron	Copper	Zinc	Phosphorus	Sulphur	Chloride*
												(mg)						(U/g)					Elements (mg)						
Parmesan	28·0	35·1	26·0	0	0·07	Tr	393	1·64	1060	0·02	0·50		0·2	0·3	0	0·9	0	1·7	1·5	760	150	1140	50	0·3	0·36	3	770	250	1110
	30·0	36·0	29·7			2·9	408	1·69			0·73	0·1	0·3			1·0						1220		0·4		4	781		
Emmental	34·9	27·4	30·5	0	0·09	3·4	398	1·67	1140	0·05	0·03	0·09	0·1	0	0·5	1·0	0	0	0	620	100	1180	55	0·9	0·13	4·6	860	0	1210
Cheddar	37·0	26·0	32·2	1·0	0·10	Tr	398	1·68	1310	0·02	0·30	0·05	0·01	0·1				0·4	0·8	610	82	750	25	0·4	0·03	4	480	230	1060
		35·0	33·5			2·1	406			0·08	0·80	0·14	0·2	0·7	0	0·8	6	2·3	1·5	700	120	800	45	1·0	0·05		520		
Edam	43·3	24·4	22·8	0	0·09	Tr	232	1·10	600	0·04	0·35	0·05	0·02	0·1				0·7		737	76	740	28	0·2	0·03	1	455	0	1640
	43·7	26·1	23·6			3·5	304	1·26		0·06	0·40	0·12	0·19	1·3	0	1·0	0	5·1	1·4	980	160	765	59	0·7	1·7	10	520		

*Variable, depends on salt content.
Above figures illustrate ranges reported in literature.
Data compiled from Paul and Southgate (1978), Anon. (1981), Scherz and Kloos (1981).

cheesemaking are similar. In principle, milk from different species of mammals could be used as the main raw ingredient, plus the addition of starter culture and coagulant. Hence, a recipe is used for the manufacture of a specific variety of cheese, and the basic stages of manufacture may include the following:

(a) Milk handling, storage and further processing.
(b) Starter cultures.
(c) Formation of the coagulum, cutting and scalding.
(d) Handling of the curd after de-wheying.
(e) Miscellaneous treatments which include milling, salting, pressing of the curd (incidentally such treatments may not occur in the same order, depending on the type of cheese produced or the mechanised system employed).

Milk Handling, Storage and Processing

Raw milk is normally received at the dairy in bulk tankers, and is cooled and stored until required. The duration and the temperature of storage of the milk prior to cheesemaking can affect the quality of the cheese. Psychotrophic bacteria can grow and survive at ordinary refrigeration temperature, and the peptide hydrolase and lipase producing species can affect both the milk clotting mechanism and the flavour of the cheese. Law *et al.* (1976) reported that cheese milk containing more than 10^7 colony forming units (CFU) ml^{-1} resulted in Cheddar cheese which was rancid after 4 months, and the cheese had a distinctive 'soapy' flavour. More recently, Al-Darwash (1983) studied the relationship between the CFU of psychrotrophic micro-organisms in refrigerated milk and the quality of Cheddar cheese, and he concluded that a temperature of 2 °C was ideal for milk stored for up to 5 days before cheesemaking.

Before further processing, the cheese milk is filtered in order to remove certain contaminants, such as cellular material, straw, hairs or soil etc. Cloth filters are the universal system employed, but centrifugal separation and/or bactofugation is used where the presence of spore-forming organisms in milk can lead to a product loss, e.g. such as during the manufacture of certain European cheese varieties. The bactofugate, i.e. the separated fraction, amounts to 2–3 per cent of the total volume of milk, and it contains both the undesirable micro-organisms and the fat content of the milk. Since the heat treatment of the cheese milk is limited to 72 °C for 15 s, the bactofugate is sterilised by live steam injection at 130–140 °C for a few seconds, and after cooling, is added back to the pasteurised cheese milk.

An alternative approach to bactofugate sterilisation is the addition of sodium or potassium nitrate ($NaNO_3$ or KNO_3 respectively) to the cheese milk in order to inhibit the growth of gas-forming butyric acid bacteria. Kosikowski (1977) reported that $NaNO_3$ was added at a rate of 50–90 g/455 litres of milk, and that Gouda and Edam cheeses may contain 23 mg $NaNO_3$/45.5 g of cheese. However, Scott (1981) reported that the addition of $NaNO_3$ (known as saltpetre) to the milk can cause colour defects in the cheese, due to the reaction of tyrosine with the nitrite (i.e. after reduction of nitrate), and hence may be limited in its use. The addition of $NaNO_3$ is controlled by statutory standards in some countries due to the reported carcinogenic effect of the reaction products of the preservative.

Seasonal variation in the chemical composition of milk has been shown to occur (see Table XII), and such variations can ultimately affect the yield and compositional quality of, for example, Cheddar cheese (Lelievre and Gilles, 1982; Banks *et al.,* 1984a,b). The standardisation of the Cheddar cheese milk to a casein to fat ratio of 0·7 is ideal (Kosikowski, 1977), and the theoretical approach to milk standardisation is mainly dependent on such factors as: firstly the type of equipment used and the efficiency of separation obtained, and secondly the control system used. However, illustrations of some of the cheese milk standardisation systems, which could be used in large dairies, have been recently reported by Scott (1981) and Tamime and Robinson (1985).

Certain additives, such as calcium chloride and colouring matter, are added to the cheese milk after pasteurisation. The former compound is important in order to provide the appropriate balance between the soluble and colloidal calcium in milk which leads to successful coagulation. Al-Obaidi (1980) has shown that a calcium concentration in milk of 140–160 mg/100 ml is ideal, and Scott (1981) has reported that no more than 0·02 per cent calcium chloride is needed for a satisfactory coagulation because larger doses can destabilise some of the casein fractions. The most popular colouring matter used in the dairy industry is the water soluble annatto, which is extracted from the fruit of *Bixa orellana.* The rate of annatto addition per 450 litres may range from:

(a) <35 ml for slightly coloured cheese;
(b) 35–55 ml for medium coloured cheese;
(c) 115–230 ml for highly coloured cheese.

The annatto is normally added to the cheese milk with the starter

A. Y. Tamime

TABLE XII
Seasonal variation in silo raw milk composition(%)

	Month	Total solids	Fat	Solids-not-fat	Protein	Casein	Casein number	Lactose	Ash	Calcium
1982	October	13·03	4·00	9·03	3·45	2·58	74·78	4·73	0·85	0·12
	November	12·65	3·95	8·70	3·20	2·45	76·56	4·67	0·84	0·12
	December	12·60	3·95	8·65	3·13	2·31	78·80	4·70	0·83	0·12
1983	January	12·60	3·90	8·70	3·09	2·31	74·76	4·79	0·82	0·12
	February	12·56	3·80	8·76	3·05	2·28	74·75	4·87	0·84	0·12
	March	12·32	3·90	8·42	3·03	2·27	74·92	4·62	0·82	0·12
	April	12·21	3·85	8·36	2·91	2·24	76·98	4·65	0·80	0·12
	May	12·31	3·60	8·71	3·20	2·38	74·38	4·68	0·84	0·12
	June	12·54	3·53	9·01	3·36	2·51	74·70	4·83	0·84	0·12
	July	12·54	3·73	8·81	3·31	2·50	75·68	4·67	0·84	0·12
	August	12·58	3·79	8·79	3·32	2·49	74·10	4·66	0·82	0·12
	September	12·73	3·83	8·90	3·34	2·50	74·10	4·76	0·81	0·12

Above values are average of two samples per month.
Each milk sample was analysed in duplicate.
Lactose was calculated by difference.
Solids-not-fat was calculated by subtracting the fat value from total solids.
The ash level was rather high due to the potassium dichromate ($K_2Cr_2O_7$) which was added to the milk sample for preservation.
After Tamime (unpublished data).

culture, or alternatively, some cheesemakers add the colouring matter 15 min before the addition of the coagulant. Swiss cheeses manufactured in stainless steel vats rather than the traditional copper kettles have been criticised to be lacking of a typical cheese flavour(s). However, in Finland such defect has been minimised by the addition of 15 parts per million (ppm) of copper sulphate to the cheese milk which is equivalent to the amount of copper absorbed by the milk when using the copper kettles (Lampert, 1975). The role of copper sulphate on the flavour of the cheese could be associated with slight fat lipolysis and/or activation of certain enzymes which are important during the maturation of the cheese.

Starter Cultures

The role of starter cultures in cheesemaking can be summarised as follows. First, the organisms produce (D), (L) or (DL) lactic acid as a result of lactose fermentation which is important during the coagulation and texturising of the curd. Second, the production of flavour compounds and, in some instances, gas, e.g. Swiss cheeses. Third, the starter culture enzymes (peptide hydrolases and lipases) play a major role during the maturation of the very hard, hard and semi-hard cheeses. Fourth, the lactic acid may contribute towards the flavour of the cheese, and the acidic condition prevents the growth of pathogens and many spoilage organisms.

The classification of dairy starter cultures has been reported by Tamime (1981) and Garvie (1984), and the type of cultures employed during the manufacture of the cheese varieties mentioned above are the following:

(a) Mesophilic lactic starters: *Streptococcus lactis, Str. lactis* subsp. *diacetylactis*, Str. cremoris* and *Leuconostoc cremoris** (*these cultures are used as aroma producing bacteria).

(b) Thermophilic lactic starters: *Streptococcus thermophilus, Lactobacillus bulgaricus, Lac. helveticus, Lac. casei,* and *Lac. plantarum* (these cultures are only used during the manufacture of high-scalded cheese, e.g. Swiss varieties).

(c) Miscellaneous bacteria: *Propionibacterium freudenreichii* subsp. *freudenreichii, globosum* or *shermanii* (these organisms are used in conjunction with thermophilic lactic cultures mainly for their ability to produce large gas holes in the cheese during the maturation period).

These sub-species of *Propionibacterium* are differentiated on the following basis:

	Ability to reduce nitrate	Acid from lactose in milk
P. freudenreichii	+	−
P. globosum	+	+
P. shermanii	−	+

Furthermore, these micro-organisms are sometimes used under the synonym of *Propionibacterium casei*, because of their significance in the dairy industry (Moore and Holdeman, 1974). Starter cultures may be preserved in one of the following forms, i.e. liquid, concentrated freeze-dried or concentrated frozen at −196°C. The latter two types can be used for direct-to-vat-inoculation (DVI) for the production of the bulk starter, or added directly to the cheese milk. The rate of acid development using an active liquid culture *vis à vis* DVI during the manufacture of Cheddar cheese is shown in Fig. 1.

The production of the bulk starter in a dairy is an important feature of a successful cheesemaking process. Over the past few decades, starter culture technology has advanced to achieve the ultimate objective of producing a pure and active culture. However, a description of some of these systems, including illustration of the equipment, has been recently provided by Tamime (1981) and Tamime and Robinson (1985).

Coagulants, Coagulum Formation and Cutting of the Coagulum

Coagulants used in the dairy industry can be classified into the following groups:

(a) Animal (rennet and pepsin).
(b) Microbial (*Mucor meihei, Mucor pusillus, Endothia parasitica* and *Bacillus subtilis*).
(c) Plant.

These enzymes are known as peptide hydrolases (proteinases), because of their activity on the casein micelle during the coagulation process of

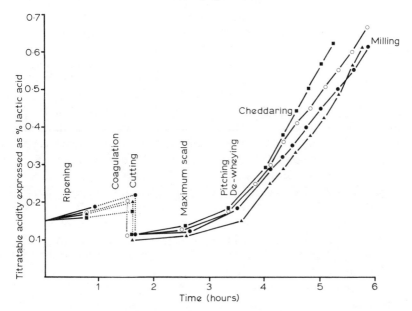

Fig. 1. The rate of acid development during the manufacture of Cheddar cheese using bulk starter culture (Auchincruive 259) and DVI (Miles M34, Hansen 850, Eurozyme MA012). (The cheese was produced in a 2273 litre vat between December 1983 and January 1984 and the curd was cheddared using the New Zealand cheddaring box; all cheeses were awarded 'First Grade' according to the scheme of the Company of Scottish Cheesemakers Ltd.) (After Tamime, unpublished data.)

the milk. Rennet, which consists of chymosin (80 per cent) and pepsin (20 per cent), is extracted from the fourth stomach of a young calf, and this enzyme is widely used in the cheese industry. Other rennet substitutes can be used, but due to the broad specific proteolytic activity of these enzymes, the finished cheese may have texture and flavour faults.

The hydrolytic activity of chymosin on the casein micelle could be divided into three phases. The primary phase of the enzyme is the destabilisation of the κ-casein by hydrolysis of the susceptible amino-acid bond at the 105-106/phenylalanine–methionine linkage. The result is the formation of two components, i.e. para-κ-casein and the casein/glycomacropeptide (Galesloot, 1958). The former fraction is insoluble, highly hydrophobic (1310 cal mol^{-1} residue), and in the presence of divalent ions (mainly calcium, magnesium, phosphates and/or citrates), it aggregates to form the coagulum which is an integral stage of cheese production. However, the casein/glycomacropeptide fraction is soluble, not markedly hydrophobic (1082 cal mol^{-1} residue), and is lost in the whey after cutting (Dalgleish, 1982a). Two aspects are still unclear regarding the specific activity of the enzyme: first, the mechanism of cleavage by the enzyme at the 105-106/Phe-Met, and second, the exact nature and kinetics of coagulum formation. A vast literature exists on the possible action of these enzymes in milk, and it is beyond the scope of this publication to review this topic in detail; however, comprehensive data in this field have been published recently by Cheeseman (1981), Fox and Morrissey (1981), Dalgleish (1982a, b), Fox (1981, 1982a, 1984) and Fox and Mulvihill (1983).

According to Davis (1965) and Fox (1982b) around 90 per cent of the coagulant is lost in the whey, and the rest is retained in the curd. Hence the secondary phase of enzyme activity is partial hydrolysis of the protein during the cheesemaking process, and the tertiary phase, where proteolysis plays a major role during the maturation period, is significant in flavour and texture development. The latter two phases may not be important in Swiss cheese, because the whey/curd mixture is heated above 50°C.

The amount of coagulant (rennet) added to milk is 28·4 ml/114 litres, and the coagulation process is normally carried out at 30°C. It is common practice to dilute the coagulant with water, i.e. 3–4 times its volume, in order to ensure homogeneous distribution of the enzyme in the milk. Factors affecting the efficiency of the coagulant have been recently published by Tofte-Jespersen and Dinesen (1979), and the usual duration of agitation of the milk/coagulant mixture is 5 min.

After the milk has achieved the desired firmness of the coagulum, the curd is cut. Stainless steel knives (vertical and horizontal) are widely used for cutting the coagulum, but recently designed cheese vats employ only vertical knives. The knives are also used as an agitation mechanism, so that the curd is cut with the sharp side of the blade, and stirring of the curd/whey mixture is with the blunt side. However, stainless steel wires, which are known as a 'cheese harp', are still used during the manufacture of Swiss cheeses employing the traditional process, and the sequence of cutting is in the shape of an 'eight'.

The duration of cutting the coagulum is around 20 min, depending on the size of the curd particle required, type of cheese manufactured and/or the size of the vat. For example, 'fine' or 'coarse' size curd particles are used for the manufacture of hard and semi-hard cheese varieties respectively. The coagulum is cut at a very low speed, and the speed is progressively increased to avoid fat and casein losses in the whey which could influence the yield of cheese.

During the manufacture of certain cheese varieties, e.g. Gouda, Edam, Fynbo and Tybo cheeses, 25–30 per cent of the whey is removed and replaced by hot water. The primary objective is to reduce the lactose content in the whey and the production of acid in the cheese.

Maximum Scald and Pitching

After the coagulum has been cut, two phases, i.e. whey and curd particles, become apparent in the cheese vat. The treatment of the curd/whey mixture at this stage varies in relation to the type of cheese produced, and a summary of such treatments, including the scalding temperatures employed, is shown in Table XIII. On completion of cutting the coagulum, the application of heat and/or the partial removal of the whey, physical and chemical changes in the curd and whey phases start to occur, and according to Czulak (1981) and Borcherds (1983) the relevant changes are as follows:

(a) A semi-permeable membrane is formed across the outer layer of the curd particle, and as the curd is stirred and the temperature is raised, the casein network continues to alter which results in the formation of a semi-rigid framework, i.e. fibrous connections. Incidentally, the milk solids retained in the curd particle constitute the main cheese components.

(b) Due to the combined action of the application of heat and the development of the network of casein micelles, syneresis (i.e. expulsion of whey entrapped in the curd particle) becomes evident, and the curd particles shrink in size.

(c) Most of the lactic acid is produced within the curd particle, because the starter bacteria are embedded in the casein framework. The excess lactic acid permeates through the membrane to the whey, so assisting in further shrinkage of the curd particles. Furthermore, the level of acid is higher inside the curd particle compared with the whey, but not the pH, due to the buffering effect of the milk solids.

(d) An equilibrium between the lactose and mineral salts in the curd particles and the whey becomes evident due to first, as the lactose is utilised by the starter bacteria, the level is reduced, and lactose from the whey permeates through the membrane to the curd particle; and second, due to the development of lactic acid inside

A. Y. Tamime

TABLE XIII
Differences in the treatment of the curd during the manufacture of certain cheese varieties

Variety of cheese	Size of cut coagulum	Temperature of scalding and other treatments
		I. High scalding temperature
Grana Padona	3 mm	Scald the curd/whey mixture to 58°C; de-whey and press the curd followed by brining and dry salting
Gruyère and Emmental	10 mm	After cutting, stir for 30 min and scald to 52–54°C in around 30 min; pitch for further 30–60 min, remove curd from whey and press; salt the cheese by brining and dry salting
		II. Medium scalding temperature
Parmesan	3–4 mm	Raise temperature from 32° to 42°C in 30 min; settle curd and de-whey at 0·2% l.a., place the matted curd in moulds and press; float cheese in brine and dry salting
Cheddar	6–8 mm	Raise temperature from 30°C to 39–40°C in 45 min and pitch for the same period; de-whey when the acidity of the pressed curd is 0·2% l.a.; form curd into 20 cm blocks, pile and press; mill at 0·65–0·68% l.a., salt at a rate of 2½–3% (estimated w/w), fill in moulds (20 kg) and press overnight
Dunlop	8 mm	Similar to Cheddar but scalding temperature is 36–37°C and less salt is added; the cheese (20 kg) is pressed overnight
		III. Low scalding temperature
Cheshire	>10 mm	Cut the coagulum coarse and scald to 32–34·5°C in 45–50 min and draw off the whey when acidity reaches up to 0·23% l.a.; cut the curd into 7·5 cm blocks, turn over and cut by hand every 20 min (*do not* pile the blocks of curd on top of each other); mill when the acidity reaches 0·65–0·7% l.a. using a peg mill and salt at a rate of 2%; press cheese overnight

Wensleydale	>10 mm	Cut the coagulum coarse and the first 15–20 min do not apply any heat; scald the whey/curd mixture to 35 °C in 15–20 min followed by idle stirring for 15–20 min; de-whey when acidity reaches 0·2, form into blocks and turn several times until acidity reaches 0·55% l.a.; cut the blocks in half, mill at 0·6% l.a. and salt at a rate of 1·8%; small size cheese is pressed for a few hours
Lancashire	>10 mm	This cheese is produced in a similar way to Wensleydale but the curd is formed into 15 cm blocks and turned continuously until the acidity reaches 0·5% l.a.; cut the blocks into smaller pieces and mill at 0·6% l.a.; salt at a rate of 2% and press overnight
Caerphilly	5 mm	Cut the coagulum and scald to 35 °C in 15–20 min, stir and de-whey when acidity reaches 0·2% l.a.; form blocks and pile along the side of vat; mill at appr. 0·3% l.a., salt at a rate of 1%, press overnight and place in a brine bath up to 24 h
Gouda/Fynbo	>12 mm	Cut coagulum, stir and de-whey 25–30% of the volume and replace with water; scald to 35 °C and drain whey when curd is consolidated, press, brine and store (curd making to pressing 60 min, pressing 90 min, brining 4–5 days); Fynbo cheese is scalded to a slightly higher temperature with brining for 2 days only
		IV. No scalding
Edam	>12 mm	Similar to Gouda but major differences are: remove only 20–25% of whey, scald to 30–31 °C, curd making to pressing 50 min; pressing time 75 min and brining for 65 h

Data compiled from: Anon. (1959), Davis (1976), Kosikowski (1977), Vries and Ginkel (1980), Scott (1981), Dijkstra (1984).

the curd particle, some of the divalent ions (i.e. bound to the casein micelles) become free and permeate to the whey.

(e) The scalding temperature may range between $>31\,^{\circ}$C and $\leqslant 55\,^{\circ}$C, and at high temperatures and acidity, the rate of syneresis from the curd particle will be reduced due to the plasticising of the curd, e.g. Swiss and Italian cheeses.

(f) Partial removal of the whey and the addition of water, for example during the manufacture of Dutch cheeses, retards syneresis, perhaps due to a lowering of the level of lactose in the whey, hence upsetting the lactose equilibrium between the curd and the whey.

(g) The continuous stirring of the curd/whey mixture during the scalding and the pitching periods exerts a physical force which expels moisture from the curd particles.

(h) High acidity at rennetting, scalding the curd/whey mixture very quickly, and cutting the coagulum when it is too firm can interfere in the rate of syneresis or moisture retention in the curd particle.

De-wheying and Curd Handling

When acid development in the curd has reached the desired level, whey is drained off in order to allow the texturising of the curd. Different treatments are employed in relation to the type of cheese produced, and some typical examples are illustrated in Table XIII. In brief, the consistency of the cheese curd could be classified as given in Table XIV, and it is evident that the temperature during scalding, level of acid in the

TABLE XIV

Nature of curd particles	Description of texture	Cheese variety
Granular	Plasticised	Gruyère and Emmental
Texturised (in some instances it is granular, e.g. American Cheddar)	Close	Cheddar and Dunlop
Slightly texturised	Coarse/Crumbly	Cheshire
Granular	Elastic	Gouda, Edam and Fynbo

cheese, method of handling the coagulum, amount of pressure applied at the pressing stage and/or the amount of moisture retained play a major role in the body characteristics of the final cheese.

Milling, Salting/Brining and Pressing

Milling of the texturised/matted curd must be carried out at the right acidity, e.g. 0·6–0·7 per cent lactic acid in Cheddar cheese; these acidities at milling are normally recommended when using DVI or bulk starter cultures respectively. Passing the mellowed curd through the mill (chip or peg type) merely cuts it into smaller pieces, so increasing its surface area so that salt can be applied more evenly. In some instances, i.e. Cheddar cheese produced in a Tebel Crockatt system, the curd is normally salted prior to milling.

Salting of the curd is achieved in two different ways: first, the addition of dry salt and second, brining, where the cheese, after pressing, is immersed in a brine solution. The function of salting could be described as follows:

— the salt acts as a preservative and flavour enhancer in the cheese;
— it helps to reduce the metabolic activity of the starter culture and assists in the liberation of their enzymes which play a major role in flavour development in the cheese;
— salting (dry or wet) helps to reduce the moisture content in the curd as a result of differences in osmotic pressure (for example Cheddar cheese curd at milling may contain <40 per cent moisture and after pressing ⩾35 per cent moisture; 60 per cent of the added salt is retained in the cheese and 40 per cent is lost in the pressed whey and on equipment (Davis, 1965). In Gouda and Edam cheeses, the moisture contents after pressing are 46·5 per cent and 51·5 per cent respectively, and after brining, the moisture levels in the cheeses are reduced to 41·8 per cent and 44·7 per cent respectively (Dijkstra, 1984);
— salt suppresses the growth of undesirable micro-organisms in the cheese;
— it helps to alter the physico-chemical characteristics of the curd;
— dry salting of the curd at high temperature can increase fat losses in the whey which can affect the yield, and the cheese may become greasy after pressing.

The duration of salting may vary from 15–20 min (i.e. dry salting of Cheddar), to a few days, for example the brining of Dutch and Swiss cheeses.

Pressing of the salted curd in moulds assists in removing more moisture from the curd, and helps to form the final shape of the cheese. The pressing of the cheese is influenced by many factors, such as:

(a) Amount of pressure applied is dependent on the cheese variety. Cheeses with low moisture require high pressures to consolidate the curd to the desired structure.

(b) Pressure should be applied gradually otherwise whey retention in the cheese will be higher due to premature rind formation. This leads to higher retention of lactose and results in a lower pH in the cheese.

(c) Duration of pressure is dependent on the type of cheese and the method of pressing system employed (see section on mechanisation).

(d) Pressing the curd at high temperature increases fat losses in the whey, reduces the rate of exudation of the whey due to rapid rind formation, and the fat is in the 'liquid' state which can impede the fusion of the milled curd.

(e) Open texture in the cheese could be associated with faulty pressing equipment, or the curd being pressed at low temperature.

(f) During pressing, better fusion of the curd particles and faster removal of the whey is more easily achieved with smaller curd particles than with larger pieces.

Miscellaneous Handling and Storage

The preparation of the 'green' cheese after pressing, for bulk storage and maturation is an important aspect of production, and one of two different approaches is usually adopted. The conventional process involves the following stages of operations:

— scalding and re-pressing of the cheese to form a harder rind;
— drying;
— bandaging/dressing;
— waxing.

The second method is to package the cheese in a barrier film which is impermeable to oxygen and moisture, or permeable to carbon dioxide.

The maturation of the cheese takes place in a controlled atmosphere, e.g. temperature and humidity, and both conditions are relevant to cheeses packaged in the conventional method. The relative humidity is maintained around 85 per cent otherwise the cheese becomes very dry due to the evaporation of moisture. If cheese is packed in a barrier pack, i.e. moisture proof, no evaporation can occur, and only the temperature requires to be controlled. The temperature of maturation of cheese is dependent upon the variety, and while, for example, Cheddar cheese and most of the UK territorial varieties are stored at 10°C, Gouda and Edam are held at 13°C and Swiss cheeses up to 20°C.

Turning of the conventionally packaged cheese in the store is essential during the early stages of maturation in order to overcome certain potential faults, e.g. deformation of shape, rotting of the cheese due to condensation between the cheese and the shelf, and/or aid even distribution of the moisture content throughout the block of cheese.

MECHANISED CHEESEMAKING

The period from 1960 could be considered as one of the most significant as regards to progress that has been achieved in the development and application of mechanised equipment for the manufacture of cheese. According to King (1966), the primary objectives of mechanisation of the cheesemaking process could be summarised as follows:

— increase productivity in a given sized factory;
— reduce the cost of manufacture, i.e. by labour savings;
— improve the working conditions and to avoid heavy manual work;
— improve, if possible, the quality of the cheese.

At present, there are many different types of cheesemaking system available on the market for the manufacture of the very hard, hard and semi-hard varieties. The reason(s) for diversification can be mainly attributed to technological advancements achieved in the manufacture of certain cheeses (i.e. Cheddar, Emmental or Edam), and to the development of different systems to produce the same variety of cheese, e.g. Cheddar. Nevertheless, Crawford (1976) has pointed out, that 'significant in the planning of mechanisation is the establishment of well-defined stages of cheesemaking process as follows':

Stage 1: curd production.
Stage 2: de-wheying, texture forming and/or handling operations.
Stage 3: milling, salting/brining and/or mould filling.
Stage 4: pressing and packaging.

Employing this approach, the recent developments of mechanised cheesemaking will be discussed.

Curd Production

Traditionally, the cheese vat (open type) was utilised for all the operations of cheesemaking, i.e. starter addition to milling and salting, with the exception of the two level system where curd handling (e.g. cheddaring, milling and salting) took place in coolers or finishing vats. However, the major developments in cheese vats have included:

(i)   enclosed vat design (vertical and horizontal);
(ii) cutting/stirring devices;
(iii) optional fittings.

Nearly a decade ago, some types of cheese vat, which had been installed in UK cheese factories (i.e. the Alfa-Laval OST-II; Golden Vale Engineering — the 'W' cheese tanks; Tebel Machinefabrieken — the Tebel Matic; the Wincanton Desco; the Sherborne vat) were described by Crawford (1976), and other cheese vats which have been widely used in different parts of the world are: the Koopmans — Curdmaker; the New Zealand — Curdmaster; and the Alfons Schwarte — Curd making tank (Scott, 1981). The latter type of vat has been recently evaluated at the Netherlands Dairy Research Institute by Vries and Ginkel (1983). Some recently developed cheese vats include:

Damrow 'Double O' vat
The Damrow Company of the USA manufacture the 'Double O' cheese vat, and this type of vat is marketed in Europe by DEC International. This type of vertical cheese vat resembles an open figure of eight, and it can be used for the production of curd for the majority of cheese varieties. The movement and mixing of the curd from one section of the vat to another is illustrated in Fig. 2, and the capacity of the vat ranges from 1000 to 35 000 litres.

The overall specifications of the vat could be described as follows:

— the vat is totally enclosed;

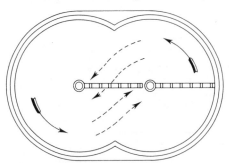

Fig. 2. Horizontal cross-section of the Damrow 'Double O' vat illustrating the direction of the cutting/stirring devices and the movement of the curd. (Bold arrows: direction of rotation, dashed arrows: curd/whey mixture movement towards centre of vat.) (Reproduced by courtesy of DEC Int. A/S — Damrow Products Division, Kolding, Denmark.)

— double blades operate in one direction for stirring and reverse for cutting; Fig. 3 shows the blade assemblies;
— the sequence of operation of the vat, which could be fully or semi-automatic, was described in detail by Crawford (1976);
— the construction of the vat offers improved cheese yields, production of consistent, high quality cheese, and minimises fat and curd 'fines' losses in the whey.

Pasilac CT 2000 cheesemaking vat

The CT 2000 cheesemaking vat is an enclosed horizontal type of tank which is manufactured by Pasilac A/S in Denmark. The capacity of this vat ranges from 6700 to 20 000 litres, and an overall schematic view is shown in Fig. 4. The cheese vat is bottom filled to minimise frothing, and only 46 per cent of the volume is utilised during production. The vat itself is provided with a heating jacket on the lower part of the shell, and a level gauge is fitted to give an exact measure of milk. A horizontal centre shaft is supplied with separate cutting/stirring tools (see Fig. 4), which operates with a pendulous movement of 180 degrees. Immediately after the addition/stirring of the milk coagulant, the shaft assembly is positioned vertically by turning the stirring paddles up and the wire cutters down into the milk. The reverse position of the shaft is achieved by turning the stirring paddles down into the curd/whey mixture after the coagulum has been cut. Both the speed of agitation, and the angle of deflection of the stirring/cutting device can be controlled to permit gentle handling of the coagulum.

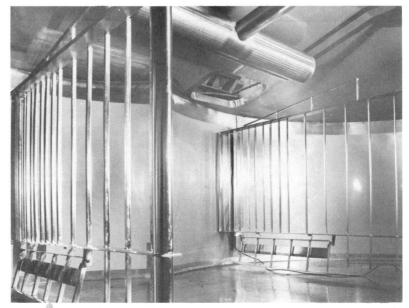

Fig. 3. Inside view of the Damrow 'Double O' vat. (Notice the vertical structure of the cutting/stirring blade assemblies attached to the top-mounted motor and the horizontal strainer is only required to replace some of the whey with water, e.g. during the manufacture of Dutch cheeses.) (Reproduced by courtesy of DEC Int. A/S — Damrow Products Division, Kolding, Denmark.)

The CT 2000 vat is mounted on four pyramidal or conical-shaped feet. The front pair are fitted with swivel joints, and the rear pair are fitted with pneumatic cylinders for tipping the vat when pumping the curd/whey mixture to the de-wheying and texture formation units.

Tebel OST-III and IV

This horizontal and enclosed cheese vat is manufactured by Tebel Machinefabrieken BV in Holland, and is suitable for the production of different cheeses. For example, the OST-CH is developed for the manufacture of Cheddar cheese, and does not contain an advanced whey strainer; this equipment is required for some types of cheese, e.g. Gouda and Edam. The tank is manufactured in five sizes ranging from 10 000 litres to 20 000 litres capacity, and the milk occupies 80–85 per cent of the volume of the tank. The volume of the tank is increased by extending the length of the tank; the diameter, height of liquid, peripheral equipment and drive-unit remain unchanged.

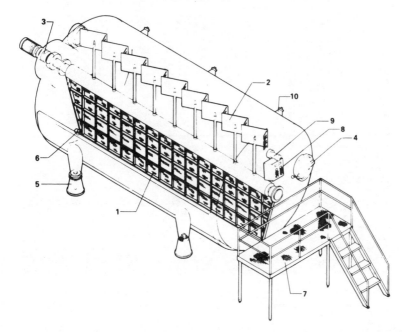

Fig. 4. Schematic illustration of the Pasilac Cheesemaking Tank type CT 2000. (1: cutting device; 2: stirring tools; 3: motor and drive; 4: manway; 5: support fitted with pneumatic bellows; 6: level gauge; 7: service platform; 8: control panel; 9: tank light; 10: spray ball.) (Reproduced by courtesy of Pasilac/ Silkeborg Ltd, Preston, UK.)

The tank is supported by four legs, and inclined at a slope of 1:40 towards the valve outlet which is 10·16 cm in diameter; the same valve is used for bottom filling of the vat.

The cutting and stirring device (see Fig. 5) is divided into sections and welded on the drive shaft. During the cutting stage of the coagulum, the shaft rotates clockwise (when viewed from the manhole) through 360 degrees, and anti-clockwise during stirring, i.e. blunt side of the knives. In order to prevent the coagulum from rotating in the tank, the cutting operation is carried out intermittently, i.e. stops twice per revolution, and only 80–85 per cent of the volume of the tank is utilised.

A permanent whey drainer can be fitted to the tank in one of its end walls and at a pre-set level. For example, if the strainer is situated at the 60 per cent tank level, the amount of whey removed is equivalent to 40 per cent. Incidentally, such a strainer is only required for the production of Dutch cheeses.

A. Y. Tamime

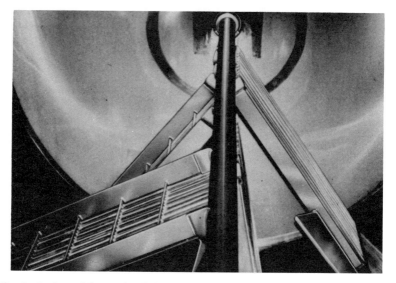

Fig. 5. A view of the cutting/stirring tools of the OST-CH vat. (Reproduced by courtesy of Alfa-Laval Cheddar Systems Ltd, Somerset, UK.)

MKT cheese vats

These cylindrical, stainless steel, cheese vats are manufactured by MKT Tehtaat OY in Finland. Model MKT-A is an open type cheese vat compared with enclosed model MKT-S, and Fig. 6 illustrates the overall similarities and construction of these vats; both vats are designed with a driving mechanism which is top mounted. The capacity of the MKT-A and S ranges from 4000 to 20 000 litres, and the emptying procedure is easy due to the conical design of the bottom of the vat. Partial removal of the whey could be achieved during agitation of the curd/whey mixture through a special, fixed, sucking device. The outer cutting mechanism (Fig. 6 — see *) is designed as a 'gate' construction, i.e. it swivels as the whey sucking device is lowered to follow the level of the whey in the vat.

APV Swiss cheese vat

This is a cylindrical cheese vat with a spherical bottom, and the interior section of the vat is normally made out of copper and the outer shell of resistant stainless steel. The vat is manufactured by APV OTT AG in Switzerland and is suitable for the production of Swiss cheeses. This cheese vat is fitted with a specially designed agitator and cutting mechanisms (see Fig. 6(b)). The latter part consists of two cheese harps,

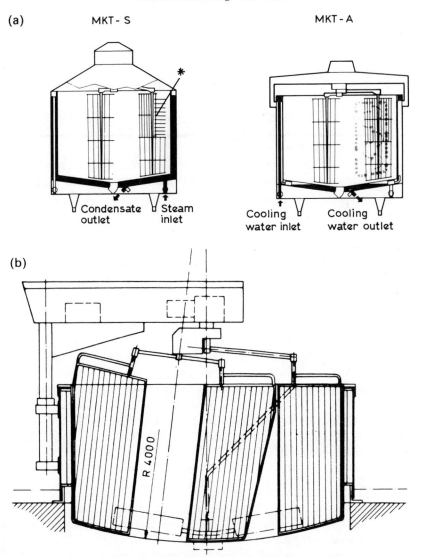

Fig. 6. Schematic illustrations of Swiss cheese vats. (a): Notice the design of the cutting mechanism and the movement of the curd during the cutting stage (reproduced by courtesy of MKT Tehtaat OY, Helsinki, Finland). (b): Notice the special design of the harp cutter and the agitator mechanism of the APV OTT cylindrical Swiss cheese vat. (Reproduced by courtesy of OTT APV AG, Worb, Switzerland.)

and the exterior one is fitted with a spherical agitator. Upon rotation, the entire content of the coagulum is cut into uniform size including the bottom part. Other specifications of the APV Swiss cheese vats include the following:

(a) The vat is fitted with an electronic device to control the speed of the agitator.
(b) The curd/whey mixture could be emptied through a pneumatic discharge valve situated on the bottom of the vat, or by a suction device over the edge of the vat.
(c) The heating of the vat content could be achieved by circulating warm water in the jacket (i.e. via two combined bottom and side wall circuits) or with steam.

However, a modification of the above vat is the oval shape type, where the bottom is flat; it has a lower filling height, but the same capacity. This cheese vat is tipped pneumatically for discharging the curd/whey mixture through a bottom valve gravimetrically or by a pump.

Performance of new-style cheese vats

Crawford (1976) has reported that 'it is very difficult to obtain factual and unbiased information on the performance of new equipment, and this is as true of the new race of cheese vats as it is of other items of machinery... The efficient functioning of the cheese vat and proper control by the cheesemaker over the first stage of the process is vital to the following stages and ultimate quality and yield of product'. However, the OST and the 'Double O' vats have been evaluated recently at the Netherlands Dairy Research Institute (Vries, 1979; Vries and Ginkel, 1980, 1984) for the production of Gouda cheese, and their overall conclusions are as follows.

Tebel OST III

(a) The fat and the curd 'fines' losses in the whey were the same and slightly lower, respectively, compared with the figures commonly obtained at the Institute.
(b) The distribution of the curd size particle was too fine, but the manufactured cheese had a satisfactory composition.
(c) The tank had no influence on the quality of the cheese, i.e. microbiological analysis at 14 days and organoleptic evaluation after 6 weeks.
(d) The curd-making tank had an adequate CIP system, as confirmed by bacteriological tests.

Damrow 'Double O'

(a) The fat and curd 'fines' losses in the whey were lower than the figures commonly obtained at the Institute.
(b) A good curd size distribution was obtained when the renneting temperature was increased from 30·5 °C to 31·0 °C.
(c) Curd production and/or handling in the vat could be properly controlled, and the composition of the cheese was good; at grading, i.e. 6 weeks after production, the quality of the cheese was good.
(d) After CIP cleaning, the bacteriological results were good.

Tebel OST IV

(a) The fat and curd 'fines' losses in the whey were lower compared with the figures normally obtained at the Institute.
(b) The difference between the calculated and the determined fat content of the second whey tended to be a little too high.
(c) The distribution of the curd size particle was somewhat fine for the manufacture of Gouda cheese, but did not affect the curd 'fines' losses in the first whey.
(d) Other comments reported were similar to those mentioned for Tebel OST III, i.e. (c) and (d).

De-wheying, Texture Forming and/or Handling Operations

The advent of mechanisation in the cheese industry has brought about some changes in the traditional process, and in most cases, the main change involves the following stages of cheesemaking: first, the de-wheying, i.e. the separation of the whey from the curd particles, and second, the texturising/handling of the curd. However, the latter aspect is of great importance, because the mechanised cheese systems have been designed for the manufacture of one or more types of cheese, and hence the available systems will be reviewed in relation to the variety of cheese.

Swiss cheese varieties

Traditional process
At the pitching stage, a heavy duty cheese cloth, which is fastened onto a metal frame, is used for the de-wheying of the curd. The curd and whey are vigorously agitated in a circular motion, and the swirling effect is

stopped by the stirring equipment which causes the curd to collect in the middle of the vat. At that moment, the cheese cloth is inserted into the vat beneath the curd, and pulled from one side to the other side. A pulley is used to lift the curd out of the copper kettle into the cheese mould. The cheese is pressed overnight in order to remove the excess moisture and to matt the curd particles. Normally, the curd weighs 100 kg which is the equivalent to the yield from each cheese vat needed to produce one cheese 'wheel'.

Damrow Swiss cheese vat

This type of mechanised vat (Fig. 7) is used for de-wheying and pressing the curd. The operation sequence could be described as follows: The curd and whey are pumped from the conventional cheese vat through a curd distributor to fill the curd evenly in the Swiss cheese vat. The curd is covered with a sanitary, re-usable woven plastic mesh (2) and the press plate is lowered into position (3). The rate of whey drainage is controlled by the cheesemaker at the whey outlet (4). Upon completion of pressing,

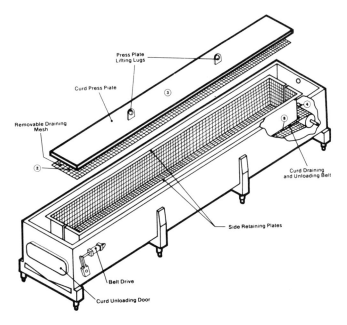

Fig. 7. Schematic illustration of the Damrow Swiss cheese vat. (Reproduced by courtesy of DEC Int. A/S — Damrow Products Division, Kolding, Denmark.)

the press plate is removed, together with the plastic mesh, and the block cutting mechanism is positioned at the curd unloading door. The bottom belt (8) continuously moves the pressed cheese to the unloading door for cutting into blocks. Finally, the cheese blocks are transferred to the brining tank.

The MKT system

This mechanised system has been developed in Finland for the production of 'block' and 'wheel'-shaped Emmental (Kiuru, 1976; Hansen, 1984a). For the manufacture of block cheeses, i.e. 84 kg in weight, the curd/whey mixture is pumped from the MKT cheese vat to a de-wheying and pressing table through a nozzle tube (see Fig. 8). The table is fitted with sieve plates along its sides to assist whey drainage, and in order to increase the removal of whey, the curd is covered with a woven, plastic sheet and press lid. After the pressing stage, the side sieve plates are un-hinged, and the pressed curd is cut to the desired size, turned and placed over a conveyor to be transported to the brining basin.

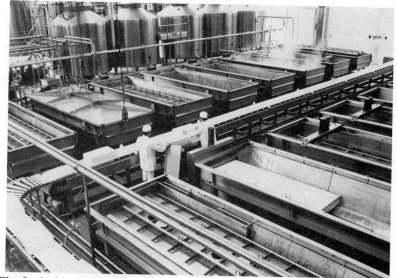

Fig. 8. A view of the MKT cheese equipment for production of block-shaped Emmental at the Keski-Pohjan Juustokunda in Toholampi, Finland. (Notice the MKT-S cheese vats in the background and some of the pressing tables are being emptied, pressed or CIP.) (Reproduced by courtesy of MKT Tehtaat OY, Helsinki, Finland.)

More recently this method of production has been modified so that the curd and whey is delivered directly to the mould (i.e. 12/pre-pressing table) through an octopus-like distributor. The accuracy of mould filling is $\pm 2\cdot 0$ kg/84 kg block of cheese. The curd is pre-pressed for 15 min, and afterwards the moulds are transferred to the tunnel press (see Fig. 23).

The production of 'wheel'-shaped Emmental is achieved using a pressing table (similar to the type mentioned above), which is furnished with cylindrical moulds formed as closed girdles made out of perforated stainless steel. The moulds are fitted with a filling tray onto which the curd/whey mixture is delivered. The curd in each mould is covered with a plastic, woven sieve and a press lid. The latter unit is fitted with a pressing cylinder actuated by compressed air at low pressure. After the pressing stage, a hydraulic lift is used to empty the moulds, and the cheese is transferred to the brining section.

Miscellaneous equipment

Hansen (1976) reported on a special pressing vat, which was designed by Perfora A/S in Denmark for the manufacture of a 450 kg 'block'-shaped Emmental. The pressing table consisted of three parts of perforated stainless steel (frame, lower and upper sheet) of the following dimensions: $1\cdot 12$ m wide $\times$ $2\cdot 8$ m long $\times$ $0\cdot 3$ m high. After the de-wheying stage, the table is tilted at an angle of 15 degrees, and the curd is turned once to remove the excess whey before pressing commences. The 450 kg cheese is later brined and portioned to 10×45 kg blocks, dried and finally packed in Cryovac® bags (registered trademark of W. R. Grace & Co.).

Cheddar cheese and related varieties

The main systems available for the manufacture of Cheddar cheese are as follows:

(a) Bell-Siro/Cheesemaker 2.
(b) Cheddarmaster.
(c) Lacto-O-Matic.
(d) Tebel-Crockatt.
(e) Alf-O-Matic Mk II.
(f) Damrow Drainage/Matting Conveyor (DMC).

Up to the 1970s, the mechanised systems employed for Cheddar cheese (i.e. (a–d) above) were widely used in the United Kingdom, Australia, New Zealand, and North America. A comprehensive review of such systems, including illustrations has been published recently by Crawford (1976), and hence only the last two systems will be discussed here.

Alf-O-Matic Mk II

This machine, which has been developed and manufactured by Alfa-Laval A/B, includes a number of different models (i.e. different number of belts — see Fig. 9), that can be used for the manufacture of separate varieties of cheese. For example, the four belt Alf-O-Matic Mk II is used for the production of Cheddar cheese, and in principle, the unit consists of:

— A fixed screen for de-wheying, and conveyor number (1) is for further drainage and moisture control in the curd.
— Conveyors number (2 and 3) are for curd matting/fusing and cheddaring respectively.
— Conveyor number (4) is for salting and mellowing of the milled curd.

TABLE XV

| Type | Dimensions (m) | | | Capacity (kg h^{-1})** |
	Length*	Height	Width*	
3 000	7·5			1 350–2 000
6 000	11·2	5·8	2·5	2 700–3 200
12 000	20·5			5 400–6 400

*Excluding the walkways.
**Depending on the residence time and treatment of the curd.

The conveyors are manufactured from perforated stainless steel and are mounted on top of each other; the whole unit is enclosed within a stainless steel casing. The Alf-O-Matic is equipped for CIP, and the specifications of the four belt model are as given in Table XV and the usual sequence of operations is:

(i) The curd/whey mixture is pumped from the cheese vat to the inlet of the Alf-O-Matic and onto a fixed de-wheying screen.
(ii) The curd particles are delivered to conveyor (1); further drainage can be achieved by stirring the curd with rotating cylinders covered with pegs. The pumping speed of the curd/whey mixture is controlled by a curd level sensor situated at the beginning of the belt, and the residence time in this section ranges between 5 and 15 min.

Full 4 Belt Alf-O-Matic
with salting for Cheddar

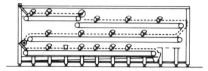

The 3 Belt Alf-O-Matic
– with salting for Cheddar types

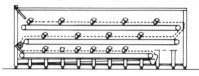

The 3 Belt Alf-O-Matic
– for curd from UF concentrated milk

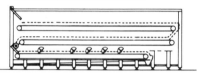

The 2/3rds Alf-O-Matic
– for use with external salting

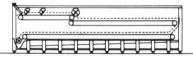

The 2 Belt Alf-O-Matic
– for stirred curd types – including salting section

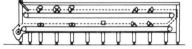

The single Belt Alf-O-Matic
– for short time Cheddar, Egmont, or to be used as a
salting belt

Fig. 9. Schematic illustration showing the different variations of the Alf-O-Matic Mark II. (Reproduced by courtesy of Alfa-Laval Cheddar Systems Ltd, Somerset, UK.)

(iii) Matting and/or fusing of the curd particles take place in conveyor (2), and the curd depth is maintained around 25–30 cm. At the end of this conveyor (e.g residence time 10–45 min), the curd mattress is inverted through a guided chute onto the cheddaring conveyor (3) where the residence time ranges between 15 and 70 min depending on the type of cheese produced.

(iv) At the end of conveyor (3), the cheddared curd passes through a mill; salt mixing and mellowing of the curd take place on conveyor (4). The mellowing stage is assisted by stirring and the residence time in this section ranges between 10 and 40 min.

(v) The salted curd is transferred to the mould filling machines by an auger mechanism or by air. The latter system is employed with the continuous 'Wincanton Block Former'.

Damrow draining/matting conveyor (DMC)

The Damrow DMC equipment is a combined unit in which the de-wheying and texture formation of the curd takes place during the manufacture of Cheddar cheese. The DMC is totally enclosed within a stainless steel chamber, and in theory, it consists of three sections, i.e. de-wheying, matting of the curd on moveable mesh belts made out of double-woven plastics, possibly polypropylene, and a milling unit. The capacity of the DMC may range from 0·1 to 6 tonnes h^{-1}, and the number of plastic belts installed in such a unit is adjusted according to the type of cheese produced. For example, a DMC unit with 2 belts is used for the manufacture of Cheddar cheese, but other cheese varieties may only require one belt, e.g. blue vein, Feta and other brine salted cheese. The sequence of operations of the DMC (see Fig. 10) can be summarised as follows:

— The curd/whey mixture is pumped from the cheese vat to the DMC inlet fitted with a special manifold for even distribution of the curd on top of the first belt.

— As the belt travels forward, the whey drains off through the belt, and by the time the curd has reached an end of this belt, matting of the curd has occurred.

— The speed of the belt and the retention guides located along and over the top of the belt ensure control of the width and depth of the curd.

— At the end of each cycle from the cheese vat, the belt continues to move on, e.g. around 120 cm, so as to separate the production of one vat from another.

— The matted curd will turn over automatically when it reaches the end of the top belt, and as it leaves the top belt, the curd is picked up (upside down) on the lower belt.
— On the lower belt, the matted curd will stretch and flow, thus simulating the traditional process of cheddaring.
— A 'chip' mill is situated at the end of the bottom belt, and the milled curd can be cut to any desired size, e.g. ranging from 13 × 10 mm to 18 × 24 mm.
— The milled curd falls into a hopper, and an auger mechanism transports the curd to the salting unit.
— The DMC is equipped with CIP.

Finishing coolers

Some degree of mechanisation can be achieved, in small cheese factories, using finishing coolers and some of these coolers have been described by Davis (1965), Crawford (1976) and Scott (1981). These coolers have been developed in the United States of America by different companies, and some examples are:

(i)   Damrow — finishing vat.
(ii) Stoelting — Stoelting table.
(iii) Kusel — Cheddar rite vat.

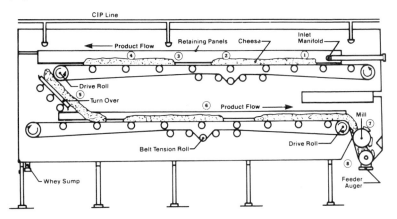

Fig. 10. Schematic illustration showing the operation sequence in the Damrow DMC. (Reproduced by courtesy of DEC Int. A/S — Damrow Products Division, Kolding, Denmark.)

The finishing coolers facilitate the following stages of cheesemaking, thus releasing the cheese vat for a second re-fill:

— the curd/whey mixture is pumped to the cooler;
— de-wheying can take place through the perforated area at the bottom of the cooler;
— cheddaring and other texture-forming treatments of the curd are carried out along traditional lines;
— a mill is fitted on the cooler at the milling stage, and the blocks of curd are delivered manually into the mill; the curd chips fall back into the cooler, and salt is applied manually.

The mechanised features of the finishing cooler could be described as:

(i) mixing tools (i.e. operated by overhead-mounted gear) during the de-wheying and mellowing stages;
(ii) mechanical unloading devices;
(iii) conveying of salted curd to the mould-filling station by the use of a spiral/screw lift elevator and/or air conveying method.

Dutch cheeses and related varieties

During the manufacture of these varieties of cheese, the curd is not 'texturised' to a great extent, and the handling of the curd/whey mixture, after leaving the cheese vat, may involve the following stages: first, partial de-wheying and pre-pressing, and second, mould filling and final pressing. Different equipment is available on the market for the production of Gouda, Edam and similar varieties, and some examples are as follows.

Pre-pressing vat

These units are manufactured by different companies, e.g. the Tebel Machinefabrieken in Holland (DBS strainer vat), Perfora A/S and Gadan A/S in Denmark. In principle, these pre-pressing vats could be considered similar in their function, and, in general, they are square-ended. The bottom and the sides are made out of perforated stainless steel, and the bottom section is designed as a wide-slat conveyor which is progressively moved forward at the end of the pre-pressing stage, so that the matted curd is cut into sections that fit the plastic moulds.

A modified version of this system is the Perfora tunnel pre-pressing vat (Hansen, 1980a), which consists of two parts, i.e. a stationary, upper section which automatically operates the press, and the mobile section which is the vat unit. The latter unit can be divided into compartments (i.e. known as the Perfora bottomless mould) ready to be filled with curd, and the whole vat is moved on rails to the press section. In some installations, a rotating, perforated drum is used (e.g Rotostrainer — Tebel Machinefabrieken) on the inlet side of the pre-pressing vat or pre-pressing tower, to separate the bulk of the whey from the curd, and the slightly drier curd is then distributed evenly before pressing commences.

'Block'-shaped Gouda cheese (e.g. 500 kg in weight) is currently produced in the Republic of Ireland (Hansen, 1981a), using the pre-pressing vat system which is 3 m long and 1·12 m wide. After the brining stage, the large block of cheese is portioned into 10 × 50 kg Gouda sections, and this method of production is somewhat similar to the system reported earlier for the production of 450 kg 'block'-shaped rindless Emmental cheese.

Pre-pressing tower

The de-wheyed curd is compacted in a tower, and a sliding plate at the base of the column allows the curd to fall into the cheese mould. This system of curd handling was primarily developed for the manufacture of Gouda and Edam cheeses as a de-wheying, pre-pressing and mould filling unit. However, the same equipment could be used for other varieties of cheese, e.g. Swiss and some Danish types. Furthermore, these machines are highly flexible for the production of 'round', 'rectangular' or 'square'-shaped Dutch cheeses, and it is adaptable for continuous and automated production lines.

Examples of some of the pre-pressing towers are:

(i) Casiomatic — Tebel Machinefabrieken BV (Hansen, 1979a, 1984b); Fig. 11 illustrates a schematic drawing of such tower.
(ii) Curd moulding — Stork Friesland. This type of pre-pressing tower has been developed as a rotary unit (Hansen 1977a, b), and recently a non-rotating version has been marketed (Hansen, 1979b).
(iii) Curd dosing — Koopmans BV. Hansen (1978a) described an installation in the Coberco cheese factory in Winterswijk for the production of 13·5 kg of Gouda; another system of curd handling and mould filling, which is supplied by Koopmans BV, is the Conomatic-R (Hansen, 1983a).
(iv) Pre-pressing tower — Pasilac/Silkeborg A/S. This unit consists of 3–5 moulding cylinders which have an hourly maximum throughput of pre-pressed cheese of 2100 kg, and the accuracy of weighing a 14 kg cheese is ± 100 g.

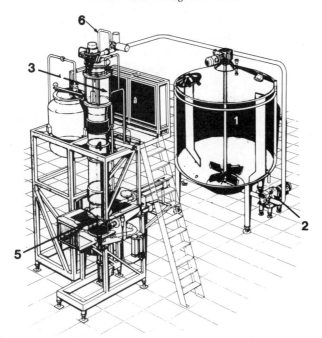

Fig. 11. Schematic illustration of the Casiomatic. 1: Buffer tank, which is situated between the cheese vat and the Casiomatic, has the following advantages: the cheese tank can be emptied rapidly; the curd/whey mixture is more homogeneous in the tank, and it is cooled which results in less moisture loss from the curd; the elimination of air entrapment from the system; fluctuations in the time scheme are overcome. 2: Positive pump. 3: Sight glass fitted with probes to control the curd and whey level in the tower. 4: Pre-press tower. 5: Mould filling station. 6: The mounting of the roto-strainer during the production of Swiss cheese varieties. The latest model is the Casiomatic III which is equipped with a special cutting device at the bottom of the tower, and coupled with multi-mould filling it is possible to produce cheese ranging from 1 to 18 kg in weight and with several shapes and sizes, e.g. round (between 11 and 42·6 cm in diameter), square (36 × 36 cm or 38 × 38 cm) and rectangular (12 × 25; 24 × 36; 28 × 35; 35 × 38 and 30 × 50 cm, respectively). (Reproduced by courtesy of Tebel Machinefabrieken BV, Holland.)

Curd Recovery Systems

Cheese whey is composed of the following milk components: fat, curd 'fines' (mainly fat and casein), whey proteins, lactose and minerals. The recovery of the fat and the curd 'fines' from the whey can, in large, centralised cheese factories, play a major role in improving the cheese yield and reducing financial losses. In principle, the recovery process is achieved in two stages: first, the removal of the curd 'fines', which is later mixed with the bulk of the cheese curd after the de-wheying stage; and second, the separation of the fat from the whey mechanically, with the cream being utilised for the manufacture of whey-butter. Some examples of the equipment which could be employed for the recovery of the curd 'fines' from the whey, are as follows.

Russell Fines Saver

This type of curd 'fines' separator is manufactured by Russell Finex Ltd in London (UK), and the overall specification of the machine is illustrated in Fig. 12(A). The filter element is made out of two polyester or nylon, mesh sleeves mounted on a stainless steel basket, which also houses the paddle-blade impellers that connect to the drive shaft of the motor. The whole unit is enclosed within a stainless steel body. As the curd 'fines'/ whey mixture enters the mesh basket, the impellers pump the liquid through the filter, and the separated curd travels up inside the basket to the outlet. By adjusting the tilt of the whole unit, the moisture content on the curd discharge side can be manipulated, i.e. the greater the angle of tilt, the drier the consistency of the curd 'fines' and vice versa. The resultant slurry of curd 'fines' can then be pumped away and distributed over the cheese curd after the de-wheying stage, or alternatively a dry 'fines' can be collected into buckets and manually spread over the cheese curd at regular intervals.

The Russell Fines Saver has an adjustable speed motor with a range from 200 to 1070 rpm, but the efficiency of curd 'fines' recovery is primarily dependent on the mesh size of the filter element. For example, a unit (40 000 litres h^{-1} throughput) fitted with a 40 μm mesh filter can recover approximately 22·7 kg of curd, but by using a finer mesh filter, e.g. 5 μm, the flow rate drops to 330 litres h^{-1}, and the weight of recovered curd will be approximately doubled.

AZO Fluid Sifter

This sifter is manufactured by Adolf Zimmermann GmbH, Osterburken in Germany, and an illustration of the unit is given in Fig. 12(B). The rotor is enclosed within a cage on which is mounted a woven, nylon screen (20–40 μm in size), and the whole unit is enclosed in a stainless steel shell. The geared motor has a variable speed ranging from 155 to 775 rpm, which helps to create a centrifugal force inside the separation chamber. This force helps to send the liquid through the screen, and the

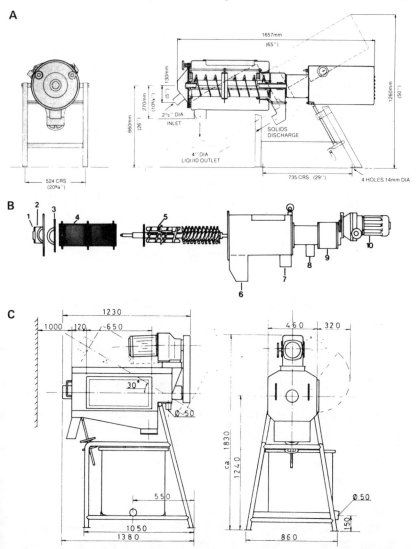

Fig. 12. Schematic illustrations of units for mechanically separating curd fines. (A) Russell Fines Saver. (Reproduced by courtesy of Russell Finex Ltd, London.) (B) AZO Fluid Sifter. (1: shaft bearing; 2: end cover; 3: cover for the sieve basket holder; 4: sieve basket; 5: rotor shaft fitted with helical screw; 6: curd 'fines' outlet; 7: whey outlet; 8: curd 'fines'/whey mixture inlet; 9: distance piece; 10: motor with variable drive.) (Reproduced by courtesy of Van den Bergh Partners Ltd, Windsor, UK.) (C) Jesma Roto-Fluid Sieve. (Reproduced by courtesy of Pasilac/Silkeborg Ltd, Preston, UK.)

'fines' are carried by a helical screw to the discharge port (see No. 6 in Fig. 12(B)). The consistency of the discharged solids is controlled by rate of throughput, size of sieve, speed of rotation and slope of the machine. Some examples of the efficiency of solids recovery by the AZO Fluid Sifter are as follows:

Size of sieve (μm)	Product throughput (litres h^{-1})	Fines recovery (%)
20	17 000	84
30	22 000	72
40	28 000	64

However, if a sieve size of 30 μm is used with a product throughput of 22 000 litres h^{-1}, by changing the rotor speed (expressed as percentage of maximum speed) from 40, 50, 60 to 70 per cent, the efficiency of 'fines' recovery would be 78, 71, 70 and 63 per cent respectively.

Jesma Roto-Fluid sieve

This is a Danish machine (Hans Jenssen Machineworks, Vejle, Denmark) which is designed for the clarification of fluids and/or the concentration of fluids containing solid particles, e.g. for the separation of curd particles from whey. The maximum capacity of the unit is 50 000 litres h^{-1}, and the mesh size ranges from 1 to 3000 μm (see Fig. 12(C)). The principal function of operation of the Jesma Roto-Fluid sieve is similar to the Russell Fines Saver or the AZO Fluid Sifter. Inside the separation chamber, the rotor flings the fluid portion through a sieving net, and the curd particles are conducted to the overflow by a screw placed around the rotor. The machine can be adjusted to a 30° angle of tilt in order to regulate the consistency of the separated curd 'fines', and/or improve the concentration of the curd particles.

Damrow Cheese Saver II

This cheese curd 'fines' recovery unit is manufactured by Damrow Company. Recovery is achieved by filtration of the whey, as compared with the other systems which rely on mechanical separation. A schematic drawing of the Cheese Saver II is given in Fig. 13. As can be observed, this unit consists of a sloped, nylon screen of 100 μm size enclosed within a stainless steel frame. The filtration capacity of the Cheese Saver II ranges between 20 and 60 m^3 of whey per hour depending on the model. The recovered curd slurry is collected into a stainless steel compartment at the bottom of the unit, and then returned to the cheese vat, or distributed over the de-whey curd by using a positive pump. Rasmussen (1980) reported the following recovery efficiency of curd solids from whey after two months trial period using whey from a Cheddarmaster drainage belt:

amount of 'fines' in whey	80	
amount of 'fines' in filtered whey	26	mg /100g whey
amount of recovered 'fines'	54	

Therefore, a cheese plant having a throughput of 100 000 litres of whey per day will save an amount of dry 'fines' equal to 54 kg $\dfrac{(540 \times 100\,000)}{1000 \times 1000}$; this is equivalent to 74 kg (54 $\times$ 1·37) of cheese containing 37 per cent moisture. Taking into consideration the cost of the unit and the price of Cheddar cheese in 1980, the capital outlay of Cheese Saver II was realised in 120 days.

AES 'Fines' recovery system

Another example of a whey filtration unit, which could be employed for the recovery of curd 'fines' from whey, is the AES system manufactured by Albaney International in the United States of America. The unit is fabricated from stainless steel and is totally enclosed (see Fig. 13(B)). The filtration screen is made out of food-grade polyester available in 100–325 μm mesh size, but it is recommended that the 200 μm mesh be used for curd 'fines' recovery from whey of most hard cheese types. The application of the AES system has been reported by Lee (1981), and, in brief, the filtration sequence is as follows:

— The cheese whey is pumped into the outer plenum (1), and surges over through the stationary mesh screen.
— The curd 'fines' are moved from the nylon screen to the curd slurry section (3) by a continuously rotating shower of filtered whey from section (2).
— The curd slurry is pumped at a rate of 7·6–11·4 litres min^{-1} to the finishing tables of the cheddaring unit(s).

Milling, Salting/Brining and/or Mould Filling

The methods available to handle the curd at this stage of production depend on the type of cheese produced, but, in general, the handling of the curd could be divided as given in Table XVI.

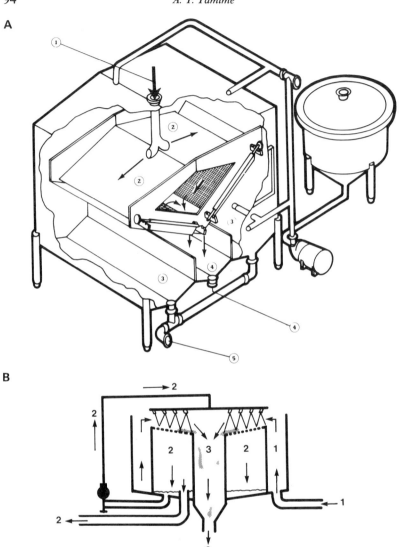

Fig. 13. Schematic illustrations of curd 'fines' filtration units. (A) 1: curd 'fines'/ whey mixture inlet from whey tanks to manifold for distribution; 2: manifold distributes cheese whey over the separation media; 3: separated whey is pumped to storage tank; 4: curd 'fines' slurry is returned to cheese vat or drained curd with a positive pump; 5: separated whey outlet. (Reproduced by courtesy of DEC Int. A/S — Damrow Products Division, Kolding, Denmark.) (B) 1: incoming unfiltered whey; 2: filtered whey; 3: curd 'fines' slurry. (After Lee (1981).)

TABLE XVI

Sequence of operation	Cheese variety
Mould filling/pressing/brining*	Parmesan/Grana Gouda/Edam Emmental/Gruyère
Milling[†]/salting/mould filling/pressing	Cheddar and related British territorials

*The Swiss and Italian varieties are rubbed with salt on the surface each day after turning in brine; however, the 'block' shaped Emmental and/or Gruyère are only brined.
[†]The salting of Cheddar is carried out prior to milling when employing the Tebel-Crockatt equipment, or during the manufacture of American Cheddar where the curd is not matted together, and hence, milling is not required.

It is evident, however, that the above division is also applicable to the method of salting the cheese curd, i.e. salting for Cheddar and related varieties and brining for Swiss, Italian and Dutch cheeses. In some of the latter cheese varieties (Italian and Swiss), the surface of the pressed curd is rubbed with dry salt in order to produce a harder rind to protect the cheese during the maturation period.

Milling equipment

Most of the mechanised equipment employed for the manufacture of Cheddar cheese includes automated milling and/or salting and mellowing facilities. Both the Alf-O-Matic Mk II and the Damrow DMC units are fitted with a curd mill at the end of the cheddaring conveyor (i.e. belt numbers 3 and 2 respectively — see Figs. 9 and 10). The matted curd is cut into curd chips of roughly 1·25–1·85 cm square section and 17·5–20·0 cm in length (Brockwell, 1981). However, the length of the curd chip is influenced by such factors as:

— depth of the matted curd;
— space between the rotating blades;
— speed of rotating cutters;
— and/or design of the mill.

An alternative type of mill, which could be employed during the manufacture of most British varieties, is the peg type; however, a different type of peg mill is used for milling Cheshire cheese curd to give 'granular' pieces which helps the finished cheese to attain the crumbly, texture characteristics.

Incidentally, the 'block'-shaped Cheshire cheese is somewhat drier, less acidic and tends to have a closer texture *vis á vis* the traditional cheese, and therefore, is more suitable for cutting and pre-packaging into retail portions.

Salting equipment

The salting should be completed in an atmosphere isolated from any humidity in order to avoid the ingress of moisture to the metering device, and thus, changes in the accuracy of salting. The salting equipment in the Alf-O-Matic Mk II operates as follows:

— Warm, dry air transfers the salt from storage to a salt metering belt.
— At the end of the belt, the salt is sucked-up into a cyclone and then falls into a 'rotary' valve.
— Hot air passing through the valve transfers the salt to a stainless steel, oscillating distribution tube to be spread onto the cheese curd. The oscillating pipe is driven by an electric, geared motor.
— The amount of salt added is controlled by a floating sensor that determines the depth of the milled curd, and actuates the salt metering device either to increase or decrease the salt dosage in line with the height of the milled curd.
— Salt mixing and mellowing take place on conveyor number 4 (see Fig. 9), where the curd chips are treated in a series of 'resting' and 'stirring' periods. The residence time ranges between 10 and 40 min.

The salting and mellowing systems developed by Damrow Company have, by contrast, been established as separate units from the DMC equipment. Two different systems are available, i.e. the Salt Retention Unit (SRU) and the Salting and Finishing Vat (SFV) (Fig. 14). In brief, the sequence of operations of the SRU is as follows:

— The curd chips are air conveyed from the DMC or other cheesemaking equipment, e.g finishing coolers, to a cyclone situated on top of the SRU (1).
— The curd chips fall to the bottom of the cyclone (2) and are delivered onto a weighing belt (3). The air is vented at the top of the cyclone.
— The curd chips enter the mixing drum (4), where the exact amount

of salt is added and mixed for a short period; (5) curd is outlet onto the mellowing top belt. The principle of using a mixing drum is similar to the Bell-Siro 'Cheesemaker' 3 system (Crawford, 1976).

— Mellowing of the salted curd takes place on two belts which are similar in operation to the DMC.

— At the end of the 2nd belt, the mellowed curd leaves the SRU through a star valve (6), and is air conveyed to a mould or hoop filling station.

The SFV unit could be used for salting untextured curd, e.g American Cheddar, or conventional cheddared and milled curd. The curd chips are conveyed from the DMC or other cheddaring/milling units to the SFV which is fitted with load cells to accurately weigh the curd. The salt is air-conveyed and added at a rate predetermined by the cheesemaker. The mellowing of the salted curd is achieved by continuous or inter-mittent stirring. The salted curd is emptied by reversing the stirrers, and discharged through the bottom, via a star valve and air-conveying system to the mould-filling station.

Both units are designed for CIP, and the salting stage could be made continuous by using an SRU of the same capacity of the DMC unit or by installing two or more SFV units (Park, 1979).

Hand salting of the milled curd is still practised in small cheese factories using finishing coolers, or in mechanised systems employing the Tebel-Crockatt equipment. The 'lawnmower' device in the Tebel strainer performs the cheddaring process, and partly mills the curd. The salt is added manually, and by passing the 'lawnmower' cutters a few times over the table, the curd is cut, turned, mellowed, and finally transferred to a peg mill (Crawford, 1976). Most of the finishing coolers are equipped with stirring devices to mix the salted curd chips, and a specially operated blade to push a portion of the curd to a conveying device leading to the mould-filling station.

In contrast to this method of salting, i.e. the curd chips, the immersion of pressed cheese in brine is often used, and examples of such varieties are the Italian, Swiss, Dutch and Danish cheeses. In general, there are two different systems which are employed to brine cheese: first, surface brining, where the cheese floats in brine tanks, and second, deep brining where the cheese is stacked on racks and lowered into brine tanks, i.e. total immersion in the salt solution.

Most brine tanks are made out of plastic, in order to minimise the

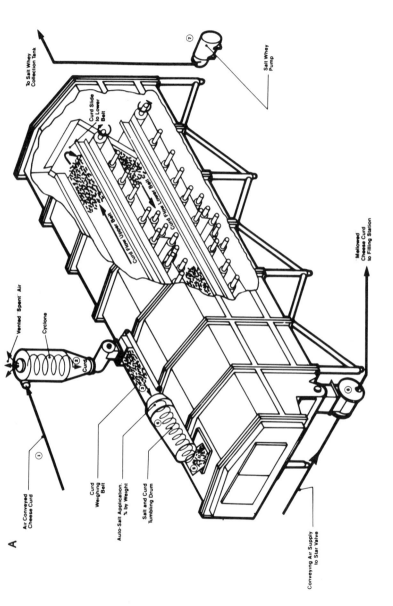

Fig. 14. Damrow enclosed salting equipment. (A) SRU salting unit.

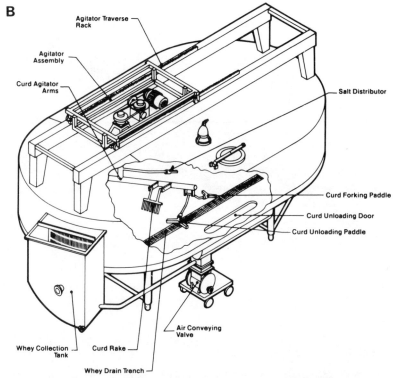

Fig. 14 — *contd.* (B) SFV salting unit (Reproduced by courtesy of DEC Int. A/S — Damrow Products Division, Kolding, Denmark.)

corrosive effects of the salt, and the construction of brining tanks has to take into account the following aspects:

(i) salt dosing equipment, i.e. dry form or in solution, so that the desired concentration of the brine can be maintained;

(ii) temperature control mechanism, so that the cheese can be brined at two different temperatures, e.g. at 18–20°C for the first few hours, followed by 14°C during the brining of Danbo cheese (Danish variety) in a Gadan/Brine-O-Matic unit (Hansen, 1979c,d);

(iii) provision of some eccentric movements of the racking unit in the brine tank, in order to overcome the problem of cheese buoyancy (i.e. top side) resting on the same bars during the brining stage;

A

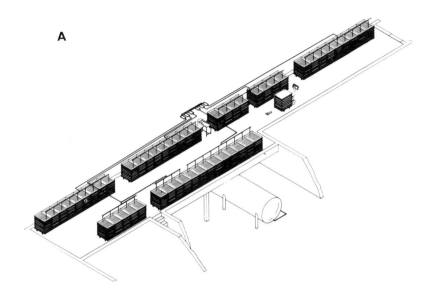

B

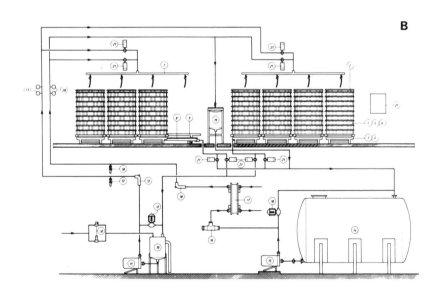

(iv) in large cheese factories, the brining process and handling of the cheese are highly automated, and some examples are illustrated in Figs. 15 and 16.

The amount of salt retained in the cheese is governed by a multitude of factors, such as:

— level or percentage of added salt;
— method of salting (i.e. addition of salt to the milled curd or brining the pressed cheese);
— duration of salting;
— moisture content of the cheese;
— type of cheese, etc.

According to Scott (1981), the salt content in the cheese may range between 1·5 and 2·5 per cent, but in some instances, the desired salt level may reach 5·0–7·0 per cent, as in Pecorino cheese (an Italian variety), or as low as 0·6 per cent salt in Emmental. Thus, the amount of added salt and/or the concentration of the brine is dependent on the variety of cheese, and Scott (1981) has reported some recommended guidelines. For example, the percentage of added salt in some of the British varieties is as given in Table XVII.

Caerphilly cheese is the only British variety where the curd is salted at a rate of 1·0 per cent. The cheese is dry salted while it is being turned in the press, and finally immersed in brine (i.e. 18·0 per cent concentration) at 15·6°C for a duration of 12–24 h.

Cheese brine is a sodium chloride (NaCl) solution, and its concentration is expressed either as percentage NaCl (w/w) (e.g. saturated brine is 35 per cent NaCl) or °Baumé. The brining condition of the Swiss, Italian and Dutch cheeses may vary slightly, and some examples are shown in Table XVIII.

Fig. 15. Gadan Cooling/Brining System type KS-600. (A) An overview of the brining system constructed at two floor levels. (B) 1: Cooling/brining trays; 2: top trays; 3: bottom trays; 4: bottom frames; 5: perforated plates; 6: cheese nets; 7: distribution pipes; 8: gutters; 9: buffer rods; 10: balance tank with valves and level control; 11: centrifugal pump for water; 12: temperature and pressurising equipment for water; 13: indicating instruments for temperature and pressure on water; 14: brine tank; 15: centrifugal pump for brine; 16: brine filter; 17: plate heat exchanger; 18: temperature and pressurising equipment for brine; 19: salt dosimeter; 20: indicating instruments for temperature and pressure of the brine; 21: electronic controller with flow-forwarding and return valves. (Reproduced by courtesy of Gadan Maskinfabrik A/S, Them, Denmark.)

TABLE XVII

Salting (%)	Cheese variety
1·6–1·8	Wensleydale
1·7–1·9	Single and Double Gloucester
1·8–2·0	Cheshire, Cotswold, Derby, Dunlop, Kingston and Leicester
2·5–3·0*	Cheddar

*According to the experience of the author, 3 per cent salt could be added to the curd chips when using direct-to-vat starter culture in order to halt acid development during the early stages of maturation, or in instances where the rate of acid increase is fast during the cheddaring stage. Incidentally, the salt content in such cheeses at one day old ranged between 1·7 and 1·8 per cent.

In practice, the brine should be monitored to maintain the following specifications:

(i) adjust the pH level to 5·2 by the addition of hydrochloric acid;

(ii) the microbiological content of the brine could be controlled by the addition of saltpetre (KNO_3) or hypochlorite (e.g. ½ litre per 1000 litres of brine), or by heat treatment of the brine (i.e. by boiling);

(iii) the appearance of a reddish discoloration on the surface of the cheese could be attributed to the high KNO_3 content in the brine, and/or to certain *Lactobacillus* spp. which have contaminated the cheese after the brining stage.

TABLE XVIII
Some examples of brining specifications of certain cheeses

Cheese variety	NaCl (%)	Temperature (°C)	Duration (days)	References
Parmesan/Grana	24	7–10	14–15	Kosikowski (1977)
Emmental	23	10	2–3	Scott (1981)
Gruyère	23	10–13	2–6	Scott (1981)
Edam	22–25	16	2–3	Scott (1981)
	22–25	12–14	3–4	
	18*	13	3	Dijkstra (1984)
Gouda	20	15	3	Scott (1981)
	18*	13	4–5	Dijkstra (1984)
Reading Yellow	20	15	0·5–1	Scott (1981)

*°Baumé (°Bé), e.g. 18, 19 and 20% NaCl is equal to 16·9, 17·8, 18·7 °Bé respectively.

Fig. 16. Salting basins for the brining of Gouda cheese in cages at DMV — Campina factory in Born. (Notice the hoist which transfers the cages stacked with Gouda cheeses ready for immersion in brine.) (Reproduced by courtesy of Tebel Machinefabrieken BV, Holland.)

Types of moulds and mould-filling equipment

The traditional multi-piece iron or tinned steel moulds have been replaced by perforated stainless steel or aluminium alloy, and recently, plastic moulds have become very popular.

In the past, cotton cheese cloth was normally used to line some types of cheese mould in order to provide a smoother finish to the pressed cheese, and to prevent the cheese from sticking to the mould. Thus, the preparation of large numbers of moulds in a cheese factory is a labour intensive process, and washing, sterilising and drying the mould liners also increases the cost of production. These problems have been overcome by the introduction of perforated, stainless steel moulds (e.g Perfora A/S in Denmark) which has made it feasible to dispense with the cheese cloths, or alternatively make use of a disposable, polyethylene liner. The specifications of one such a cheese mould liner manufactured by Smith & Nephew Plastics Ltd (UK) are as follows:

— tear strength 70–110 g;
— burst strength $127 \cdot 53$–$156 \cdot 96$ kN m^{-2};
— porosity 40–70 s;
— gauge 100–150 μm.

Plastic moulds, which are manufactured by Crellin BV (Kadova) and Arend BV (Lauda) in Holland, are widely used for semi-hard cheese varieties. These moulds are made out of polyethylene, do not require a cheese liner, and are suitable for use on a mechanised cheese line, i.e. mould filling, pressing, de-moulding, washing and re-filling. Basically, the moulds are constructed of two pieces, and they are produced in a multitude of shapes. However, the block-type cheese mould, which is suitable for Cheddar cheese, consists of three pieces constructed within a stainless steel frame to withstand higher pressure. Some of the technical specifications of the Kadova and Lauda cheese moulds have been reported by Anon. (1976), Hansen (1978b), Anon. (1980a) and Dijkhuizen (1981), and some of these plastic cheese moulds could be supplied as multicheese moulds (Hansen 1981b).

Mould filling equipment is primarily designed to deliver a certain amount of curd to a mould, and hence to produce cheeses, after pressing, of roughly the same weight. Such equipment could be incorporated either as part of the curd handling system, or in a position dependent on the variety of cheese produced. The following are examples.

Emmental — traditional 'wheel'-shaped cheese
The entire curd content from each vat is delivered to a mould after de-wheying, and the cheesemaker has to alter the volume of the milk used in order to overcome the effects of seasonal variation on yield; thus, producing cheeses of roughly the same weight.

Cheddar and related varieties
The sequence of filling the salted, curd chips into moulds is as follows:

   (i)   tare-off weight of mould;
   (ii) first stage filling — up to 70 per cent or more of the final weight
         of the cheese is delivered into the mould;
   (iii) pre-press the curd;
   (iv) second stage filling of the mould to the desired weight, e.g. 20 kg;
   (v)   transfer cheese moulds to the press.

Two stage filling ensures that the pressed cheeses' are uniform in weight, and Hansen (1981c) describes a filling sequence of Cheddar cheese chips into Lauda moulds in a large factory at Waitoa in New Zealand.

A different approach to the filling of Cheddar cheese moulds is the 'Press-n-Fill' Mk IV (see Fig. 17), which is a volumetric hoop filler manufactured by Wincanton Engineering Ltd in Dorset (UK). The salted curd is pre-pressed into a small, rectangular tower/chamber of similar dimensions to a 20 kg 'block'-shaped hoop. The lightly pre-pressed curd is discharged over an awaiting cheese mould. While the discharge door is in the closed position, a horizontal guillotine cuts the curd, and an ejector ram pushes out the block (i.e. after the discharge door has opened) and gives the curd a second pre-press in the hoop. During the mould filling stroke, the mould liners are held in position by a clamp frame which also prevents spillage of the curd. The rate of filling is 3 × 20 kg moulds min^{-1}.

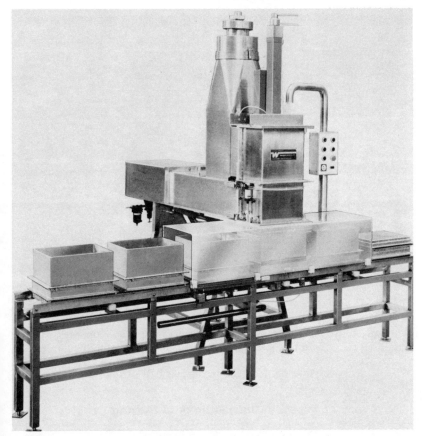

Fig. 17. The Wincanton Mark II 'Press-n-Fill' Unit. (Reproduced by courtesy of Wincanton Engineering Ltd, Dorset, UK.)

Semi-hard cheese varieties

The de-wheying and curd handling equipment, e.g. pre-pressing vat or tower, are designed as mould fillers, and the accuracy of filling is very good (see Fig. 18).

The cheese moulds (metal, i.e. aluminium alloy, or plastic, e.g. Lauda and Kadova) can be easily cleaned in a tunnel washing system. A special detergent should be used when cleaning the metal moulds to avoid black discoloration, and the polyethylene type should be cleaned with an alkaline and/or acid detergent at 50–70 °C. However, it is recommended that an acid treatment should follow the alkaline wash in order to neutralise the latter compound.

Fig. 18. A Gouda cheese factory (DMV-Campina) in Born in the Province of Limburg. (1: Six Casiomatics capable of delivering 264 pre-pressed and moulded 12·5 kg Gouda cheeses per hour; 2: mould cleaning equipment; 3: Kadova cheese moulds ready to be conveyed to tunnel presses. (Reproduced by courtesy of Tebel Machinefabrieken BV, Holland.)

Pressing, Packaging and Storing

Pressing equipment

The very hard, hard and semi-hard varieties of cheeses are pressed in either vertical or horizontal presses under normal atmospheric conditions and/or under vacuum. The pressing parameters, i.e. application of the right amount of pressure for a specific duration of time, gives the cheese its final shape, produces a cheese with a firm and smooth surface, and helps to lower its moisture content to the desired level. The pressure applied is dependent on the type of cheese, and it can be achieved using one of the following:

— spring-loaded or screw mechanism;
— hydraulic presses;
— pneumatic systems;
— vacuum presses.

The former type is manually operated, and hence the other systems are more common in mechanised cheese factories. One important aspect, which must not be overlooked, is the difference between the 'air-line' pressure and the 'actual' pressure applied to the cheese; the former pressure is normally lower. Since the pressure is defined as mass per unit area, and the calculated 'actual' pressure on the cheese takes into account:

— the air-line pressure;
— the diameter of the cylinder head;
— the surface area of the cheese.

Therefore, different cheese varieties are pressed at different pressures and a summary of some relevant data is shown in Table XIX. However, high pressures are used when pressing 'block'-shaped cheeses, and some examples are given in Table XX.
In general, the cheeses are pressed in individual moulds (i.e. in single or multi-row) or in bulk, and examples of some cheese pressing systems are as follows.

Horizontal 'Creeping' or Gang Press
The 20 kg metal mould is placed on its side between two guided side rails, and when the press is full, the pressure is achieved pneumatically using compressed air. The press head section moves forward gradually, and the pressure is maintained all the time. In order to avoid too long a pressing cylinder, the bottom base of the press has metal notches welded to it, and as the press moves forward, it interlocks at certain intervals, which prevents it from moving backwards. This type of press is not mechanised, and the cheese moulds are placed manually in the press.

TABLE XIX

Pressures applied and treatment of some traditional cheese varieties during pressing

Cheese variety	Size (cm) Diameter × height	Weight (kg)	Pressing treatment
Caerphilly	25·0 × 6·0	3·0–4·0	Press at 12 kN m^{-2} for 10–20 min; turn, rub with salt and re-press overnight at 49·8kN m^{-2}
Cheddar	30·5 × 35·5	18·0–22·5	First day press at 75 kN m^{-2} for 12–16 h; second day turn, pull cloth, scald (optional) and re-press at 200 kN m^{-2} for 24 h
Cheshire	30·5 × 25·5	up to 27·2	First day press at 24·9 kN m^{-2} for 2 h, turn and re-press at 49·6 kN m^{-2} overnight; second day increase pressure to 99·6 kN m^{-2}
Cotswold	Variable	0·45–1·0	First day press with 13 kg per 1 kg of cheese; second day turn into fine cloth and press at 24·9 kN m^{-2} for 24 h
Derby	41·0 × 11·5		First day press at 24·9 kN m^{-2} for 2 h, turn and re-press at 49·8 kN m^{-2} increasing to 74·7 kN m^{-2} overnight; second day turn and press at 124·5–149·4 kN m^{-2} for 24 h
Dunlop	35·0–40·0 × 20·0	13·6–15·8	Press lightly for 15 min increasing to 50 kN m^{-2} for 3 h; turn and re-press at 150 kN m^{-2} overnight
Edam	30·0 × 13·5 25·0–11·0	4·0–4·5 2·0–2·5	Press cheese in *groups* at 980·0–1470·0 kN m^{-2} for 3 h; turn and re-press at 1470·0–2450·0 kN m^{-2} (number of moulds not given)

Emmental	70·0–100·0 × 13·0–25·0	100·0–110·0	Press at 7·0 kN m^{-2} for 5–15 min, and press overnight at 30·0–60·0 kN m^{-2}
Gloucester (double)	45·7 × 12·6	18·0–20·0	First day press gently increasing to 24·9 kN m^{-2} for 30 min, and turn and press at 49·8 kN m^{-2} for 16 h; second day turn and press at 74·7 kN m^{-2} for 24 h
(single)	38·0 × 7·6	6·8–8·0	Press at 34·0 kN m^{-2} for one day increasing to 133·0 kN m^{-2} on the second and third day
Gouda	24·0–25·0 × 6·5–12·0	up to 20·0	Press lightly increasing to 96·5–193·0 kN m^{-2} for 5–8 h
Gruyère	40·0–64·0 × 8·0–13·0	35·0–40·0	Press and turn cheese for 2–3 days at 60·0–70·0 kN m^{-2}
Kingston	—	0·45–2·2	Press at 15·0 kN m^{-2} for 4 h minimum
Lancashire	—	18·0–22·5	First day apply no pressure; second day press at 24·9 kN m^{-2} increasing to 99·6 kN m^{-2}; third day scald at 60°C for 30 s and press at 124·6 kN m^{-2}, fourth day bandage and re-press at 49·8 kN m^{-2} for 12 h

TABLE XIX (*Contd.*)

Cheese variety	Size (cm) Diameter × height	Weight (kg)	Pressing treatment
Leicester	45·0 × 11·0	18·0–22·5	First day press at 24·9 kN m^{-2} increasing to 49·8 kN m^{-2} in 2 h and turn and re-press at 74·7 kN m^{-2} overnight; second day turn and press at 99·6–149·4 kN m^{-2} for 24 h
Parmesan	35·0–45·0 × 17·0–22·0	30·0	Press at 12 kN m^{-2} for 1 h, turn and re-press for 12–24 h at same pressure
Wensleydale (White)	—	1·5–5·0	First day apply no pressure; second day press at 24·9 kN m^{-2} for 2 h, bandage and re-press at same pressure for 3 h

Adapted from Scott (1981).

TABLE XX

Cheese	Size $(L \times W \times D\ cm)$	Weight (kg)	Pressure $(kN\ m^{-2})$
Emmental	$66 \times 40 \times 20$	38–45	2845–4315
Cheddar			300– 500
Cheshire	$36 \times 28 \times 18$	18	200– 300
Dunlop			150– 300

Corrosion of the metal frame is a major problem, and in some factories in the United Kingdom, two presses are built on top of each other in order to reduce the floor area required.

Vertical press
This type of press comes in different designs, and the simplest, manually operated type consists of up to four vertical units. Each unit is divided into four sections which can hold, for example, 2 × 20 kg block cheese moulds. The pressing cylinder is mounted on the top, and the base of each section is guided by two vertical guide rails. An illustration of such a press was provided by Anon. (1980b).

The development of a vertical press, which is around 10 m high and suitable for 20 kg block moulds, has been reported by Crawford (1976). The press is filled and emptied using an automatic stacking and de-stacking mechanism, and it can hold around 42 moulds. The press is enclosed with plastic sheeting, and is suitable for CIP operation. A similar press (Towerpress) is manufactured by Tebel Machinefabrieken BV in Holland, and a combined vacuum and tower press suitable for Cheddar cheese is shown in Fig. 19.

A different type of vertical press, i.e. the 'Pallet' Press has been reported by Scott (1981), and it has a pressing capacity of 14 tonnes of cheese. The press is divided into two main sections, and each section holds 36 × 20 kg block moulds stacked in six layers. This press is highly mechanised, and can also be used under partial vacuum. Incidentally the 'Pallet' Press is not manufactured anymore.

Table, trolley, conveyor and/or tunnel presses
These types of presses have been mainly developed for pressing semi-hard cheese varieties, and the pressure is applied from above (see Fig. 20). These presses consist of a number of rows which can be filled manually

A. Y. Tamime

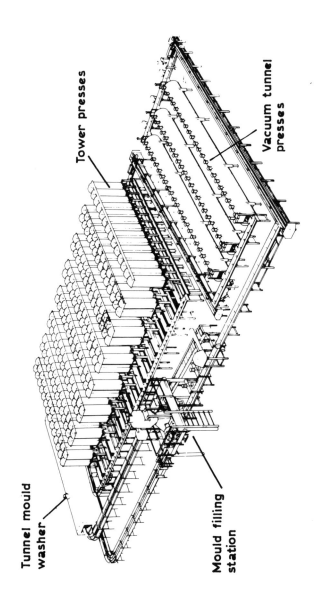

Tower presses

Vacuum tunnel presses

Tunnel mould washer

Mould filling station

Fig. 19. A Cheddar cheese pressing system. (Reproduced by courtesy of Tebel Machinefabrieken BV, Holland.)

or mechanically, and pressing commences whenever a row is filled. For a highly mechanised cheese plant, the latter two types of press are widely used (Scott, 1981). In some instances, cooling of the pressing area is recommended during the manufacture of Dutch cheeses and most equipment suppliers provide such facilities. Furthermore, some of these presses are automatically programmed to increase the pressures when required. The economics and efficiency of three of the above presses are illustrated in Fig. 21.

A recent development of a conveyor press is the 'air cushioning' press by Koopmans BV. The pressing cylinder does not press directly onto the cheese mould, but instead the pressure is applied on a pressure box which is tightly packed within a pressure tube. This tube adjusts any irregularities of the positioning of the lids and all the cheeses are pressed under the same pressure. Such developments facilitate pressing cheeses of different sizes at the same time (Hansen, 1984c).

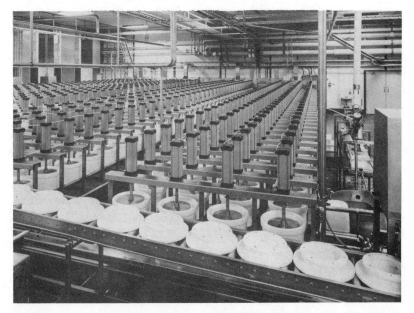

Fig. 20. Four conveyor presses in a cheese factory in Europe. (The Dutch variety of cheese is pressed in a Kadova plastic mould.) (Reproduced by courtesy of Tebel Machinefabrieken BV, Holland.)

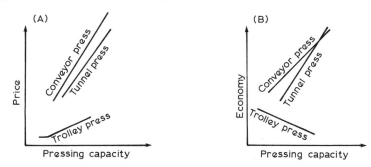

Fig. 21. Comparison of efficiencies and economics of different presses manufactured by Tebel. (In (B) the following aspects are taken into account: depreciation, space, labour time, number of moulds and maintenance.) (Reproduced by courtesy of Tebel Machinefabrieken BV, Holland.)

Rotary presses

Gadan Cassette® Press. This type of press (Fig. 22) consists of 8–12 arm-brackets which hold the cassettes. Both ends of the cassette are open, and the cheese moulds are loaded mechanically. Pressing commences after filling each cassette, and as the arm-brackets rotate, the cassette pivots on both brackets, and hence the moulds remain upright. Any shape of individual cheese moulds (round, rectangular or loaf) up to 20 kg capacity can be handled in this Gadan pressing system.

Hermann Walder. This rotating cheese press consists of guided rails for receiving the moulds, and a clamp to hold each mould under the pressing cylinder. The whole unit is enclosed within a stainless steel frame which is cylindrical in shape and 3 m in diameter (Hansen, 1979e). Pre-pressing commences when each horizontal section of the press is filled with cheese moulds, and the press rotates 1/12 of the circumference of the rotor before full pressure is applied. As a result, the cheese moulds rotate during the pressing period. The process can be made continuous by installing two presses in parallel, and furthermore, as the filled press can rotate, it is possible to empty the pressed moulds and re-fill with moulds to be pressed simultaneously.

MKT tunnel press. This type of press (Fig. 23(A)) has been developed by MKT Tehtaat OY for pressing an 84 kg 'block'-shaped Emmental

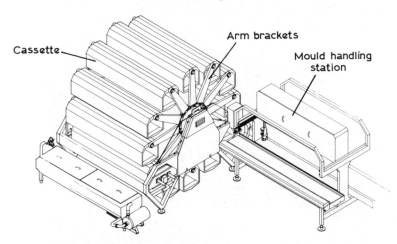

Fig. 22. The Gadan Cassette® Press. (The arm brackets are designed in such a way as to deliver the compressed air supply required to the top of each cassette.) (Reproduced by courtesy of Gadan Maskinfabrik A/S, Them, Denmark.)

cheese. The pre-pressed cheese is moulded and transferred by conveyor to the tunnel presses. Each press holds 12 × 84 kg cheese moulds, and inside the press, the lids are placed automatically. Pressing commences as each press is filled, and the duration of pressing is 20 h. During the pressing period, the moulds are rotated slowly in order to ensure a uniform moisture content in the cheese. The rotary action of the press simulates the traditional method of turning the Emmental cheese several times a day during pressing.

APV cheese press. The cheese moulds are constructed on both sides of a supporting structure which is rotated by a motor up to 180 degrees (see Fig. 23(B)), and this type of press has been developed by APV OTT AG in Switzerland. This type of press is suitable for the production of Gruyère or Sbrinz cheeses.

The presses mentioned above are designed for a CIP cleaning system.

Vacuum presses
The development of the vacuum press was primarily aimed at reducing the time required to press a cheese, but it also helps to cool the cheese during the pressing operation. Examples of such presses are as follows.

Fig. 23. Presses used during the manufacture of Swiss cheeses. (A) MKT presses Yhteisjuustola Emmental Cheese Factory in Finland, and notice the MKT-S cheese vats in the background. (Reproduced by courtesy of MKT Tehaat OY, Helsinki, Finland.) (B) APV OTT cheese presses which are used during the production of Gruyère and Sbrinz cheeses. (Reproduced by courtesy of APV OTT AG, Worb, Switzerland.)

Tebel tunnel press. This press is shown in Fig. 19 for pressing Cheddar cheese, and the first stage of pressing takes place in a tunnel under vacuum, and the second stage of pressing is carried out in a Tower Press. It is possible to suggest that, in certain instances, the cheese could be packaged directly after the vacuum pressing stage, but alternatively, the cheese could be retained in another pressing system until the packaging shift commences work the following day.

Gadan CH-2000 Cassette® Press. This press consists of a cassette rack system which is designed for pressing Cheddar cheese in 'block'-shaped, plastic moulds (6 × 20 kg). The cassette is made out of stainless steel capable of withstanding vacuum and a pressure of 98 kN m^{-2}. An illustration of the Gadan CH-2000 Cassette® Press is shown in Fig. 24, and it can be observed that the mould filling, stacking, pressing, de-stacking, mould washing and re-filling is highly mechanised.

The inside top of each cassette is fitted with a high strength, flexible, plastic tube which can be filled with compressed air to exert the desired pressure on the cheese. The rack model is supplied with rails in front of the cassettes to carry the mould transport unit. Thus, the sequence of operation is as follows:

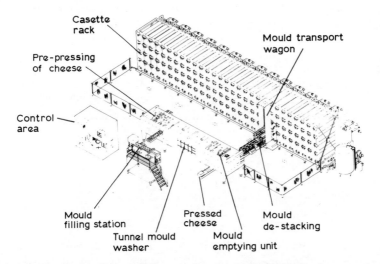

Fig. 24. The Gadan Cassette® Vacuum Press type CH-2000. (In the control area the system is fully computerised and what takes place in each cassette could be viewed on a closed circuit TV set. (Reproduced by courtesy of Gadan Maskinfabrik A/S, Them, Denmark.)

— stack cheese moulds in a cassette;
— close the door of the cassette;
— apply vacuum in the filled cassette;
— fill plastic tube with compressed air;
— pressing time for Cheddar cheese is 2 h (see also Hansen, 1982a,b).

Wincanton Block Former. This machine has been designed as a continuous system for the fusion of the salted, curd chips under vacuum, so avoiding the need for cheese moulds and presses. The Wincanton Block Former resembles a rectangular tower around 6–7 m high, and recently a round tower has been developed for the production of 'wheel' or 'cylindrical'-shaped cheeses. By mid-1984, the total number of Wincanton Block Formers installed all over the world amounted to 207, and one 'round' was also in operation.

The salted curd is air-blown to the tower, and depending on the condition of the curd being fed to the tower, the filling time is around 20 min; the residence time of the curd in the tower is approximately 35 min. In theory, the weight of the curd mass in the tower, and the vacuum condition, is able to produce a block of cheese every 1½ min, once the tower has been filled with salted curd. The sequence of block discharge is as follows:

(i) the elevator platform, which is enclosed at the bottom of the tower, is raised beneath the guillotine;
(ii) the guillotine is withdrawn and the column of curd (i.e. supported by the platform) is lowered to the height of a standard block of cheese; then the guillotine moves forward to sever a block from the column of curd;
(iii) the door opens and an ejector discharges the block of cheese into a bag-loader;
(iv) the door closes and the cycle is repeated; the overall sequence of operation has been reported by Wegner (1979) and Hansen (1982c).

Carousel® Press. The Carousel® Press has been developed by D. C. Norris & Co. (Engineering) Ltd, (UK) for the pressing of salted curd chips in plastic moulds under vacuum. This cheese press was first installed in the South Caernarvon Creameries. The press consists of 24 stacks of cylindrical, stainless steel canisters which are supported by a box-section frame rotating on a circular track on nylon wheels. Filled moulds pass through a curd levelling unit before the lids are placed on. The moulds are stacked in each canister (i.e. to hold six or three moulds in alternating stacks) via a vertical feed conveyor. When each canister is fully loaded the Carousel® moves slowly around to the next position, and the mechanically controlled door is closed (Pope, 1984). Each canister is fitted with specially guided rails suitable for holding 'block' or 'wheel'-shaped moulds. Over each mould, a pneumatically operated cylinder presses the cheese while the vacuum is drawn to the required level.

The pressing specifications for a 20 kg block Cheddar cheese are: vacuum 80–85 kN m^{-2}, air line pressure 275 kN m^{-2} and the duration of pressing is 1–2 h. After pressing the cheese, each canister is emptied, the moulds are transferred to a de-lidding and knock-out station, and the cheese is packaged; the moulds are washed to be re-filled with curd.

Bulk presses

Salted curd can be pressed in bulk, and this development was aimed at eliminating the use of a large number of individual moulds. Examples of such pressing systems are as follows.

Large Hoop or 'Ton' Press. This type of press was developed roughly two decades ago in New Zealand for Cheddar cheese pressing, and the specifications of the system have been reported by Crawford (1976) and Scott (1981). In brief, the press consists of a rectangular chamber fitted with a piston from the bottom end. The piston is hydraulically operated to both compress the curd, and extrude the pressed curd. The sequence of operations of the 'Ton' Press are as follows:

   (i)   fill the chamber with salted curd;
  (ii)   close the door of the press;
 (iii)   apply vacuum, e.g. 95 kN m^{-2} for 20 min;
 (iv)   maintain vacuum and apply low pressure at 6200 kN m^{-2} for 10 min;
  (v)   release vacuum and press at 9300–10 340 kN m^{-2} overnight (during peak season, the pressing time is only 8 h);
 (vi)   after pressing, the cheese is pushed to the height of a 'block'-shaped cheese and guillotined, and then further cut to give 4 × 18 kg cheese at each extrusion.

This type of press incorporates a micro-processor control system for operation, fault warning and the CIP programme.

Block-System (290 kg) Press. This is a Damrow pressing method, and the complete processing line is illustrated in Fig. 25. As can be observed, the cheese, e.g. Cheddar, is pressed under normal atmospheric conditions followed by pressing under vacuum in specially designed chambers. The sequence of operations of the Damrow Block-System including the time required for each hoop pressing (in parentheses) are:

— Station 1: assemble the hoop on a special cart fitted with four castors (3 min).
— Station 2: pre-fill the hoop with salted curd by using a cyclone filler, let the curd settle and finally fill the hoop to the desired weight on scales (2½–4 min).
— Station 3: apply no pressure and allow whey to drain freely (20 min).
— Station 4: pre-press each hoop (3 min).
— Station 5: main pressing area (120 min).
— Station 6: vacuum chamber where the hoop is retained for a minimum period (30 min).
— Stations 7 and 8: turn over the hoop and carry out the necessary capping and banding (3 min).
— Station 9: before transferring the hoop to this section, weigh the cheese (1½ min) and store cheese at 4–7 °C (4–5 days).
— Station 10: remove hoop.
— Stations 11, 12 and 13: bulk packaging of cheese.

A slightly modified Damrow pressing line for Cheddar cheese has been reported by Hansen (1982d) where vacuum pressing is not employed, and after storage in a cold room (3–4 days at 2–4 °C) the cheese is cut into 18 kg blocks and packaged. It is recommended that around 300–400 hoops are required for a cheese plant processing 250 000 litres of milk per day into Cheddar cheese.

Cheese pressed in individual moulds is transferred to a de-moulding machine or 'knock-out' station before any further handling of the cheese is carried out. This latter equipment removes the lid from the mould, inverts the mould and removes the cheese. In instances where mould liners are used, they must be removed (manually) before the cheese is packaged. However, 'traditional' cheeses are retained in their cloth bandages until the cheese is fully matured.

Cheese (e.g. Gouda and Edam) pressed in plastic moulds is easy to remove from the mould compared with cheeses requiring high pressure, i.e. the pressing of Cheddar cheese in Lauda moulds. However, the

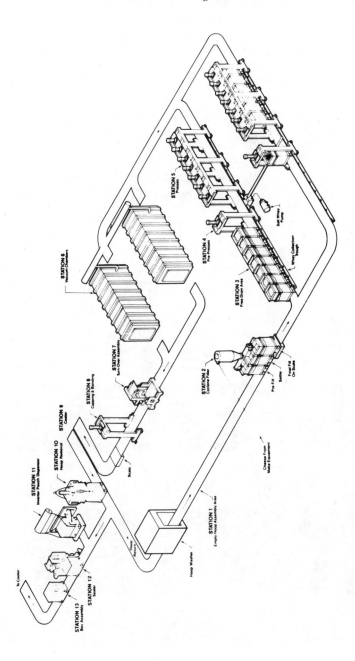

Fig. 25. The Damrow Block-System pressing. (Reproduced by courtesy of DEC Int. A/S — Damrow Products Division, Kolding, Denmark.)

pressed Cheddar cheese is easily removed from the mould using compressed air, and such a system has been described and schematically illustrated by Hansen (1978b).

Bulk packaging the cheese

After the pressing stage, the handling and packaging of the cheese is dependent on its variety, and examples may include the following treatments.

Emmental cheese

A pressed 84 kg 'block'-shaped Emmental cheese is wrapped in a cling film and stacked on a specially designed pallet (see Fig. 26), which can be mechanically turned during the ripening/storage period (Hansen, 1984a). At present, the packaging operation is carried out manually, but mechanisation of the process is under investigation for future development.

The smaller 'block'-shaped Emmental could be packaged mechanically in a Cryovac-BK bag. This type of packaging material is a laminate of

Fig. 26. Mechanical handling for turning large 'block'-shaped Emmental during the maturation period. (Reproduced by courtesy of MKT Tehtaat OY, Helsinki, Finland.)

different layers of plastics, and the sequence of operations using this system is as follows.

— The bag loader (e.g Cryovac Model BX 14) indexes one of the taped bags into position, opens it, and then automatically loads the block of cheese before depositing the bagged product onto the in-feed conveyor of the vacuum chamber.
— An operator straightens the neck of the bag to ensure that a perfect, wrinkle-free, heat-seal closure is obtained after air has been removed from within the machine's chamber.
— A second operator carries out visual inspection as the cheese passes through a check/weighing station and a weight label is applied.
— The cheese passes through a shrink tunnel and is placed in a plastic box or cardboard container.

The principle behind the Cryovac* packaging system (Registered trade mark of W. R. Grace & Co.) ensures that the cheese is matured under the following conditions:

(a) in an atmosphere free from oxygen, thus preventing surface mould growth;
(b) the shrinking of the packaging material provides a tight wrapping of the cheese free from surface wrinkles;
(c) the packaging material prevents moisture loss, physical damage and contamination of the cheese;
(d) the Cryovac-BK bag is impermeable to oxygen, but permeable to carbon dioxide; this latter property is necessary to prevent gassing during the secondary fermentation stage of Emmental cheese.

Cheddar and related cheese varieties
Different systems have been employed for the packaging of Cheddar cheese and other British territorials. These may include the following.

Pukkafilm®. This type of packaging material consists of a waxed cellulose laminate which is supplied by DRG Flexible Packaging (UK). First, the cheese block is wrapped with the laminate; second, it is over-wrapped with waxed cellulose; and third, the cheese is placed in a chamber for sealing by the application of heat and pressure (Hansen, 1975; Crawford, 1976).

Unibloc® System. The pressed cheese is wrapped with a plastic film, e.g. Saran® which is manufactured by the Dow Chemical Co. Ltd, and over-wrapped with a layer of paper prior to packaging within six wooden slats (British Patent 937441). The cheese is compressed within the slats by a specially designed machine, and the pressure is maintained by placing four metal straps around the cheese. In some instances, the wrapped cheese is placed within a thin cardboard box before final packaging. This box serves as a dispatch unit when the cheese leaves the factory, and the wooden slats are retained on the premises.

The film wrapping of the cheese could be mechanised, but in many cases, the process is carried out manually. However, a comparative costing of different packaging equipment, including the Unibloc® System to wrap 18 kg blocks of cheese at a rate of 5–6 min⁻¹ has been reported by Gray (1975).

A new development in the Unibloc® System is the use of lightweight plastic tray/lids (Kenkaps® or Plastic Cheese Kaps® — UK Patent pending) in conjunction with four, wooden, side slats. This development provides rigidity, and the telescopic effect maintains the shape of the block (i.e. overcomes the tapered edge), reduces labour requirements to a minimum, and the steel bands for strapping are not required. Incidentally, the cheese block is packaged, in a vacuum pouch or heat-shrink bag and not in a Saran® film.

Storpac®. The packaged cheese, e.g. in a vacuum pouch or heat-shrink bag, is wrapped in a thin cardboard box (optional), and is placed in a wooden box with a loose cover (UK Patent 1433361). The latter piece is held onto the box using a plastic band for strapping. On dispatch, the cheese is removed from these boxes which are retained in the factory.

Heat-shrink bags. An example of such a bag is the Cryovac — BB1* bag which consists of three main layers: polyolefine/PVDC barrier layer against oxygen and moisture/cross-linked polyolefine. The sequence of operations of packaging Cheddar cheese in this system is similar to the method described for packaging Emmental cheese. For 18 kg blocks, the bag is heat-sealed after the vacuum stage; however, a perfect seal could be achieved using a metal clip when packaging small Cheddar cheese truckles, or 'baby' Gouda and Edam. The clipping machine, e.g. Tipper Clipper, is supplied by Ben Langen (UK) Ltd, and the machine is equipped with a bag trimming facility after the sealing stage; however, an alternative machine, which is widely used in the industry for Cheddar cheese truckles or 'baby' Gouda and Edam, is the Cryovac VC10. Gas production in Cheddar cheese during the maturation period is considered a serious problem, and a quick remedy is to package the cheese in a carbon dioxide permeable material, e.g. Cryovac-BK bag.

Vacuum pouches. Different types of plastic film laminates could be used to package Cheddar cheese, and such pouches should provide a barrier against oxygen ingress and moisture loss. One such example is th Diolon® pouch which consists of 20 μm nylon (polyamide) and 60 μm polyethylene, and is supplied by Transparent Paper plc in Lancashire (UK). The sequence of operations can be summarised as follows:

— place the 18 kg block of cheese in the pouch;
— transfer to vacuum chamber;
— remove the air and heat-seal;
— package cheese in a shipping container, e.g. Unibloc®, plastic box or cardboard container.

Gouda, Edam and related cheeses

After the brining stage, the cheese is plasticised twice to prevent mould growth during the ripening period, and this process is repeated several times if the cheese is to be stored for long periods. Prior to dispatch, the cheese is washed, dried and coated with paraffin wax and overwrapped with a red cellophane film (the latter packaging material is optional). The mechanical handling of the cheese in the store, and the waxing equipment are discussed below.

An alternative approach for the packaging of the 'loaf', 'block' or 'round' Dutch cheeses is to wrap the product in a heat-shrink bag which is either sealed by heat or with a metal clip.

It is evident that different types of packaging materials are used in the cheese industry, and some relevant technical specifications of these laminates are shown in Table XXI. Cheese packaging line(s) can be highly mechanised, and Fig. 27 illustrates the packaging of 18 kg block Cheddar cheese in the United Kingdom.

Storage of the Cheese and Miscellaneous Handling

The quality of any cheese variety is dependent on many factors such as the quality of the milk, the activity of the starter cultures, and the

TABLE XXI
Some properties of cheese packaging materials

Film	Laminate	Thickness (μm)	Moisture vapour transmission	Oxygen permeability[†]	Carbon dioxide permeability[†]
Saran®	PE, PVC, PP, polyester, foil and paper	25	3·1 g m⁻²/24 h/38°C/ 90% RH	12-17 cm³ m⁻²/24 h/23°C	59-93 cm³ m⁻²/24 h/23°C
Diolon®	Nylon/PE	20/60 50/70 80/100	6·5 6·0 4·2	70 35 20 } cm³ m⁻²/24 h/75% RH/25°C	NA NA NA
Diomex®	Polyester/PE	12/50	5·0 g m⁻²/24 h/38°C/ 90% RH	100 cm³ m⁻²/24 h/0% RH/22°C	NA
Metallised film	Polyester/foil/low density ethylene butene copolymer	12/12/50	<1·0	<1·0 cm³ m⁻²/24 h/75% RH/25°C	NA
Pukkafilm®	Cellulose/wax/ Pukkacote	NA	<0·5 g m⁻²/24 h/25°C/ 75% RH	<20 cm³ m⁻²/24 h/23°C	NA
Novaflex II®	Nylon/polythene	30/50	2·3-2·6 g m⁻²/24 h/25°C/ 75% RH	30-40 cm³ m⁻²/24 h/50% RH/23°C	NA
Cryovac — BB1*	Polyolefine/PVDC/ polyolefine	60	Max. 15 g m⁻²/24 h/ 38°C/100% RH	Max. 40 cm³ m⁻²/24 h/ 50% RH/23°C	Nominal 150 cm³ m⁻²/ 24 h/50% RH/ 23°C
— BK1	As above			0-200	Min. 500
— BK5	As above			0-375	Min. 1400

PE = polyethylene. PP = polypropylene. PVC = polyvinyl chloride. PVDC = polyvinylidene chloride polymer. NA = not available. ® = registered trade mark. *see text.
[†]Pressure applied during testing is 1.01×10^2 kN m⁻².
Note that the permeabilities are at different test conditions and only if the expression is the same can the figures of each packaging material be compared.

manufacturing stages; however, other criteria, for example, the bulk handling of the cheese in the store and the conditions provided during the maturation period are important. The latter aspects can influence the biochemical changes in the curd during the storage period, and can ultimately affect the quality of the cheese. The handling and storage conditions of cheese may vary from one variety to another, and some typical examples include the following.

Swiss cheeses

The Emmental and Gruyère type cheeses undergo a secondary fermentation during the early stages of storage, and 'eye' formation takes place in the cheese due to the metabolic activity of the propionic acid bacteria. Thus, the 'wheel'-shaped Swiss cheeses are matured under controlled temperature and relative humidity (RH) conditions, and examples of the handling of the cheese in the store are as follows.

Fig. 27. Bulk packaging (5–6 blocks min^{-1}) of Cheddar cheese at a factory in North Tawton, UK. (1: overhead 'ski' type or J-hanger conveyor transferring cheese from the de-moulding station to the packaging department; 2: placing the block of cheese in a heat shrink bag; 3: vacuum and heat seal chamber; 4: shrinking tunnel.) (Reproduced by courtesy of Express Dairy (UK) Ltd, Middlesex, UK, and W. R. Grace Ltd, London, UK.)

Gruyère
Store the cheese at 10 °C for 3 weeks, and then at 15–20 °C for 2–3 months in an atmosphere of 90–95 per cent RH followed by storage at 12–15 °C and 85 per cent RH for a duration of 8–12 months (Scott, 1981). The cheese is turned regularly and rubbed with a damp cloth (i.e. soaked in brine solution), which aids the growth of those bacteria that provide a red-brown smear coat.

Emmental
Store the cheese at 10–15·6 °C for 10–14 days at 90 per cent RH then at 20–24 °C for 3–6 weeks at 80–85 per cent RH follwed by storage at 7·2 °C or less for 6–12 months at 80–85 per cent RH. The cheese is also turned and the surface is wiped with a cloth soaked in brine (Scott, 1981).

Prior to dispatch, the 'wheel'-shaped cheeses are cleaned, stamped, waxed and packaged in a shipping container.

'Block'-shaped Swiss cheeses
Since the cheese is packaged in a barrier-type material, only the storage temperature is taken into account and humidity control is not necessary. For example, the handling of 84 kg 'block'-shaped Emmental in a factory in Finland (Fig. 26) is as follows: store the packaged cheese at 8 °C for 4 weeks, then at 23–24 °C for 6 weeks and then 6 °C for a few weeks. The cheese is ripened only for 3 months (Hansen, 1984a). According to Scott (1981) a 'block'-shaped Emmental (e.g. 18–45 kg) is handled as follows: dry the cheese after brining for 4 days at 7 °C; package in barrier film material; store at 17–18 °C for 10–15 days until 'eye' holes have formed (1·3–1·9 mm in diameter) and finally store at 8–12 °C until the cheese is matured. Turning of small size 'block'-shaped Emmental in the store is not necessary.

As mentioned elsewhere, temperature control during the pressing, secondary fermentation and storage of large size Swiss cheeses is critical, and an extensive study on the effect of temperature variation in cheese on the rate of development of D(−) and L(+) lactic acid has been recently reported by Gehriger (1979); however, an extensive review on Emmental type cheeses, which covers published data over the last 10 year period, has been reported by Kasprzyk *et al.* (1983).

British cheeses
The traditional type of British cheeses are stored at different temperatures (e.g. 7·2–18·3 °C) depending on the variety (Scott, 1981), and most likely

at 85 per cent RH to prevent dryness and crack formation in the rind. It is necessary that the cheese is turned during the maturation period, and such a process is carried out manually in small cheese factories or by using rotating shelves. The latter type is illustrated by Davis (1965). 'Block'-shaped cheeses are normally matured at around 10°C, and plastic laminates (i.e. impermeable to oxygen and moisture) are used as the packaging material. Therefore turning of the cheese in the store is not necessary. The stacking of cheese blocks in the maturation room is dependent on factors such as: shipping or outer packaging container, daily production and/or mechanised system employed. Some examples of handling 'block'-shaped cheeses are:

(a) Cheeses wrapped in cardboard boxes are normally stacked manually 2 × 18 kg blocks high on wooden or metal shelves to avoid deformation in the shape of the block. The system could be mechanised by using specially designed metal pallets, and one such type is constructed with angle-iron uprights with indentation feet for stacking the pallets on top of each other. The pallet contains two folding shelves where cheese blocks, i.e. three on top of each other, can be stacked on the metal base and the shelves. In total, the pallet holds 72 blocks of cheese (12 cheeses in each layer), and the stacking of these pallets is carried out using a forklift truck.

Cheddar cheese packed in cardboard boxes could be stacked mechanically or manually by using a 'Boxpallet' (Hansen, 1982e) where the cheese is distributed on four levels with 12 cheeses on each level and has a capacity of 48 cheeses.

(b) Cheeses wrapped in Unibloc® or Storpac® are stacked on wooden pallets where each pallet holds 45 cheeses.

(c) Cheese wrapped in plastic containers is stored in metal pallets similar to the type mentioned above without the folding shelves, and the capacity of each pallet is 60 cheeses.

It is evident that different systems could be used for storing 18 kg Cheddar and related cheese varieties, and some examples are illustrated in Fig. 28.

(d) Cheese, which is not wrapped in any type of an outer/shipping container, can be stored in an OR-Maturing System developed by Olavi Räsänen OY in Mikkeli in Finland (see Fig. 29). All the different sections are made out of wood, and no cartoning of the cheese is required, and no metal strap handling is necessary

Fig. 28. Different illustrations of modern cheese stores in the United Kingdom. (A) Cheese wrapped in cardboard boxes at the SMMB Galloway Creamery. (Reproduced by courtesy of the Scottish Milk Marketing Board, Paisley, Scotland, UK.)

when stacking. Each pallet holds six different decks of 8 × 18 kg block cheeses. The collar supports the frameworks and construction of each deck, and in this system, every block of cheese is subjected to an individual pressure.

The OR-Maturing System is employed in one cheese factory in the United Kingdom, where after 3–4 weeks, the collar section is

Fig. 28 — *contd.* (B) A view of cheese packaged using the Unibloc® system at Sorbie Creamery. (Reproduced by courtesy of Sorbie Cheese Ltd, Wigtownshire, Scotland, UK.) (C) The storage of Cheddar cheese in Storpac® at Maelor Creamery. (Reproduced by courtesy of Dairy Crest Creameries, Milk Marketing Board, Thames Ditton, UK.)

Fig. 28 — *contd.* (D) The Express method ('Blue Boxes') for maturing Cheddar cheese at Priestdykes Creamery. (Reproduced by courtesy of Express Dairy (UK) Ltd, Middlesex, UK.)

removed for constructing new units for stacking 18 kg blocks of Cheddar cheese. Prior to dispatch, the cheeses are wrapped with thin cardboard box, and the wooden parts are retained in the creamery.

Dutch cheeses
Gouda, Edam and other related cheese varieties are stored for at least 28

days after manufacture at 12-13 °C and 85-90 per cent RH (Dijkstra, 1984), and then at a temperature of 10 °C and 85-90 per cent RH until the cheese is matured (Scott, 1981). The cheese is turned many times during the maturation period, and this process is highly mechanised in large cheese factories (Elten, 1979, 1981; Weenik, 1975; Hansen, 1981d,e,f, 1982f). The control of air in the cheese store is important, and some technical aspects have been reported by Bouman (1979) and Viswat (1979).

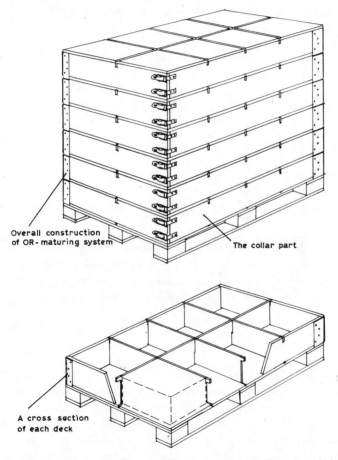

Overall construction of OR-maturing system

The collar part

A cross section of each deck

Fig. 29. The OR-Maturing System. (The construction of each unit is strong enough for 6 pallets to be stacked on top of each other.) (Reproduced by courtesy of Olavi Räsänen, Mikkeli, Finland.)

The brined cheese is plasticised a few times to prevent mould growth, and in order to prevent soiling of the shelves in the store; the cheese is plasticised from one side only, so that when it is turned, the dried section of the cheese is in contact with the shelf. At the end of the maturation period, the cheese is washed, dried, coated with paraffin wax (red or yellow) at 90–120°C and possibly over-wrapped with a cellophane material before it is dispatched to market. Mechanical handling of Dutch cheeses has been reported by Anon. (1979, 1982b) and Hansen (1980b, 1983b). In addition, Dutch cheeses could be packaged in an impermeable material, for example, by employing the Cryovac* packaging system.

Retail Packaging of Cheese

Consumer packaging of cheese, e.g. 250 g–1 kg can be carried out in cheese factories, or by specialised dairy packers. The sequence of operations involves the following:

— un-wrapping;
— portioning;
— packaging;
— weighing and labelling;
— packaging the cheese in shipping containers.

It is necessary to ensure that the handling of the cheese should be carried out under controlled atmospheric conditions, i.e. air should be filtered, and if possible, the whole area should be under positive pressure to minimise airborne contamination. Different types of plastics are used for the packaging of retail portions of cheese, and the same protective measures used for bulk packaging of cheese (e.g. protection against moisture loss and oxygen and carbon dioxide permeabilities) are also enforced.

Un-wrapping of the cheese blocks
Removal of the shipping container is normally carried out manually in an area isolated from the rest of the retail packaging line. The second stage of operation is the removal of the 'bulk' packaging material, and in some instances, cleaning the surface of the cheese from any mould growth.

It has been argued, however, that certain mould species produce mycotoxins which can penetrate the cheese, and it is recommended that a 'slice off' the whole side should be removed, rather than a mere scraping, in order to ensure that the retail portions do not contain any toxic substance(s).

Portioning of the cheese

As mentioned elsewhere, the very hard, hard and semi-hard cheese varieties are produced in different shapes and sizes, and the retail portioning of the cheese is dependent on factors such as:

— original shape of the cheese;
— consumer size and shape required;
— minimising wastage (i.e. off-cuts) during the portioning of the cheese.

In general, cheese is portioned into different shapes, e.g. wedge, rectangular or square, and Fig. 30 illustrates how some cheese varieties are cut. Different cheese portioning machines are available on the market, and some examples are as follows.

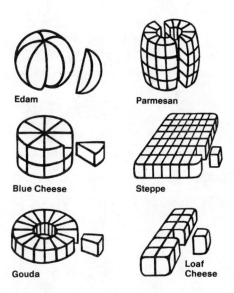

Edam

Parmesan

Blue Cheese

Steppe

Gouda

Loaf Cheese

Fig. 30. The portioning of different cheeses. (Reproduced by courtesy of Alpma Hain & Co. KG, Rott/Inn, West Germany.)

Alpma KSA 75 AV

This cutting machine is designed for handling the 'block' and 'wheel'-shaped 80 kg Emmental cheese (see Fig. 31). The cheese is cut into bars which are then fed through a high speed portioning machine type Alpma CUT 21. The latter machine has a capacity of 30–80 cuts min^{-1} depending on the variety of cheese.

Alpma HT II

This is a hydraulic cheese cutter suitable for portioning Edam, Gouda and all hard cheese types including Parmesan and Grana cheeses. The latter two varieties are normally cut by using special wires, and thus the portions have smooth surfaces which are not broken.

Wright Pugson cheese cutting machines

Different models of cheese cutting machines are manufactured by

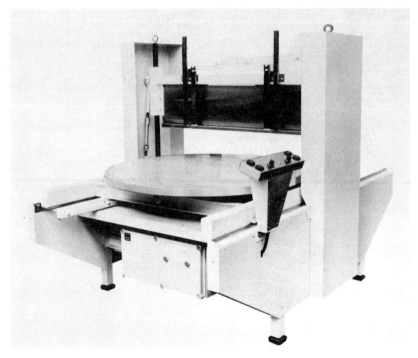

Fig. 31. An automatic machine type KSA 75 AV for cutting 'block' and 'wheel'-shaped Emmental up to a weight of 80 kg. (Reproduced by courtesy of Alpma Hain & Co. KG, Rott/Inn, West Germany.)

Wright Pugson Ltd, in Dorset (UK). These machines are entirely pneumatic in operation, constructed in stainless steel with polypropylene cutting heads, and some examples of these portioning machines are:

— Models RB9 MK IV and V for cutting 18 kg blocks of cheese into 4 × 5 kg or 8 × 2½ kg in approximately 20 s.
— Models 20 and 20A are capable of cutting cheese blocks into 4 × 4·5 kg or 6 × 3·2 kg and incorporate accurate height sensing to ensure precise halving of the height of the cheese.
— Model DC3 is an ideal high speed cutter and the rate of production is up to 800 × 225 g random weight pieces per minute.

Codat 400 and 500 range machines

These machines (Fig. 32) are designed to cut 18 kg blocks of cheese into retail size portions and place them automatically in form/fill/seal type packaging machines. This type of cutting/loading machines is produced by Codat Systems Ltd, in Dorset (UK), and was developed in conjunction with the National Research Development Corporation.

The motions of the machine are powered by pneumatic cylinders, except the cutting plate which is operated by a hydraulic cylinder, and the sequence of operations is as follows:

— the blocks of cheese are loaded manually onto the roller accumulator which has a capacity up to 20 blocks;
— each block is released and raised automatically to the level of the cutting plate and then centred on the machine axis;
— an electronic micro-processor measures the length and the height of the block before raising the block to a pre-selected thickness through the first cut frame wire and adjustable side trimmers;
— the portions are formed one layer at a time by using a retractable single wire slicer;
— the cheese portions are lifted by vacuum pick-up heads, spaced into dimensions to match the packet spacing of the form/fill/seal machine and lowered and placed accurately into the pockets.

The Codat machine was first installed at Maelor Dairy Crest Creamery in the United Kingdom for portioning 18 kg 'block'-shaped Cheddar cheese (Hansen, 1980c), and it was reported that by operating four cheese cutting machines (e.g. the wrapping throughput 200 g × 84 packages per minute), it was estimated that 1680 × 18kg blocks of Cheddar cheese could be packaged per 7 h working day.

Fig. 32. The Codamac 402/12 for cutting 18 kg blocks of cheese and loading the portions into packaging machines. (Reproduced by courtesy of Codat Systems Ltd, Dorset, UK.)

Equipment for cheese packaging

A multitude of high speed cheese portions packaging machines are available on the market. It would be impractical, of course, to discuss all the different types of machines in detail, but from a technical point of view certain important specifications must not be overlooked, for example:

— capital cost;
— type of packaging material used;

— proposed method of packaging, e.g. over-wrapping, vacuum, heat-shrink, and/or gas flushing;
— sanitary standards;
— versatility and reliability of the machine;
— power and labour requirements;
— degree of automation and handling;
— other specifications such as: availability of pricing, date marking and safety measures.

Some examples of cheese portions packaging machines are as follows.

Vacuum packaging machines

The majority of high speed, vacuum packaging machines are known as form/fill/seal, and one of the leading manufacturers is the Multivac Verpackungsmaschinen in West Germany. The packaging material is delivered to the dairy in large reels of plastic sheet, and two reels are fed into the machine. The base reel, which is a thermoplastic material, e.g. Diolon® 50/70 μm or 80/100 μm (see Table XXI) is warmed, and vacuum or compressed air stretches the film down into cooled multi-raw moulds which transforms the sheet into the desired shape or die. The cheese portions are placed into the die, and as the cheese container moves forward to the vacuum and heat-sealing chamber, it is covered by the plastic sheet from the top reel, e.g. Diomex® 12/50 μm or 20/60 μm (see Table XXI). The following stage consists of removing the air from the chamber, heat-sealing the thermoplastic materials to each other, and finally longitudinal and cross cutting units separate the cheese portions before transferring them to the weighing and labelling section. An illustration of such a machine is shown in Fig. 33 where 'block' and 'wedge'-shaped cheese portions are packaged.

The shelf-life of cheese packaged in this system (i.e. kept under refrigeration) is up to three months.

Gas-flushing machines

The replacement of air from the cheese package by gas (carbon dioxide or nitrogen) is known as gas-flushing and it prevents the spoilage of the product by oxidation.

In practice, carbon dioxide is widely used, compared with nitrogen as a gas-flushing agent, because it reacts with the moisture in the cheese to produce carbonic acid (H_2CO_3), and gives the product a neat, 'snug' appearance similar to a vacuum pack.

Fig. 33. The packaging of cheese portions in a form/fill/seal machine under vacuum. (1: thermoforming section; 2: bottom reel; 3: vacuum and heat sealing section; 4: cutting section; 5: top reel.) (These packaging machines are the Alpma VAC 470 and 471 models which are developed in conjunction with Multivac; however, a larger throughput machine is the Multivac 7000.) (Reproduced by courtesy of Alpma Hain & Co. KG, Rott/Inn, West Germany.)

Some examples of gas-flushing cheese packaging equipment are the Multivac, Alpma/Hayssen RT 118 and 318 and Rose Forgrove®. The latter machine is the RF 250 N Flowpak (Fig. 34) which has a capacity of 16–85 packs min^{-1} depending on the product, feeding arrangement and wrapping material. The plastic film (e.g. polyester/polyethylene laminate) is fed into the machine from a reel, a pillow-type pack with fin-sealed longitudinal and end seams is produced. As the wrapping material is formed into a tube in the folding box, a portion of cheese is delivered and, before fin-sealing the edges, the oxygen is replaced by gas; the residual oxygen is estimated at less than 1 per cent.

Over-wrapping machines

These machines merely over-wrap the cheese portion with a plastics film, e.g. Saran® or polypropylene laminate, and the estimated shelf-life of the product is up to 3 weeks. The object of packaging the cheese by this method is to provide the customer with a 'fresh-look' appearance to the cheese.

Fig. 34. The Rose Forgrove® (RF 250 N Flowpak) gas-flushing machine for packaging retail portions of cheese. (Reproduced by courtesy of Rose Forgrove Ltd, Seacroft, Leeds, UK.)

A typical example is the wrapping machine Model 670 which is manufactured by Carrington Packaging Ltd in Lancashire (UK) (Fig. 35), and the sequence of wrapping operations is as follows:

— The infeed conveyor comprises moving pushers and static base rail, and the cheese is simply transferred to the elevators position.

— A sheet of wrapping material is cut from a reel, and it is delivered to a position above the elevator and below the metal die box (i.e. a machine casting tailored to the shape and size of the cheese portion).

— As one elevator rises carrying the cheese into the wrapping material through the die box, the movement of the cheese is restricted by a spring-loaded top pressure plate at the top of the stroke and by product clamps on either side; however, the outer elevator withdraws, leaving the cheese supported on the smaller inner elevator and allowing two side tuckers to move from opposite sides and fold the trailing end of the packaging material under the cheese.

— The inner elevator withdraws immediately and the cheese is supported on the side tuckers; however, a third tucker, which is known as bottom folder, moves in from the rear and folds the wrapping material under the cheese.

— As the cheese is moved forward on to the discharge belt, the final trailing edge of the wrapping material is folded under the product.

— The discharge belt, which is made out of thin PTFE, carries the wrapped cheese over a heated plate to seal the folds of the wrapping material.

Incidentally, while the wrapping is tight enough, in instances where there is considerable variation in the dimensions of the cheese portion, it is recommended that the packaged cheese is passed through a heated tunnel to obtain a tighter final pack.

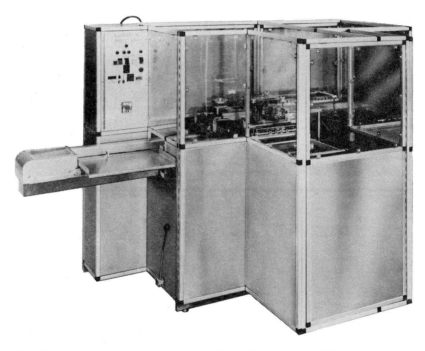

Fig. 35. Cheese portions wrapping machine. (This is a Model 670 which has the capacity of up to 80 packs per minute.) (Reproduced by courtesy of Carrington Packaging Ltd, Lancashire, UK.)

Heat-shrink packaging machines

In principle these machines are similar to the type used for bulk packaging of cheese. In Scotland, Cheddar and Dunlop cheeses produced on different islands (i.e. in 'square' block or 'wheel'-shapes of 200–450 g) are packaged in heat-shrink bags, e.g. Cryovac*, and normally the closure used is a metal clip.

Vacuum-skin packaging machines

The most recent development in packaging cheese portions is the use of a vacuum skin packaging technique, and a typical example is the Cryovac VS 44 Darfresh* machine (see Fig. 36). The product is placed on the lower web, which may be left flat or formed to suit the product if desired, before being indexed into the chamber of the machine. The top web, which has been pre-heated to improve its flexibility, is draped over the cheese portions using the minimum pressure differential, so that air is removed and the product is fully encased in a protective 'second skin'. The top and bottom webs adhere properly with each other, but may be peeled apart easily when required by the customer without damaging the product. Incidentally, a labelling device could be installed on the VS 44 machine before the cutting tool (see Fig. 36).

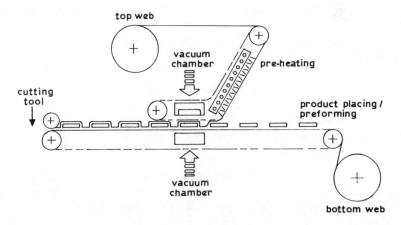

Fig. 36. The Cryovac VS 44 Darfresh* machine. The capacity of this machine is 10 cycles per minute, giving up to 80 packs per minute on small portions. (Reproduced by courtesy of W. R. Grace Ltd, London, UK.)

Miscellaneous equipment

Most of cheese portions packaging machines are equipped with a weighing and labelling unit. The latter machine provides the customer with price per kg, and the actual price of the cheese portion, the expiry date, the type of cheese packaged and place of origin (e.g. English, Scottish, Irish, New Zealand or Canadian Cheddar). The final handling of the retail cheese pack involves placing it into shipping containers ready for dispatch to the retail shops.

MECHANISATION OF CHEESE PRODUCTION AND PLANT DESIGN

As the scale of cheese production increases or becomes more centralised, the use of mechanisation to handle the milk, curd, whey, pressing and packaging of the cheese becomes inevitable. A wide range of equipment is available for the manufacture of different cheese varieties, but the final choice of any mechanised cheese system is governed by:

— scale of production;
— type of cheese(s) produced;
— degree of mechanisation;
— degree of automation, e.g. 'push button' or fully integrated system;
— labour saving;
— capital cost.

In view of the different equipment which could be used for the manufacture of cheese, the plant may be divided into the following sections in order to achieve a high degree of mechanisation and/or automation.

Area/Department 1

In this area milk handling takes place which may include:

— milk reception and storage;
— bactofugation and/or standardisation of the milk may be required during the manufacture of certain cheese varieties;
— heat treatment of the milk.

Area/Department 2

In this section, the following functions are carried out:

— production of the coagulum;
— de-wheying and curd handling;
— mould filling.

Area/Department 3

The pressing and brining of the cheese is carried out here, and different methods could be employed to provide the necessary processing systems.

Area/Department 4

In this department, bulk packaging and storage of the cheese is carried out.

Area/Department 5

The handling of the whey is carried out here, and the functions involved may include:

— recovery of curd 'fines';
— fat separation;
— miscellaneous treatment(s) of the whey depending on its utilisation.

Areas/Departments 6, 7 and 8

Cleaning-in-place station, the preparation of starter culture and effluent treatment, are carried out in these departments respectively; refer to Tamime and Robinson (1985) for further details.

Area/Department 9

In this section retail packaging of the cheese could be carried out (i.e. optional).

Area/Department 10

In this area plant maintenance and production of energy (e.g. hot water and steam) takes place.

Numerous mechanised and automated cheese systems are available on the market, and these can either be 'custom built' or obtained as a modular unit (see Figs. 37–40). Whatever system is chosen trained personnel are still required to operate and look after the plant, and the real attraction of automation could be summarised as:

(a) improved efficiency of the operator under more congenial conditions;
(b) improved quality and productivity;
(c) better utilisation of floor processing area.

The design and layout of a cheese factory play a major role in order to achieve the ultimate objectives mentioned above, and in an ideal situation the construction of a cheese factory and plant installation are carried out simultaneously. But the permutations available, particularly within existing buildings, mean that each cheese plant design has to be considered in its own right. In theory, and in practice, the cheese production line should take into account factors such as:

(i) the throughput of the milk processing plant should take into account the need for a continuous flow of milk to the cheese vats;
(ii) the capacities of, and duration of emptying of, the cheese vats should be synchronised for the continuous flow of the cheese curd to the curd handling equipment;
(iii) the number of working shifts required during a normal processing day;
(iv) the time required to wash the cheese vat before re-filling in order to avoid the diversion of milk;
(v) the de-centralisation of CIP station is recommended, so that different equipment could be cleaned when required rather than cleaning all the plant at one time;
(vi) the provision of an ample supply of hot water and steam in order to avoid an unnecessary shut down of the plant;
(vii) the systems of cheese pressing and brining could influence the degree of automation adopted, and for the former system, whether or not cheese moulds are required;
(viii) the area available for storage is also critical in order to ensure adequate space available, especially if long maturing cheeses are produced, and the degree of automation employed to handle the cheese in the store;

Curd / whey mixture inlet

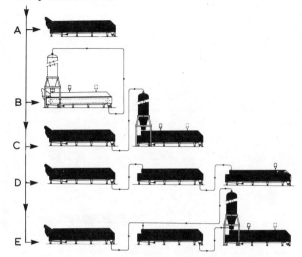

Fig. 37. Cheddarmaster installations for the production of different cheese varieties. (A: Feta or soft cheeses; B: Cheddar or Mozzarella; C: Cheddar or unwashed Colby; D: stirred Cheddar or British territorials; E: Cheddar or stirred British territorials.) (Reproduced by courtesy of Van den Bergh and Partners Ltd, Windsor, UK.)

    (ix)   the design of any cheese plant should accommodate some flexibility for future increase in production without incurring high capital investments or tremendous changes in the existing equipment.

It is evident that a cheese plant is a highly complexed structure, and Fig. 41 illustrates the layout of a factory for the production of Emmental and Edam cheeses in Finland

FUTURE DEVELOPMENTS AND CONCLUSION

The process of cheesemaking (i.e. very hard, hard and semi-hard varieties) is an effective method to preserve the nutritional components of milk for a long period of time. It is likely, however, that the future developments in this field, based on current scientific research work, will aim to fulfil some of the following requirements.

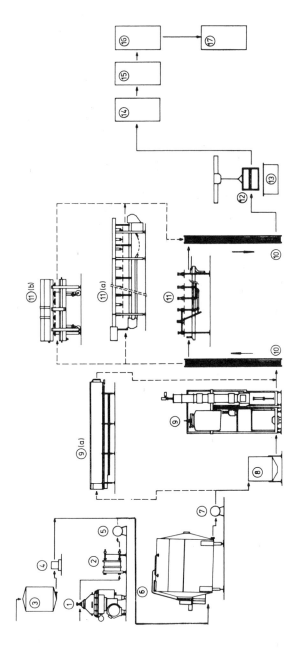

Fig. 38. Alfa-Laval production lines for Swiss, Gouda and Edam cheeses. (1,2: Equipment for preliminary treatment of milk; 3: bulk starter tank; 4: starter proportioning pump; 5: centrifugal pump; 6: cheese vat OST IV; 7: special pump for curd/whey mixture; 8: buffer tank; 9: Casiomatic; 9(a): DBS strainer vat; 10: conveyor; 11: conveyor press; 11(a): tunnel press; 11(b): trolley press; 12: container; 13: basin for brine; 14: green store; 15: surface treatment; 16: ripening store; 17: slicing and packaging.) (Reproduced by courtesy of Alfa-Laval Cheddar Systems Ltd, Somerset, UK.)

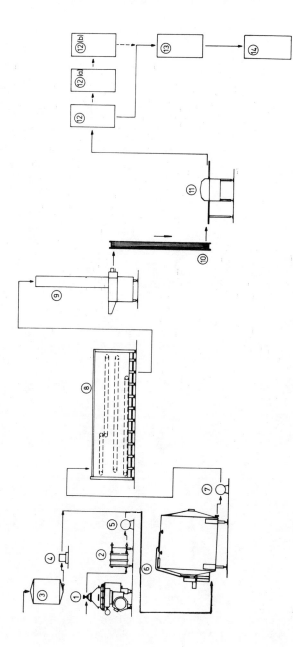

Fig. 39. Alfa-Laval Cheddar cheese production line. (1–7: see Fig. 38; 8: Alf-O-Matic; 9: Wincanton Block Former; 10: conveyor; 11: vacuum packaging; 12: packaging — wrap around carton; 12(a): packaging — shoe box carton; 12(b): packaging — Unibloc®; 13: cheese store; 14: slicing and packaging.) (Reproduced by courtesy of Alfa-Laval Cheddar Systems Ltd, Somerset, UK.)

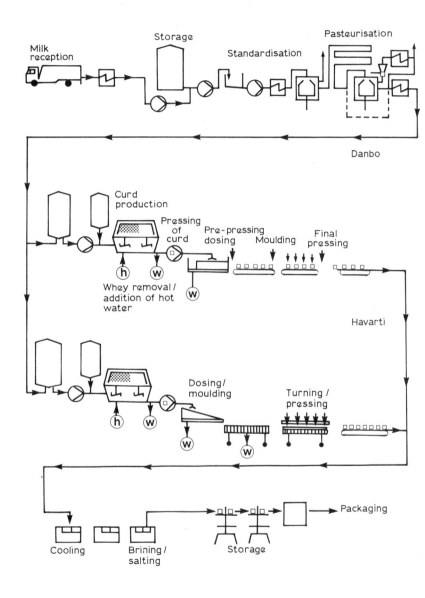

Fig. 40. Pasilac/Silkeborg production lines for manufacture of Danish cheeses. (Reproduced by courtesy of Pasilac/Silkeborg Ltd, Preston, UK.)

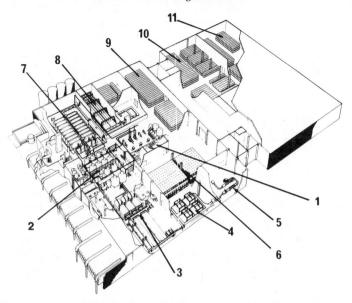

Fig. 41. Layout of the Yhteisjuustola cheese factory in Lapinlahti, Finland. (1: milk treatment; 2: MKT-S cheese vats. Edam section — 3: de-wheying, mould filling and pressing equipment; 4: brining tanks; 5, 6: packaging and maturation rooms. Emmental section — 7: MKT tunnel presses; 8: brining tank; 9, 10, 11: maturation rooms at 8°C, 23–26°C and 4–6°C, respectively, 12: cheese wrapping and packaging section.) (Reproduced by courtesy of MKT Tehtaat OY, Helsinki, Finland.)

First, Milk as a Raw Material

The variation in the chemical composition of milk, which ultimately affects the composition and quality of milk, can be overcome by standardisation of the casein and fat components to a desired fixed ratio. The concentration of milk by ultrafiltration is orientated towards better utilisation of the cheese milk (i.e. improved productivity, reduced cost and increased yield), but more research is required to optimise the cheesemaking process.

Second, the Starter Culture

Continuous selection and identification of cheese starter culture strains are likely to increase, in order to overcome certain problems during manufacture (e.g. attack by bacteriophages), and to produce cheese of similar flavour characteristics within desired and well defined flavour parameters. The understanding of the physiology of these cultures may help to improve their function during cheesemaking by the application of genetic manipulation, and up-to-date information in this field has been reported by Gasson and Davies (1984). However, the direct acidification of milk by acids (e.g. lactic or phosphoric) is likely to be applicable only to non-ripened cheese varieties.

Third, Milk Coagulants

Since the enzyme rennet (i.e. a mixture of chymosin and pepsin) is most widely used in the cheese industry and in view of its world-wide shortage, the major development in this area will be the search for an alternative and acceptable replacement, or to manipulate genetically certain micro-organisms to produce chymosin more economically.

Fourth, Mechanisation of Cheesemaking

Mechanisation will continue with a view to improving productivity and reducing labour costs. The latter aspect has already greatly improved, and according to Wilbrink (1982), the labour requirements to produce 100 kg of Edam cheese (in 2 kg size) and the same amount of Gouda cheese (in 10 kg size) in 1965 were 149 and 94 man-minutes respectively; however, in 1981 these figures were reduced to 42 and 12 man-minutes respectively. A high degree of automation, including the corporation of micro-processor control, has been achieved in the cheese industry over the past two decades — but there are still areas in cheesemaking where process control by instrumented measurement could be an advantage.

For example:

— an *in-tank* instrument for the measurement of the milk gel rigidity before cutting of the coagulum commences;
— an *in-tank* pH measurement and recorder of the acid development during cheesemaking;
— an instrument for measuring the ripening indices during the maturation of cheese in the store.

Fifth, Accelerated Ripening of Cheese

Different methods have been used to accelerate the ripening process of Cheddar cheese, and a comprehensive report has been compiled by Law (1978, 1984). What is highlighted from current research is the need for a better understanding of the biochemistry of the maturation of cheese, including the mechanisms of flavour development, in order to overcome obstacles, e.g. undesirable flavours, discoloration, and body and texture defects.

Sixth, New Method(s) of Making Cheese

The manufacture of cheese in a shorter period helps to contain the rising cost of production and helps to increase throughput of existing plants. A typical example is the 'Short Method' of Cheddar cheese manufacture which was developed in Australia. Hammond (1979, 1982) has reported such a process, where the manufacturing times (i.e. from cutting → de-wheying and from de-wheying → milling) based on the conventional method to produce Cheddar cheese requires 140 and 90 min respectively *vis à vis* the 'Short Method' 120 and 30 min respectively. Thus, the manufacturing time is reduced by 1 h and 20 min; however, the starter cultures consisted of mesophilic lactic acid bacteria and an active strain of *Streptococcus thermophilus* (TS3 or SD1).

Seventh, Packaging Materials and Equipment

Packaging technology of cheese has developed greatly over the past few years, and the future trend is to increase the rate of packaging (i.e. bulk and/or retail portion) and to improve the specifications of the packaging materials at a reduced cost.

Eighth, Flavouring of Cheese

Flavouring of natural cheeses with herbs have been produced for some time, and a recent development is the addition of alcoholic beverages which is aimed towards increase of cheese consumption. Tamime (1984) concluded that the addition of concentrated wine flavour to Cheddar cheese did not affect the quality and such cheese has a potential market.

154　　　*A. Y. Tamime*

ACKNOWLEDGEMENTS

The author acknowledges Dr R. J. M. Crawford and Miss Janet H. Galloway for their critical review of the manuscript, and Dr J. H. Dijkstra (NIZO) for providing some technical data on Dutch type cheeses. The author is also grateful for the enthusiastic response and assistance of all the companies in providing extensive data required, to Mr E. C. McCall (WSAC) for the preparation of the figures, and Mrs C. McInnes (WSAC) for typing the manuscript.

REFERENCES

Al-Darwash, A. K. (1983). In: Changes in the Characteristics and Properties of Milk from Production to Consumption — Cheese Manufacture and Quality, PhD thesis, University of Glasgow, Scotland, UK.

Al-Obaidi, G. Y. (1980). In: A study of the Use of Coagulants in Cheddar Cheesemaking, PhD thesis, University of Glasgow, Scotland, UK.

Anon. (1959). In: *Cheesemaking,* Bulletin No. 43, HMSO, London.

Anon. (1976). *North European Dairy Journal,* **42**, 248.

Anon. (1979). *North European Dairy Journal,* **45**, 264.

Anon. (1980a). *North European Dairy Journal,* **46**, 36.

Anon. (1980b). In: *Dairy Handbook,* Alfa-Laval A/B, Lund, Sweden.

Anon. (1981). In: *Geigy Scientific Tables, Vol. 1,* 8th edn, (Ed. C. Lentner), Ciba-Geigy Ltd, Basle, Switzerland.

Anon. (1982a). In: *Yearbook of Agricultural Statistics — 1978/1981,* Brussels, Belgium.

Anon. (1982b). *North European Dairy Journal,* **48**, 103.

Banks, J. M., Muir, D. D. and Tamime, A. Y. (1984a). *Journal of the Society of Dairy Technology,* **37**, 83.

Banks, J. M., Muir, D. D. and Tamime, A. Y. (1984b). *Journal of the Society of Dairy Technology,* **37**, 88.

Borcherds, K. B. (1983). In: *Miles Jaarlikse Kaasmakers Symposium — Cape Town,* Miles Lab. (Pty) Ltd, Cape Town, South Africa.

Bouman, S. (1979). *North European Dairy Journal,* **45**, 4.

Brockwell, I. P. (1981). In: *Proceedings of the 2nd Biennial Marschall International Cheese Conference,* Wisconsin, USA.

Cheeseman, G. C. (1981). In: *Enzymes and Food Processing* (Ed. G. G. Birch, N. Blakebrough and K. J. Parker), Applied Science Publishers Ltd, London.

Crawford, R. J. M. (1960). In: A Study of Factors Affecting the Activity of Lactic Acid Producing Cultures in Cheesemaking, PhD thesis, University of Glasgow, Scotland, UK.

Crawford, R. J. M. (1976). *Journal of the Society of Dairy Technology,* **29**, 71.

Czulak, J. (1981). In: *Proceedings of the 2nd Biennial Marschall International Cheese Conference,* Wisconsin, USA.

Dalgleish, D. G. (1982a). In: *Food Proteins* (Ed. P. F. Fox and J. J. Condon), Applied Science Publishers, London.
Dalgleish, D. G. (1982b). In: *Developments in Dairy Chemistry — 1 Proteins* (Ed. P. F. Fox), Applied Science Publishers, London.
Davis, J. G. (1965). In: *Cheese, Vol. I,* J. & A. Churchill Ltd., London.
Davis, J. G. (1976). In: *Cheese, Vol. III,* Churchill Livingstone, London.
Dijkhuizen, G. (1981). In: *Proceedings of the 2nd Biennial Marschall International Cheese Conference,* Wisconsin, USA.
Dijkstra, J. H. (1984). Personal communication.
Eck, A. (1984). *Dairy Science Abstracts,* **46,** 242.
Elten, G. J. van (1979). *North European Dairy Journal,* **45,** 98.
Elten, G. J. van (1981). In: *Proceedings of the 2nd Biennial Marschall International Cheese Conference,* Wisconsin, USA.
FAO (1972). In: *Production Yearbook, Vol. 26,* FAO, Rome, Italy, pp. 208–14.
FAO (1982). In: *1981 FAO Production Yearbook, Vol. 35,* FAO, Rome, Italy, pp. 229–33.
FAO (1984). In: *1983 FAO Production Yearbook, Vol. 37,* FAO, Rome, Italy, pp. 243–9.
FAO/WHO (1972). In: *Recommended International Standards for Cheeses and Government Acceptances,* CAC/C1-C25, FAO, Rome, Italy.
FAO/WHO (1978). In: *Joint Committee of Government Experts on the Code of Principles Concerning Milk and Milk Products,* CX 5/70-19th Session, FAO, Rome, Italy.
FAO/WHO (1984). In: *Code of Principles Concerning Milk and Milk Products, International Standards for Milk Products and International Individual Standards for Cheeses,* Codex Alimentarius Vol. XVI, 1st Edition, FAO, Rome, Italy.
Fox, P. F. (1981). In: *Proteinases and their Inhibitors, Structure, Function and Applied Science* (Ed. V. Turk, and L. J. Vitale), Pergamon Press, Oxford.
Fox, P. F. (1982a). In: *Use of Enzymes in Food Technology* (Ed. P. Dupuy), Technique et Documentation Lavoisier, Paris.
Fox, P. F. (1982b). In: *Proceedings of XXI International Dairy Congress, Vol. 2,* Mir Publishers, Moscow.
Fox, P. F. (1984). In: *Developments in Food Proteins — 3* (Ed. B. J. F. Hudson), Elsevier Applied Science Publishers, London.
Fox, P. F. and Morrissey, P. A. (1981). In: *Enzymes and Food Processing* (Ed. G. G. Birch, N. Blakebrough and K. J. Parker), Applied Science Publishers, London.
Fox, P. F. and Mulvihill, D. M. (1983). In: *Proceedings of IDF Symposium — Physico-Chemical Aspects of Dehydrated Protein — Rich Milk Products,* International Dairy Federation, Brussels.
Galesloot, Th. E. (1958). *Netherlands Milk and Dairy Journal,* **12,** 130.
Garvie, E. I. (1984). In: *Advances in the Microbiology and Biochemistry of Cheese and Fermented Milk* (Ed. F. L. Davies and B. A. Law), Elsevier Applied Science Publishers, London.
Gasson, M. J. and Davies, F. L. (1984). In: *Advances in the Microbiology and Biochemistry of Cheese and Fermented Milk* (Ed. F. L. Davies and B. A. Law), Elsevier Applied Science Publishers, London.

Gehriger, G. (1979). In: *Proceedings of the 1st Biennial Marschall International Cheese Conference,* Wisconsin, USA.
Gray, B. E. (1975). *Journal of the Society of Dairy Technology,* **28**, 11.
Hammond, L. A. (1979). In: *Proceedings of the 1st Biennial Marschall International Cheese Conference,* Wisconsin, USA.
Hammond, L. A. (1982). *Australian Journal of Dairy Technology,* **37**, 71.
Hansen, R. (1975). *North European Dairy Journal,* **41**, 301.
Hansen, R. (1976). *North European Dairy Journal,* **42**, 127.
Hansen, R. (1977a). *North European Dairy Journal,* **43**, 88.
Hansen, R. (1977b). *North European Dairy Journal,* **43**, 93.
Hansen, R. (1978a). *North European Dairy Journal,* **44**, 138.
Hansen, R. (1978b). *North European Dairy Journal,* **44**, 195.
Hansen, R. (1979a). *North European Dairy Journal,* **45**, 66.
Hansen, R. (1979b). *North European Dairy Journal,* **45**, 241.
Hansen, R. (1979c). *North European Dairy Journal,* **45**, 86.
Hansen, R. (1979d). *North European Dairy Journal,* **45**, 274.
Hansen, R. (1979e). *North European Dairy Journal,* **45**, 153.
Hansen, R. (1980a). *North European Dairy Journal,* **46**, 126.
Hansen, R. (1980b). *North European Dairy Journal,* **46**, 167.
Hansen, R. (1980c). *North European Dairy Journal,* **46**, 152.
Hansen, R. (1981a). *North European Dairy Journal,* **47**, 191.
Hansen, R. (1981b). *North European Dairy Journal,* **47**, 208.
Hansen, R. (1981c). *North European Dairy Journal,* **47**, 199.
Hansen, R. (1981d). *North European Dairy Journal,* **47**, 22.
Hansen, R. (1981e). *North European Dairy Journal,* **47**, 62.
Hansen, R. (1981f). *North European Dairy Journal,* **47**, 106.
Hansen, R. (1982a). *North European Dairy Journal,* **48**, 7.
Hansen, R. (1982b). *North European Dairy Journal,* **48**, 223.
Hansen, R. (1982c). *North European Dairy Journal,* **48**, 137.
Hansen, R. (1982d). *North European Dairy Journal,* **48**, 70.
Hansen, R. (1982e). *North European Dairy Journal,* **48**, 317.
Hansen, R. (1982f). *North European Dairy Journal,* **48**, 19.
Hansen, R. (1983a). *North European Dairy Journal,* **49**, 218.
Hansen, R. (1983b). *North European Dairy Journal,* **49**, 253.
Hansen, R. (1984a). *North European Dairy Journal,* **50**, 138.
Hansen, R. (1984b). *North European Dairy Journal,* **50**, 112.
Hansen, R. (1984c). *North European Dairy Journal,* **50**, 202.
Hynd, J. (1984). Personal communication.
IDF (1981). In: *IDF Catalogue of Cheeses,* Doc. 141, International Dairy Federation, Brussels.
IDF (1982a). In: *Consumption Statistics for Milk and Milk Products,* Doc. 144, International Dairy Federation, Brussels.
IDF (1982b). In: *The World Market for Cheese,* Doc. 146, International Dairy Federation, Brussels.
IDF (1983). In: *Consumption Statistics for Milk and Milk Products,* Doc. 160, International Dairy Federation, Brussels.
IDF (1984a). In: *Consumption Statistics for Milk and Milk Products,* Doc. 173, International Dairy Federation, Brussels.

IDF (1984b). In: *Memento — Containing the Reports of the 67th Annual Session of IDF,* 3–8 July, 1983 in Oslo, Norway, International Dairy Federation, Brussels.

Kasprzyk, P., Michel, J. F. , Seuvre, A. M. and Mathlouthi, M. (1983). In: *Maturation des Fromages a Pate Pressee Cuite de Type Emmental,* No. 331, Actualites Scientifiques et Techniques dans les Industries Agro-Alimentaires, Department Biologique Appliquée, IUT, Dijon, France.

King, D. W. (1966). *VII International Dairy Congress,* D, 723.

Kiuru, V. (1976). *North European Dairy Journal,* 42, 44.

Kosikowski, F. (1977). In: *Cheese and Fermented Milk Foods,* 2nd edn, Edward Brothers Inc., Michigan, USA.

Lampert, L. M. (1975). In: *Modern Dairy Products,* 3rd edn, Chemical Publishing Company Inc., New York.

Law, B. A. (1976). IDF Doc. No. 108, International Dairy Federation, Brussels.

Law, B. A. (1984). In: *Advances in the Microbiology and Biochemistry of Cheese and Fermented Milk* (Ed. F. L. Davies and B. A. Law), Elsevier Applied Science Publishers, London.

Law, B. A., Sharpe, M. E. and Chapman, H. (1976). *Journal of Dairy Research,* 43, 459.

Lee, B. J. (1981). In: *Proceedings of the 2nd Biennial Marschall International Cheese Conference,* Wisconsin, USA.

Lelievre, J. and Gilles, J. (1982). *New Zealand Journal of Dairy Science and Technology,* 17, 69.

Lindsay, R. C., Hargett, S. M. and Bush, C. S. (1982). *Journal of Dairy Science,* 65, 360.

Marth, E. H. (1953). *Milk Products Journal,* 44 (10), 31.

Moore, W. E. C. and Holdeman, L. V. (1974). In: *Bergeys Manual of Determinative Bacteriology,* 8th edn (Ed. R. E. Buchanan and N. E. Gibbons), The Williams & Wilkins Co., Baltimore, USA.

Park, W. J. (1979). In: *Proceedings of the 1st Biennial Marschall International Cheese Conference,* Wisconsin, USA.

Paul, A. A. and Southgate, D. A. T. (1978). In: *McCance and Widdowson's — The Composition of Foods,* 4th edn, HMSO, London.

Pederson, C. S. (1979). In: *Microbiology of Food Fermentations,* 2nd Edn, AVI Publishing Co., Connecticut, USA.

Pope, V. J. (1984). *Dairy Industries International,* 49 (6), 31.

Rasmussen, V. (1980). Personal communication.

Schelhaas, H. (1982). In: *Proceedings of XXI International Dairy Congress, Vol. 2,* Mir Publishers, Moscow.

Scherz, H. and Kloos, G. (1981). In: *Food Composition and Nutrition Tables 1981/82,* 2nd edn, Wissenschaftliche Verlagsgesellschaft GmbH, Stuttgart, West Germany.

Scott, R. (1981). In: *Cheesemaking Practice,* Applied Science Publishers, London.

Speckmann, E. W. (1979). In: *Proceedings of the 1st Biennial Marschall International Cheese Conference,* Wisconsin, USA.

Tamime, A. Y. (1981). In: *Dairy Microbiology, Vol. 2* (Ed. R. K. Robinson), Applied Science Publishers, London.

Tamime, A. Y. (1984). *Dairy Industries International,* **49** (7), 30.

Tamime, A. Y. and Robinson, R. K. (1985). In: *Yoghurt: Science and Technology,* Pergamon Press, Oxford.

The Cheese (Scotland) Regulations (1966). No. 98-S.8; Amendment (1967) No. 93-S.8; Regulation (1970) No. 108-S.4; Amendment (1974) No. 1337-S.115; Amendment (1984) No. 847-S.84, Statutory Instruments, HMSO, Edinburgh, Scotland.

The Colouring Matter in Food (Scotland) Regulations (1973). No. 1310-S.100; Amendment (1975) No. 1595-S.229; Amendment (1976) No. 2232-S.184; Amendment (1979) No. 107-S.12, Statutory Instruments, HMSO, Edinburgh, Scotland.

The Emulsifiers and Stabilisers in Food (Scotland) Regulations (1980). No. 1888-S.175; Amendment (1982) No. 514-S.65; Amendment (1983) No. 1815-S.171, Statutory Instruments, HMSO, Edinburgh, Scotland.

The Food Labelling (Scotland) Regulations (1981). No. 137-S.20, Amendment (1982) No. 1779-S.192, Statutory Instruments, HMSO, Edinburgh, Scotland.

The Miscellaneous Additives in Food (Scotland) Regulations (1980). No. 1889-S.176; Amendment (1982) No. 515-S.66: Statutory Instruments, HMSO, Edinburgh, Scotland.

The Preservatives in Food (Scotland) Regulations (1979). No. 1073-S.96, Statutory Instrument, HMSO, Edinburgh, Scotland.

The UK Food Additives and Contaminant Report (1982). In: *Report on the Review of Enzyme Preparations,* FAC/REP/35, HMSO, London.

Tofte-Jespersen, N. J. and Dinesen, V. (1979). *Journal of the Society of Dairy Technology,* **32**, 194.

USDA (1978). In: *Cheese Varieties and Descriptions,* revised Edn, Handbook No. 54, United States Department of Agriculture, Washington, DC.

Viswat, E. (1979). *North European Dairy Journal,* **45**, 13.

Vries, E. de (1979). In: *Test of a Curd-Making Tank, Type OST III with Capacity of 10,000 Litres Manufactured by α-Tebel,* NIZO Report No. 110, Ede, The Netherlands.

Vries, E. de and Ginkel, W. van (1980). In: *Test of a Curd-Making Tank, Type Damrow Double O with Capacity of 16,000 Litres,* NIZO Report No. 113, Ede, The Netherlands.

Vries, E. de and Ginkel, W. van (1983). In: *Beproeving van een Gesloten Wrogebereider met een Inhoud van 16,000 l Fabrikaat Alfons Schwarte GmbH,* NIZO Report No. 118, Ede, The Netherlands.

Vries, E. de and Ginkel, W. van (1984). In: *Test of a Curd-Making Tank, Type OST IV with Capacity of 10,000 litres Manufactured by Tebel BV,* NIZO Report No. 120, Ede, The Netherlands.

Weenik, A. J. H. (1975). *North European Dairy Journal,* **41**, 408.

Wegner, F. (1979). In: *Proceedings of the 1st Biennial Marschall International Cheese Conference,* Wisconsin, USA.

Wilbrink, A. (1982). In: *Proceedings of the XXI International Dairy Congress, Vol. 2,* Mir Publishers, Moscow.

Chapter 3

Modern Cheesemaking: Soft Cheeses

M. B. Shaw

Dairy Crest Foods, Research and Development Division, Telford, Shropshire, UK

CLASSIFICATION OF SOFT AND SEMI-SOFT CHEESE

Legislative Designations

The UK Cheese Regulations describe compositional standards for some 29 cheese varieties which are listed in a Schedule. These standards are expressed as minimum fat in the dry matter (FDM), and maximum moisture content in the cheese. All cheeses other than those in the Schedule, are categorised in the Regulations as either 'soft' or 'hard', depending on whether or not they are 'readily deformed by moderate pressure' (sic). Soft cheese must bear one of the descriptions given in Table I, depending on the fat and moisture content. Soft cheese may also be described as 'Cream cheese' if it contains not less than 45 per cent milk fat, or 'double cream cheese' if it contains not less than 65 per cent milk fat.

Legislation in other Member States of the EEC also draws distinctions between hard and soft cheeses. In France, five categories are identified: pâtes fraîches (fresh or unripened soft cheese), pâtes molles (ripened soft cheese), pâtes pressées non-cuites (hard pressed cheese, unscalded), pâtes pressées cuites (hard pressed cheese, scalded) and pâtes persillées (blue veined cheese). Similarly, in West Germany, legislation distinguishes between the following categories: hard, semi-hard, ripened soft, unripened soft and sour milk cheese.

Distinction is also drawn between 'soft' and 'hard' cheeses with regard to permitted ingredients. For example, in the UK Cheese

TABLE I
Legal designations for soft cheese (UK Cheese Regulations)

Description	Milk fat	Water
Full fat soft cheese	Not less than 20 per cent	Not more than 60 per cent
Medium fat soft cheese	Less than 20 per cent but not less than 10 per cent	Not more than 70 per cent
Low fat soft cheese	Less than 10 per cent but not less than 2 per cent	Not more than 80 per cent
Skimmed milk soft cheese	Less than 2 per cent	Not more than 80 per cent

Regulations, hard cheeses may contain: 'common salt (sodium chloride), starter, coagulant, and various permitted miscellaneous additives such as calcium chloride and a range of permitted colouring matters including carotene and annatto'.

Soft cheeses may contain: 'the ingredients mentioned above, together with flavourings, starches (whether modified or not), permitted emulsifiers and stabilisers, such as alginates, carrageenan, guar gum and locust bean gum, and the permitted miscellaneous additives lactic acid, citric acid, acetic acid, hydrochloric acid, orthophosphoric acid, and D-glucono-1,5-lactone'.

Functional Classification

While many cheese varieties will be encompassed by the terms 'hard' and 'soft' within the legal designations, quite often it will be found that the true nature of the product is not adequately described, particularly in the area of semi-hard and semi-soft cheese, as in the case of Edam and Saint Paulin for example. Thus, a number of classification systems have evolved where a cheese will be placed in a category based on consideration of such factors as consistency, fat content, moisture content, cooking (scalding) temperature and method of ripening (see Table V of Chapter 2).

Generally, the terms semi-soft and soft, when applied to consistency, are attributed to products having a moisture in fat free cheese (MFFC) of 61–69 per cent, and greater than 69 per cent respectively. Figures 1 and 2 illustrate the range of soft and semi-soft cheese covered in this chapter, sub-divided into groups according to the method of ripening.

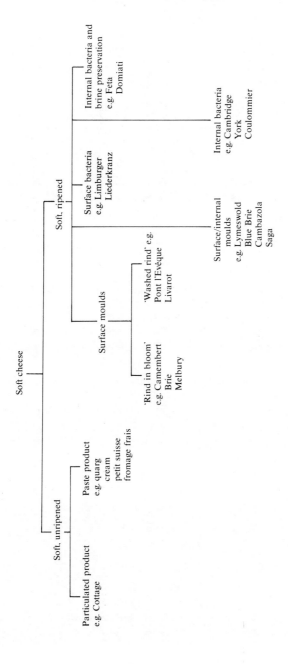

Fig. 1. Classification of soft cheese.

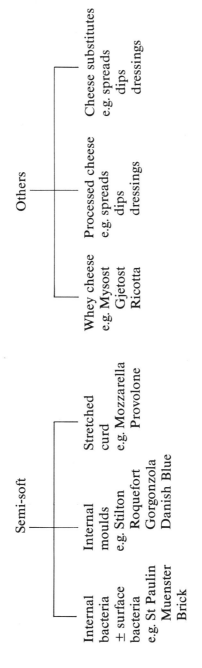

Fig. 2. Classification of semi-soft and other soft/semi-soft cheese.

FUNDAMENTALS OF SOFT AND SEMI-SOFT CHEESE MANUFACTURE

Process Principles

The manufacture of cheese is basically a means of preserving milk over the short to medium term, with the essential characteristics being the lowering of pH and water activity (a_w). Milk, with a water activity of 0·995 and a pH in the range 6·5–6·7 may be considered easily perishable, while cheeses will become more shelf-stable, that is less perishable, as the water activity and pH are progressively reduced. The relationship between pH and a_w for a number of cheeses is shown in Fig. 3.

During the initial stages of cheese manufacture, a_w is about 0·99 and this value will drop, depending on the processing regime applied for each particular cheese variety as described later in this chapter. Generally, the harder the cheese (lower MFFC per cent), the lower the a_w value and, conversely, the softer the cheese (higher MFFC per cent), the higher the a_w value. The a_w value is important when considering the ripening/maturation characteristics and subsequent shelf-life of cheese varieties. Generally, after salting and during maturation, the a_w values are lower than the optimum for the growth of starter bacteria, and hence the a_w value exerts a control over their metabolic activity and multipli-

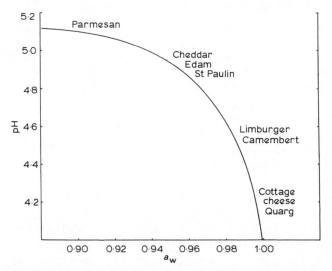

Fig. 3. Relationship between pH and water activity (a_w) of various cheeses.

cation, and over the subsequent maturation rate and expected shelf-life of the cheese. Thus, cheese such as Parmesan, with a relatively low a_w value in the order of 0·90 will mature slowly, have a relatively high pH value, and may have a shelf-life in excess of 2 years.

At the other end of the spectrum, a soft cheese, such as quarg or cottage cheese, will have a high a_w value in the order 0·99, and consequently the metabolic activity of the starter bacteria will be extremely high, resulting in a fairly low pH, and a comparatively short shelf-life in the order of 2 weeks. Intermediate products, such as Camembert, for example, which have an a_w value in the order of 0·98 and initial pH in the region of 4·6, will mature in a relatively short time (4–8 weeks) and have a maximum shelf-life in the order of 8–12 weeks.

Soft and semi-soft cheeses generally have a_w values in the range 0·96–0·99 and pH values in the range 4·3–5·0 with shelf-lives (excluding processed cheese) in the range 2–12 weeks at refrigeration temperatures. Thus, they are regarded as short shelf-life products, and require a complete 'cool chain' from the point of manufacture until consumption.

The controlled transformation of milk into product is achieved by a number of stepwise procedures which are illustrated in Fig. 4. Obviously, the manufacture of any particular cheese variety might not involve every step shown, and mechanisation and new technologies, such as membrane processing, may have an influence on the unit processes carried out during manufacture.

The fundamental principle of soft/semi-soft cheese manufacture involves a reduction in pH and water activity brought about by a controlled lactic fermentation, accompanied by subsequent drainage of whey and salting of the curd. Milk coagulation is achieved by a combination of enzymatic action by a coagulant protease (which may either be an animal rennet or an alternative microbial coagulant), and the production of lactic acid by lactic starter bacteria. The acidification profile during the various stages of manufacture, such as milk ripening, coagulation, curd draining and subsequent cheese maturation, is of prime importance.

The rate and quantity of lactic acid produced affects the degree of solubilisation of the calcium in the milk, and consequently, the rheological properties of the coagulum; it affects the extent of whey drainage, as well as the mineral composition and final texture of the cheese. In addition to the acid producing role, starter bacteria are also important during the maturation stage (if any), where they contribute to

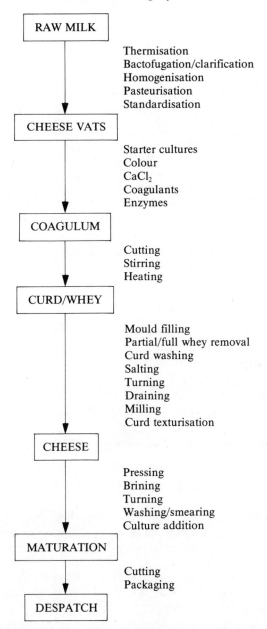

Fig. 4. Steps in soft/semi-soft cheese manufacture.

the overall flavour characteristics of the product. The degree of product acidification can be controlled by adjusting the amount of lactose available for fermentation. A number of methods are available such as curd washing, ultrafiltration and use of thermophilic starter cultures. The degree of specificity of proteolysis achieved by the coagulating enzymes is also of paramount importance in influencing the rheological properties of the coagulum, its subsequent whey drainage ability, the yield of product and the cheese maturation profile.

Where a ripening or maturation period is required for a particular soft/semi-soft cheese, this can be of variable duration dependent on the final flavour characteristic required. The time varies from zero for such cheeses as cottage, quarg and York which are consumed fresh, to over 3 months for full-flavoured products, such as Stilton, Camembert and St Paulin. The degree of maturation and type of flavour developed will depend on a number of factors, which include: the moisture and fat content of the cheese, the action of proteases and lipases derived from the coagulant, the species of starter bacteria, the surface or internal microflora of the cheese, and the ripening humidity, temperature and time.

Cultures Used in Soft and Semi-soft Cheese Manufacture

Starter cultures

The selection, maintenance and use of starter cultures is perhaps the most important aspect of cheesemaking, particularly when considering the modern, mechanised processes where predictability and consistency are essential.

The reasons for using starters can be summarised:

1. It ensures consistent acidity development at a controllable rate.
2. Acid production aids rennet action and subsequent coagulum formation.
3. Acid aids expulsion of moisture (i.e. whey) from the curd.
4. Starters govern the flavour, body and texture of the cheese.
5. Growth of undesirable bacteria in the curd is suppressed.

The organisms selected for use as cheese starters are shown in Table II, and the selection of particular cultures depends on the cheese variety being manufactured, for example:

1. To produce a fresh, acid cheese, such as quarg or cottage cheese,

one would use a starter containing homofermentative, fast acid-producing strains of *Streptococcus lactis* and *Str. cremoris*.

2. To produce a cheese with some openness in texture, such as Stilton, one would use strains of *Str. lactis* and *Str. cremoris* together with *Str. lactis* sub-sp. *diacetylactis* for gas and flavour production.
3. To produce a sweet, washed curd cheese, such as St Paulin, one might use slower acid-producing strains of *Str. lactis* and *Str. cremoris*, together with a proportion of the flavour-producing *Leuconostoc* species.
4. To produce a high-scald, stretched-curd cheese, such as Mozzarella, one would use *Str. thermophilus* and *Lactobacillus bulgaricus* or possibly in addition, *Str. durans* or *Str. faecalis* for flavour development.

Over the years, a number of systems have evolved whereby an active starter culture can be introduced into the cheese vat.

1. Mother Culture and Bulk Starter Inoculation:
   (a) Sub-culturing of starters from freeze-dried cultures.
   (b) Sub-culturing of starters from frozen liquid cultures.
2. Direct Bulk Starter Inoculation:
   (a) Using deep frozen concentrated cultures.
   (b) Using freeze-dried concentrated cultures.

TABLE II
Cheese starter organisms

Type	Function
Streptococcus lactis *Streptococcus cremoris*	Acid production
Streptococcus lactis sub-sp *diacetylactis*	Acid, gas and flavour production
Leuconostoc cremoris *Leuconostoc dextranicum*	Gas and flavour production
Streptococcus thermophilus *Lactobacillus bulgaricus* *Lactobacillus helveticus*	Acid production in high-scald cheese
Streptococcus durans *Streptococcus faecalis*	Acid and flavour production in high-scald cheese
Propionibacterium shermanii	Gas and flavour production

3. Direct Vat Inoculation (DVI):
    (a) Using deep frozen concentrated cultures.
    (b) Using freeze-dried concentrated cultures.

The cultures, as supplied by commercial starter manufacturers, are usually in one of the following forms:

1. single strains;
2. defined multiple strains;
3. mixed strains.

Future developments in the starter area will undoubtedly revolve around the production of bacteriophage resistant starter strains using DNA recombination techniques, while the increasing costs of bulk starter production may result in a greater use of direct-to-the-vat (DVI) starter systems, using freeze-dried, bacteriophage resistant, single-strain starters. Alternatively, there may be an increase in the use of direct acidification procedures for, in particular, soft unripened cheese manufacture, but as more knowledge of cheese flavour chemistry is acquired, even the manufacture of ripened cheeses could be considered.

Other cultures

Non-starter organisms often used in the manufacture of soft and semi-soft cheeses are listed in Table III. The most commonly used are the blue and white moulds, *Penicillium roqueforti* and *P. candidum* respectively. These strains were classically, naturally occurring moulds found in cheese maturation stores, but nowadays they are available as pure strain cultures from commercial suppliers. Both blue and white moulds serve to give the final cheese a pleasing appearance and contribute to the flavour, and various strains are available, with more or less proteolytic and lipolytic activity, resulting in a range of flavour intensities. White moulds are also available which give different surface growth characteristics in relation to time, and blue moulds are available which give a range of colours from a pale green through to dark blue. The other cultures listed are mainly involved in colour and flavour development; *Brevibacterium linens*, for example, is partly responsible for the colour and flavour of such cheeses as Muenster, Livarot and Limburger, and the reddish-brown coloration on the surface of some Brie and Camembert.

Generally, it can be said that cheese flavour is a result of the production of a complex mixture of chemical compounds arising from

TABLE III
Other organisms used in soft/semi-soft cheese manufacture

Type	Function
Penicillium roqueforti	Blue mould veining, flavour production
Penicillium candidum *Penicillium camemberti*	White mould surface growth
Penicillium album	White/blue mould flavour production
Brevibacterium linens *Brevibacterium erythrogenes*	Colour and flavour production
Micrococcus varians	Flavour production
Debaryomyces hansenii *Candida utilis*	Yeasts involved in flavour production
Rhodosporidium infirmominatium (precursor to *B. linens*)	Yeasts involved in surface colour production
Geotrichum candidum Mycoderm	Moulds involved in flavour production
Bifidobacterium	Used for product stabilisation

the enzymic interaction of a number of micro-organisms present in the curd. Thus, it is difficult to credit any individual strain with the development of a particular chemical compound, and the organoleptic qualities of any cheese will result from:

1. The lactic fermentation: acids (lactic, acetic, butyric), ketones (diacetyl), esters.
2. Lipolysis (fat breakdown): acids (butylic, caproic, capric), methylketones.
3. Proteolysis (protein breakdown): peptides, amino acids, ammonia.

It can be seen, then, that complex flavours can develop in cheeses as a result of the activities of a range of organisms, including bacteria, yeasts and moulds. Blends are available, especially in France, of one or more of these micro-organisms for the production of a uniquely flavoured cheese. For example to produce a farmhouse-style Pont l'Evêque with a greyish-white and orange coat, one would use a mixture of *Brevibacterium erythrogenes*, cheese smear micrococci, Mycoderm, *Penicillium camemberti* and *Geotrichum candidum*. The cultures are generally added to the milk, and/or sprayed onto the cheese surfaces.

Rennet and Rennet Substitutes

The general conception of cheese curd is that it results from the action of a proteolytic enzyme which cleaves the milk protein, casein, thus rendering it insoluble, so forming a coagulated mass which encloses the other milk components such as the fat.

The most commonly used enzyme is an acid protease (EC 3.4.23.4), designated as rennin or chymosin, which is extracted from the abomasum of the suckling calf. The casein complex in milk has been shown to comprise four moieties: $\alpha, \beta, \kappa, \gamma$, and it is the κ-casein, which exerts a stabilising influence against coagulation. The rennet enzyme, chymosin, cleaves the phenylalanine-methionine bond (105–106) in the κ-casein molecule which effectively destabilises the casein complex. It is a two stage process, the first being enzymic:

$$\kappa\text{-casein} \xrightarrow{\text{enzyme}} \text{para-}\kappa\text{-casein} + \text{macropeptide}$$

and the second, non-enzymic stage which occurs concurrently:

$$\text{para-}\kappa\text{-casein} \xrightarrow[\text{pH } 6 \cdot 0 - 6 \cdot 4]{Ca^{++}} \text{dicalcium para-}\kappa\text{-casein}$$

This stage requires the presence of calcium ions, and is the reason cheesemakers often add calcium chloride to the cheese vat. There is a third stage where proteolysis of the casein continues at a low level during cheese maturation, so making a contribution to flavour development. Some 6 per cent of the chymosin added to the milk will be retained in the curd, while the remainder is found in the whey.

Over the years, a number of alternative coagulants have been evaluated for cheesemaking, and these have been screened in order that they satisfy the following criteria:

1. Must be produced from raw materials available in sufficient quantities and at an acceptable price.
2. Toxicologically safe.
3. Suited to different cheese types without changes in cheesemaking procedure.
4. Must not adversely affect yield and/or product quality.
5. Must be similar to calf rennet, particularly with regard to chemical composition.
6. Extraction and manufacture on an industrial scale must be possible to a high microbiological standard.

Some of the sources of alternative coagulants are given in Table IV. It has been found that the bacterial and plant enzymes are too proteolytic for cheesemaking, although they are still used in some parts of the world. Work with pig and bovine pepsins and mixtures of these with standard or high chymosin rennet preparations has produced encouraging results in cheesemaking trials. The pepsins tend to coagulate milk more slowly but exhibit greater proteolytic activity at the lower pH values associated with maturing cheese.

Rennet extracted from a young calf will contain between 88 and 94 per cent chymosin and between 6 and 12 per cent pepsin, while extracts from the older bovine animal will contain 90–94 per cent pepsin and only 6–10 per cent chymosin. Thus the ratio of chymosin:pepsin depends on the age of the calf at slaughter. With commercial pressures on suppliers to produce adequate quantities of 'Standard Rennet' at a realistic price, there has been a tendency over the years to supply product with a reduced proportion of chymosin. Presently, most standard rennets contain between 75 and 80 per cent chymosin, mixtures of standard and bovine rennet 40–50 per cent chymosin, and bovine rennet 25 per cent chymosin.

It was this pressure on the suppliers that led to the development of the fungal-derived, microbial rennet enzymes. The majority of microbial rennets available on the market are derived from the fungi, *Mucor miehei* and *Mucor pusillus*, and are marketed under the names Rennilase®,

TABLE IV
Sources of alternative coagulants

Group	Source of enzyme
Bacteria	*Bacillus polymyxa*
	Bacillus subtilis
	Bacillus mesentericus
Fungi	*Mucor miehei*
	Mucor pusillus
	Endothia parasitica
Plant	Paw Paw (papain)
	Pineapple (bromelain)
Animal	Calf (chymosin)
	Ox (pepsin)
	Pig (pepsin)
	Chicken (pepsin)

Hannilase®, Marzyme®, and Emporase®, for example. The products have been shown to have similar proteolytic specificity to chymosin, and are supplied as standardised activity solutions with recommended usage rates for cheesemaking.

In 1981, more than one-third of all cheese produced world-wide utilised microbial rennets, and they have found exclusive use in the production of vegetarian, Halal and Kosher cheeses.

A problem highlighted some time ago was that of residual coagulant activity in whey, particularly when it was destined for use in formulated dairy products, baby foods and dietary aids. The residual coagulant was shown to cause proteolysis during whey processing, resulting in off-flavour development or even coagulation of milk containing foods.

The most recent development in the alternative coagulant area has come as a result of intense activity by newly-formed, biotechnology companies. DNA recombinant technology would seem to offer the potential for the large-scale and economical production of the chymosin enzyme, and already small-scale, cheesemaking trials are being carried out using the product. The application of biotechnology, and in particular the genetic engineering techniques afforded by DNA recombination, will almost certainly have a considerable effect on the selection of starters and coagulants which will be made available in the next decade for cheesemaking.

MANUFACTURING PROCESSES FOR SOFT AND SEMI-SOFT CHEESE

Figures 1 and 2 illustrate a functional classification of soft and semi-soft cheese varieties, and Table V lists some typical chemical compositions of representative cheeses from each category. Certain countries may set compositional standards for specific varieties, and not withstanding these, there will be natural variations in levels of, for example, protein and fat due to seasonal variations in the milk supply. Often the fat in dry matter of a cheese may be adjusted or 'standardised' by varying the fat content of the whole milk to be processed. This may be achieved by:

1. removing milkfat using centrifugal separators;
2. adding liquid skimmed milk;
3. adding skim-milk powder; or
4. by using processes such as ultrafiltration.

TABLE V

Typical chemical composition of soft and semi-soft cheese

Category	Variety	Moisture	Fat	FDM	Protein	Salt
Soft, unripened particulated	Cottage, creamed	79·9	4·0	19·0	14·0	1·0
Soft, unripened paste	Quarg	79·0	0·2	1·0	15·0	0·7
Soft, ripened surface moulds	Camembert	50·0	23·0	45·0	20·0	1·5
Soft, ripened surface/internal moulds	Lymeswold	42·0	39·0	68·0	14·0	1·6
Soft, ripened surface bacteria	Limburger	46·0	27·0	50·0	21·0	1·7
Soft, ripened internal bacteria, brine preservation	Feta	58·0	21·0	50·0	20·0	4·0
Semi-soft, surface bacteria	St Paulin	48·0	26·0	50·0	20·0	1·7
Semi-soft, internal moulds	Stilton	42·0	31·0	53·0	22·0	1·8
Semi-soft, stretched curd	Mozzarella	58·0	17·0	40·0	21·0	1·5
Whey cheese	Ricotta	72·0	10·0	36·0	12·5	1·5

Variations in moisture, salt content, and pH of the cheese may also be experienced due to differences in processing conditions: for example, the rate of addition of starter culture, scald temperature and profile, and the time/temperature relationship of post whey-off operations.

Not all cheese is manufactured from cow's milk and, in certain countries, substantial quantities are made from sheep's, goat's and water buffalo's milk; Table VI compares the average chemical composition of each milk. Both soft and semi-soft varieties may be found, while nowadays it is becoming increasingly common to manufacture product using cow's milk, with the addition of lipase, for example, in an attempt to mimic natural flavours found in cheese made from goat's milk. Varieties made from goat's milk, or mixtures of goat's and sheep's milk, include Feta, Lightvan, Domiati and a large range of French soft,

TABLE VI
Average chemical composition of milk from various mammals

Mammal	Moisture (%)	Fat (%)	Protein (%)	Lactose (%)
Sheep	80·6	8·3	5·4	4·8
Goat	87·8	3·8	3·5	4·1
Buffalo	82·4	7·4	4·7	4·6
Cow	87·3	3·7	3·4	4·8

mould-ripened cheeses, including local delicacies such as Crottins de Chavignol, Sainte' Maure, Tome de Romans and Levroux. Sheep's milk is used to manufacture the classic blue cheese, Roquefort, and water buffalo's milk has long been used in Southern Italy to produce Mozzarella. Sheep's, goat's and buffalo's milk all lack carotene, and consequently cheese made from them, unless artifically coloured, will be white. The major difference between goat's milk and cow's milk is the higher level of capric, caprylic and caproic fatty acids, which give the cheese a sharp, pungent flavour.

The following sections review the manufacturing processes involved for each category of soft and semi-soft cheese as shown in Figs. 1 and 2. Where appropriate, processing variations are included, together with a review of the application of equipment and process developments aimed at mechanising unit operations and for increasing manufacturing efficiency. Figure 4 illustrates the range of steps encountered in soft and semi-soft cheese production, although the manufacture of a particular variety may not involve every operation shown.

SOFT CHEESE MANUFACTURE

Soft Unripened Cheese

Soft unripened fresh cheese may be characterised by the following:

1. High moisture content (up to 80 per cent but varies with fat content).
2. Mildly acidic to bland flavour (dependent on fat content).
3. Short shelf-life.
4. Little rennet used.
5. No pressing of curd.
6. Product ready for consumption immediately (no maturation period).

Cottage cheese contains discrete curd particles as a result of its method of manufacture, while the majority of the unripened cheeses are of the paste-type. Variation is due mainly to the fat content, and a wide range of products, both plain and with sweet or savoury additives, may be encountered. Quarg (or quark) is of West German origin, while in France, the 'fromage frais' category includes such examples as Fromage Blanc, Petit Suisse, Gervais and Neufchatel. In the USSR, Tvorog is found, and in the UK, Bakers, Lactic and Cream cheese are all in this category.

Cottage cheese

Manufacturing process
Cottage cheese is manufactured from pasteurised skimmed milk by way of an acid coagulation stage, followed by cooking the cut curd particles in whey, washing the curd particles with water, draining and blending the curd with a cream dressing. The unit processes for manufacturing cottage cheese are shown in Fig. 5. The illustrated process is known as a 'short set', in which the coagulum is produced at 31–32 °C in about 5 h using a 5 per cent bulk starter culture, and the total process time is between 9 and 10 h. This method is now favoured over the more traditional 'long set', in which the coagulum is produced at 22 °C in about 14–16 h using 0·5–1·0 per cent starter addition. Although traditionally an acid set product, it is now usual to add very low rennet levels of 1–3 ml per 1000 litres skimmed milk. The rennet addition allows a higher pH at cutting, the curd tends to be more resilient during cooking (scalding), and whey fines are reduced.

The coagulum is cut near to the milk protein isoelectric point of pH 4·6 (usually in the range 4·40–4·85), and the size of the cut is varied depending on whether small, medium or large curd particles are required. After cutting and a quiescent period of some 15–20 min, stirring is commenced, and the whey temperature is progressively raised from 31–32 °C to 50–55 °C. This cooking process is critical to cheese quality and yield, and in the initial stages, the scald profile must not exceed a 1 °C rise evey 5 min or curd clumping will be experienced. At a whey temperature of 40 °C, the increments may be increased to 2 °C every 5 min. At 50–55 °C the curd particles have the desired moisture content and firmness. To prevent the curd particles clumping at this temperature, it is necessary to introduce wash waters which also serve to remove lactic acid and lactose from the cheese grains, these losses give rise to the characteristic bland flavour. The chilled water is often

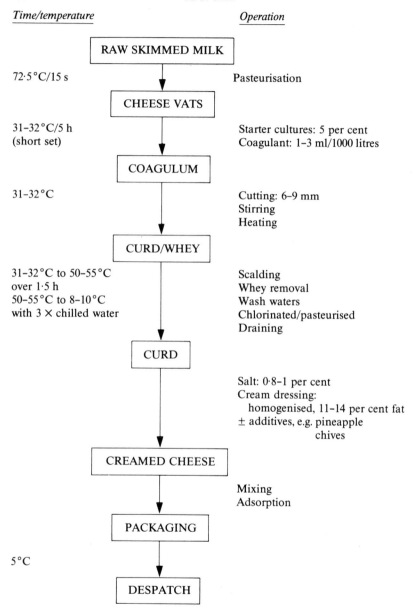

Time/temperature   *Operation*

RAW SKIMMED MILK

72·5°C/15 s   Pasteurisation

CHEESE VATS

31–32°C/5 h   Starter cultures: 5 per cent
(short set)   Coagulant: 1–3 ml/1000 litres

COAGULUM

31–32°C   Cutting: 6–9 mm
                                 Stirring
                                 Heating

CURD/WHEY

31–32°C to 50–55°C   Scalding
over 1·5 h   Whey removal
50–55°C to 8–10°C   Wash waters
with 3 × chilled water   Chlorinated/pasteurised
                                 Draining

CURD

                                 Salt: 0·8–1 per cent
                                 Cream dressing:
                                     homogenised, 11–14 per cent fat
                                     ± additives, e.g. pineapple
                                                       chives

CREAMED CHEESE

                                 Mixing
                                 Adsorption

PACKAGING

5°C

DESPATCH

Fig. 5. Steps in cottage cheese manufacture.

introduced into the cheese vat as the whey is drained off and in this way the temperature is reduced to 8–10°C. The procedure is repeated 2–3 times, and it is essential to have chilled water of very good microbiological quality. To this end, the water is often chlorinated (5–25 ppm) and/or pasteurised. Also the pH of the wash water is often adjusted to the pH value desired in the final product.

After the washing stage is completed, the curd is drained and blended with a formulated cream dressing. The dressing formulation will vary from manufacturer to manufacturer with the fat content adjusted to give 4·0 per cent in the final product. Often, sorbates are added to the dressing at 0·1 per cent to extend the shelf-life, stabilisers to prevent syneresis occurring, and salt to give 0·8–1·0 per cent in the product. Starters may also be added to the dressing in order to reduce the pH, and in some cases, to produce carbon dioxide, both aimed at extending the shelf-life of the product. Finally various additives may be used in order to produce a range of products, and these may include pineapple, chives, or pimentoes, all of which must be of good microbiological quality before adding to the cottage cheese.

After the curd and dressing have been blended and some 30–40 min allowed for adsorption, the cottage cheese is ready to be packaged. The packed product should be immediately transferred to a cold room, and after 6–7 h, the product temperature should be less than 5°C. During this time a further adsorption of free cream will take place. The cheese is now ready for sale with a shelf-life of 2–3 weeks.

Equipment and process developments

Traditionally, cottage cheese has been manufactured in rectangular, open vats, and while some manufacturers still utilise this method, a number of equipment developments have been introduced in order to maximise process efficiency (see Fig. 6).

The majority of equipment developments have evolved in the USA where there is a well developed market for cottage cheese. Various systems are available from manufacturers such as Grace, Stoelting, Rietz and Damrow, who have engineered equipment aimed at mechanising the following process stages.

Curd manufacture. Conventional cottage cheese vats employed steam in the vat walls, but circulated water is now used in order that a gentler, more uniform heating can be achieved. This change was

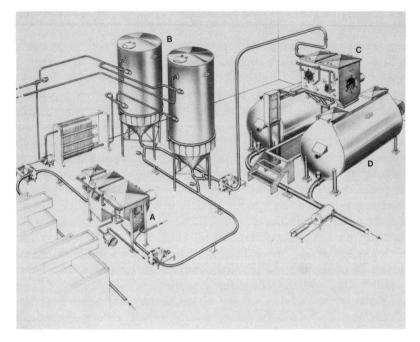

Fig. 6. Equipment for the manufacture of cottage cheese. (A: whey drainer; B: curd washers/coolers; C: curd drainer; D: creamers.) (Courtesy of Bepex Corp.)

necessary to obviate problems with curd 'burn-on' and differential heat treatment, and this type of vat is often referred to as a 'spray-vat'.

As the smaller, traditional vats of 4000–5000 litres began to be replaced by larger vats upwards of 10 000 litres, the need to achieve a constant heat distribution during the 55–60°C cook became more critical. Thus, the cooking process referred to as 'jet cooking' was introduced. The basis for this method is the introduction of culinary steam into the whey layer formed on the top of the vat after cutting. This development went 'hand-in-hand' with the development of more efficient stirring systems necessary to ensure good heat distribution, due to the fact that the warm whey naturally tended to remain at the vat surface. The Stoelting 'verti-stir' system uses agitators with large blades which gently scoop the curd from the bottom of the vat into the warm whey stream, so that all the curd particles receive a uniform heat treatment.

Further development of the 'jet cooking' system has resulted in the use of a concentric tube heat exchanger, which allows the whey to be heated without the dilution effect experienced with steam injection systems.

Washing/draining. Amongst the early cottage cheese equipment developments, the removal of the curd washing, draining and creaming operations from the cheese vat contributed much to overall process efficiency and plant utilisation. A number of manufacturers offer combined drainers/creamers, and the washing process is performed in the cheese vat followed by draining and addition of the cream dressing in a separate, tared tank system. Damrow and Rietz for example, offer whey drainer systems where undiluted whey can be recovered, followed by separate washer/cooler tanks and curd drainers. Stoelting have developed a high speed, continuous drainer called 'Flexi-press', which separates curd from the final wash water at rates in the order of 100 kg min^{-1}. Damrow also have a belt system which is made up of four sections. In the first, the whey is drained from the curd particles which are spread as a thin layer on the belt. Washing is carried out in the second and third sections, with the chilled water sprayed onto the third section being recovered and sprayed onto the second section — this guaranteeing rapid cooling with an overall low water consumption. Finally, the fourth section permits drainage of the curd and is equipped with pressure rollers which can be adjusted to control the final moisture content of the curd.

Creaming. A number of batch curd creamers are available which are equipped with gentle agitation systems and sprays for the addition of curd dressing. Load cell systems are often used to ensure a correct and accurate ratio of curd to dressing.

Recent developments which can be applied to the manufacture of cottage cheese are reviewed in a later section, and these include the application of ultrafiltration, the use of direct acidification procedures, and continuous curd manufacture, as exemplified by the Alpma continuous coagulation system.

Quarg and other fresh or unripened soft cheeses

Manufacturing process
The basic manufacturing process involves an acid coagulation assisted

by the addition of small amounts of rennet, a coarse cut of the coagulum, separation of the curds and whey by various methods, cooling of the curd, followed by packaging and sale. Figure 7 illustrates the process steps used in the manufacture of a low fat quarg from skimmed milk. The illustration is for a 'long set', in which the coagulum is produced at 22–23 °C in about 16–18 h using 1–2 per cent of a bulk starter culture. The incubation proceeds until a pH of 4·7 is attained (0·55–0·6 per cent titratable acidity as lactic acid). The incubation time may be reduced to 5–6 h ('short set') using a 5 per cent starter inoculum at 30 °C. The incubation time is generally chosen to fit in with production scheduling. The cultures used are generally homofermentative strains of *Streptococcus lactis* and/or *Str. cremoris*, and the small amount of coagulant added allows the formation of a firm coagulum at pH 4·7–4·8, that is above the protein's isoelectric point of pH 4·6. The coagulum treatment depends on the methods used for whey separation. Traditionally curd drainage is carried out in muslin bags, and the coagulum is cut very coarsely into approximately 20 cm cubes. The introduction of centrifugal whey separators led to the coagulum being slurried up using mechanical agitators prior to feeding the separators (see later).

After whey drainage/separation, the curd is cooled either in muslin in cold rooms or, more usually, using a tubular heat exchanger/cooler. The chilled curd is then blended with salt and, in some cases, the addition of cream and/or condiments, such as herbs and fruits, may be carried out. The product is then filled into pots or tubs and distributed immediately. The shelf-life of the product will be in the region of 14 days.

Quarg may be bulk packed and used for the manufacture of other soft cheeses, such as medium and full fat varieties, where butter or cream is blended in; the blend is heat treated, often up to 80°C, a stabiliser is added to reduce serum separation, and various additives such as herbs may be introduced. The final product has the advantage of an extended shelf-life in the order of 10–12 weeks.

Equipment and process developments

Traditionally whey separation is achieved by cutting the coagulum and transferring the curd mass to linen or muslin bags, where the whey drainage proceeds until the optimum curd moisture content is reached. This method is cumbersome and labour-intensive, and the product can be subjected to aerial contamination from yeasts and moulds. A further development of this system, which is used extensively in France, is the

Time/temperature   *Operation*

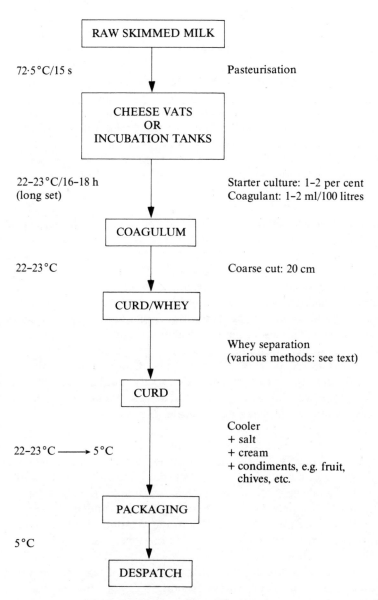

Fig. 7. Steps in Quarg manufacture.

Berge process, where whey drainage takes place in a series of cloths held in a rack which allows the curd to be slowly compressed over a defined time period. Even with this semi-mechanised process, whey drainage times can still be up to 8 h, and difficulties may be encountered in obtaining consistent cheese quality.

As the demand for quarg and quarg based products increased, it became obvious that the cloth drainage systems were not suitable for large-scale, sanitary operations. This commercial pressure led to the development of centrifugal separators and ancillary equipment in the late 1950s and early 1960s. The two major manufacturers of these separator systems are Westfalia and Alfa-Laval.

In the Westfalia KDA quarg separator, for example (see Fig. 8), coagulated skimmed milk flows through a central feed tube into a set of discs revolving at about 5500 rpm. Whey flows to the top of the separator bowl and is discharged foam-free under pressure. The cheese curd is ejected from the separator bowl by nozzles, and the hood, cooled by chilled water circulation, guides the quarg vertically into a curd collecting trough. Four, slowly rotating, scraper blades then move the quarg to the discharge outlet from which it slides into a quarg catcher. Separator capacities are typically up to 2000 kg curd h^{-1}. The cheese curd is transferred from the quarg catcher by a variable speed, positive displacement pump through a tubular cooler and a quarg mixer into a holding tank. The quarg mixer is used for the manufacture of creamed quarg (medium and full fat products). Mixing takes place in a cylinder fitted with three pairs of vanes, and a piston proportioning pump controls the amount of cream added to the quarg. Blended quarg is passed to a silo tank where a positive displacement pump delivers the chilled quarg to the packing machine.

Separators can also be used for the manufacture of full fat cheese, although the bowl design is significantly different from that in the quarg separator. Typically, full fat cheese is made from pasteurised, standardised milk with a butterfat content of 11 per cent. This milk is homogenised at 45–80 °C, and the resultant fat–protein complex will be less dense than whey (0·93 compared to whey density of 1·025), and it is this fact that dictates the design of the bowl (as in the Westfalia KSA separator, for example), which continuously discharges the lighter, full fat cheese solids concentrate.

A range of products of various fat contents may be obtained by blending different proportions of quarg (0·2 per cent fat) and full fat cheese (33 per cent fat) from the two designs of separators.

Fig. 8. Quarg separation line showing Westfalia KDA separator, tubular cooler and Quarg mixer. (Courtesy of Westfalia Separator, UK.)

Following on from the introduction of complete separator production lines for quarg, there have been a number of developments aimed at increasing the yield of quarg from a unit volume of skimmed milk. Currently, the two most successful processes are as follows.

The Thermoquarg process. Using a standard, separator-based quarg production line, the whey has a protein content in the order of 0·8 per cent; these valuable whey proteins are lost from the quarg and reduce yield. Westfalia have developed a Thermo-process which allows the recovery of approximately 50 per cent of the whey proteins, resulting in an increased quarg yield of some 10 per cent. The process is as follows: skimmed milk is heated to between 85 and 90°C for 2–3 min, followed by cooling to normal ripening temperature when the additions of starter and rennet are made. Following formation of the coagulum, the bulk is further heat treated at, typically, 60°C for 3 min to allow further whey protein denaturation and complexing with the casein. The coagulum is then cooled to separation temperature, and the curd discharging from the separator will contain the whey proteins, thereby increasing the yield.

Ultrafiltration. The use of membrane processing, and in particular ultrafiltration, for the manufacture of soft cheeses has received much attention, and ultrafiltration (UF) may be used for quarg manufacture where the skimmed milk is concentrated to the total solids content of the final product, i.e. 17–20 per cent. The UF concentrate is inoculated with starter culture and rennet, and the coagulated product can be homogenised and packaged without whey separation. Very little protein is lost during ultrafiltration and the need for a separator is eliminated, since the de-watering is achieved in the UF plant where it is lost as permeate. Increases in cheese yield of up to 40 per cent have been reported, although, at high levels of whey protein incorporation, the product may have different organoleptic qualities from conventional quarg.

A number of process developments have taken place aimed at extending product shelf-life and improving process efficiency. These include:

1. Direct acidification.
2. Specific heat treatments of the product.
3. Use of bifidobacteria.

Specific heat treatments are used to extend the shelf-life of a number of fresh, unripened soft cheeses. The base cheese is blended in a batch cooker with a suitable stabiliser (usually proprietary brands based on guar gum, locust bean gum or carboxymethyl cellulose) and heated by indirect or direct injection to between 60 and 80 °C for 2–5 min with continuous stirring. Additives, such as fruits and herbs, may also be mixed in at this stage prior to product packing. Hot filling and/or the addition of preservatives, such as potassium sorbate, also assist in extending shelf-life up to 12 weeks. Products are often aerated or whipped by incorporating nitrogen gas under pressure to produce stable foams with up to 200 per cent overrun, and these may be used in layered products with fruit jellies and/or cream resulting in a range of attractive, value added, retail products.

The use of specific cultures in soft cheese starter mixtures has been investigated as a means of extending product shelf-life.

A number of cultures are now available which have a limited acidification rate below pH 4·8–5·0. For example, Biogarde® (West Germany) produce a culture containing *Str. thermophilus, Str. lactis, Str. lactis* subsp. *diacetylactis, Leuconostoc cremoris, Lactobacillus acidophilus* and *Bifidobacterium bifidum* for the manufacture of fresh, unripened cheeses. It is the presence of the bifidobacteria which has the major effect on product stabilisation against over-acidification, and shelf-life in a cold chain can be usefully extended from 2 to, at least, 3 weeks.

Soft Ripened Cheeses

Figure 1 shows the range of ripened soft cheese classified according to the major ripening agent, that is surface or surface/internal moulds and surface or internal bacteria.

Surface mould-ripened cheese

Manufacturing process
Surface mould-ripened soft cheeses, or pâtes molles as they are known in France, are characterised as cheeses that have undergone other fermentations as well as a lactic fermentation. The curd is not scalded and the cheeses are not pressed, but they are matured. The French recognise two sub-categories of the mould-ripened cheese; those with a 'rind in bloom' (croute fleurie) such as Camembert, Brie and Melbury, and those with 'washed rinds' (croute lavée), such as Pont l'Evêque and Livarot.

Of the 'croute fleurie' or surface mould types, Camembert and Brie are undoubtedly the best known, and the 'rind in bloom' characteristic is due to the growth of a white mould species, *Penicillium candidum*. During the production of the 'croute lavée' varieties, such as Pont l'Evêque and Livarot, the white mould growth is removed during cheese maturation by 'smearing' or washing, and the surface growth of a red-pigmented bacterium, *Brevibacterium linens*, is encouraged.

Over the years, a number of methods have evolved for the manufacture of soft mould-ripened cheese, and the technological development of various mechanised systems will be discussed later. As the systems have evolved, slight changes have been made to the manufacturing regime, and various 'recipes' may be found showing variations in time/temperature relationships at different stages, amount and type of starter culture, and/or coagulant addition.

Fundamentally, two types of manufacture exist: the traditional and industrial, although in practice a spectrum of systems can be found with varying degrees of technological innovation offering good examples of how a traditional method has evolved into highly sophisticated, mechanised plants having a large milk throughput. Figure 9 shows the typical steps for manufacturing Camembert by an industrial process.

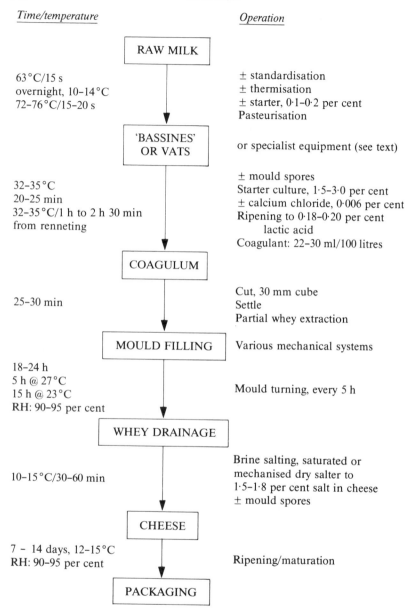

Time/temperature   *Operation*

RAW MILK

63°C/15 s   ± standardisation
overnight, 10–14°C   ± thermisation
72–76°C/15–20 s   ± starter, 0·1–0·2 per cent
                                            Pasteurisation

'BASSINES'
OR VATS   or specialist equipment (see text)

                                            ± mould spores
32–35°C   Starter culture, 1·5–3·0 per cent
20–25 min   ± calcium chloride, 0·006 per cent
32–35°C/1 h to 2 h 30 min   Ripening to 0·18–0·20 per cent
from renneting   lactic acid
                                            Coagulant: 22–30 ml/100 litres

COAGULUM

                                            Cut, 30 mm cube
25–30 min   Settle
                                            Partial whey extraction

MOULD FILLING   Various mechanical systems

18–24 h
5 h @ 27°C
15 h @ 23°C   Mould turning, every 5 h
RH: 90–95 per cent

WHEY DRAINAGE

                                            Brine salting, saturated or
10–15°C/30–60 min   mechanised dry salter to
                                            1·5–1·8 per cent salt in cheese
                                            ± mould spores

CHEESE

7 – 14 days, 12–15°C
RH: 90–95 per cent   Ripening/maturation

PACKAGING

Fig. 9. Steps in Camembert manufacture (industrial).

Traditional manufacture

The essential points of the traditional manufacturing process can be summarised as follows:

1. Raw milk is used for manufacture, and acidification results from the natural bacterial flora. In cases where this limits the rate of acidity development, a mesophilic, mixed strain starter culture may be added at reduced levels (0·05 per cent).
2. The raw milk is filled into 100 litre tanks or 'bassines' after temperature adjustment to 30–32 °C. Calcium chloride may be added at the rate of 6–10 g per 100 litres milk to aid coagulum formation.
3. Acidity development proceeds until a level of 0·20–0·25 per cent lactic acid is achieved.
4. Rennet, or other suitable coagulant, is added at the rate of 15–20 ml per 100 litres milk.
5. The coagulum is ready to be moulded approximately 1 h 10 min to 1 h 30 min after renneting.
6. No cutting of the curd occurs, and it is ladled by hand from the bassine to a series of moulds, which are open-ended cylinders with whey drainage holes, placed on drainage mats in trays which allow turning of the filled moulds.
7. Whey drainage is allowed to proceed in the moulds for some 24–28 h in a room having a temperature of 28 °C and relative humidity (RH) of 95–100 per cent. To assist whey drainage, the moulds are turned approximately every 5 h.
8. The cheeses, after removal from the moulds, are salted by sprinkling the surfaces with dry salt.
9. The cheese is ripened or matured at a temperature of between 11 and 13 °C and relative humidity of 90–95 per cent for from 3 weeks to 1 month.
10. The cheese is then packaged and stored at between 4 and 8 °C prior to distribution.

Industrial manufacture

The essential points of the industrial manufacturing process can be summarised as follows:

1. Milk is often standardised for fat content to give a cheese of known 'fat in dry matter'. Milk is pasteurised at 72–76 °C for 15–20 s,

and between 1·5 and 3 per cent of a mesophilic mixed strain starter culture is added. Heterofermentative cultures are generally used comprising strains of *Streptococcus lactis, Str. cremoris. Str. lactis* sub-sp. *diacetylactis* and *Leuconostoc cremoris.* In certain situations, thermisation of the milk may be carried out on reception (63 °C for 15 s) followed by an overnight prematuration of the milk by inoculating the milk with 0·1–0·2 per cent starter culture at between 10 and 14 °C. Addition of calcium chloride is generally carried out at between 5 and 20 g per 100 litres milk.

2. The pasteurised, standardised (and perhaps thermised and pre-ripened) milk, after dosing with starter, is filled into 100 litre bassines (or specially constructed vats or equipment) with the temperature being adjusted to 32–35 °C.

3. Acidity development proceeds until a level of some 0·18–0·20 per cent lactic acid is achieved (20–25 min).

4. Rennet addition at the rate of 22–30 ml per 100 litres milk is carried out.

5. Coagulum forms and is ready for moulding at approximately 1 h 00 min to 2 h 30 min after renneting.

6. The curd is cut into 30 mm cubes and may be allowed to settle in the vat for some 10–15 min followed by partial extraction of whey using a pump. The cut curd is then transferred manually or mechanically to various types of mould systems where it is distributed as evenly as possible amongst the moulds.

7. Whey drainage is allowed to proceed in the moulds for approximately 18–24 h: 5 h at 27 °C followed by 15 h at 23 °C with relative humidity between 90 and 95 per cent. To assist whey drainage, the moulds are turned approximately every 5 h.

8. The cheeses, after removal from the moulds, are salted either in a saturated brine solution at 10–15 °C for 30–60 min, or by means of a mechanical dry salter. Salt content of between 1·5 and 1·8 per cent is considered acceptable in the finished cheese.

9. The cheese is ripened or matured at a temperature of between 12 and 15 °C and relative humidity of 90–95 per cent for from 7 to 14 days.

10. The cheese is then packaged and stored at between 4 and 8 °C prior to distribution.

The characteristic white mould coat found on Camembert and other soft ripened cheeses was traditionally formed by a naturally occurring mould strain found in cheese ripening rooms. Nowadays this surface mould growth is controlled by the use of pure cultures of *Penicillium candidum*, a strain having pure white coat forming ability and characterised by its high salt tolerance and highly aerobic nature.

The cultures of *P. candidum* may be inoculated in three ways:

1. By addition of the mould culture to the milk.
2. By spraying a solution of the mould culture onto the cheese surfaces.
3. By applying dry mould spores along with the salt to the cheese surfaces.

Ripening or maturation of the cheese is brought about by the action of both the lactic bacteria and the *P. candidum*. Camembert characteristically ripens from the outer surface as the *P. candidum* releases proteolytic enzymes. As the hydrolysis continues, casein is progressively broken down to ammonia, the body becomes smooth, and the breakdown of fat gives characteristic flavours. The central white, pasty layer gradually diminishes as ripening continues, and eventually the cheese will become over-ripe and liquefy with a characteristic aroma of ammonia. Traditional Camembert ripens differently from the industrial varieties, and this is attributed mainly to differences in the bacterial flora in the cheese.

Other varieties of soft ripened cheese are manufactured using similar methodology, but by varying the speed of whey drainage, ripening conditions and mould size, a different finished product results. For example, Brie manufacture is very similar to that for Camembert. However, slight differences in production and the dimensions of the cheese mean that the internal ripening, and hence characteristic flavour and aroma, is different.

Equipment and process development

In view of the high proportion of the production cost attributable to raw materials (that is, the milk), and the fact that the cheeses are sold as small units (usually 250 g pieces), it is essential to minimise the variation observed in unit weight, and attempt to reduce the mean milk volume used per cheese. In traditional systems, 90–100 litre bassines or vats are used for producing the curd which is subsequently filled into a constant number of moulds. Consequently, irregularities of filling the bassines will result in variations in cheese weights, and certain lines of equipment development have been aimed at reducing this variability.

Distribution of curd into the moulds has also been a problem with a variation of weights being observed.

The net effect of the above two points is shown in Fig. 10, which illustrates the influence of mechanisation on cheese weight dispersion and volume of milk per cheese. Essentially, during traditional manufacture, the problem of cheese weight scatter results in the manufacture of cheeses of average weight 270 g from an average volume of 2·0 litres of milk per cheese, in order that the minimum number of cheeses is downgraded because of short weight. By introducing mechanical processes, the weight scatter observed has been substantially reduced, and now cheeses of average weight 250 g can be manufactured from an average volume of 1·85 litres of milk without fear of producing an excessive number of under-weight cheeses.

Thus, the primary objective of mechanisation of soft cheese manufacturing processes has been the reduction of weight dispersion due to irregularities in vat filling and distributing the curds among the moulds, together with the improvement in productivity. Other economies have been made in the handling and packaging stages, as these two items, traditionally of a very labour intensive nature, were also prime candidates for mechanisation.

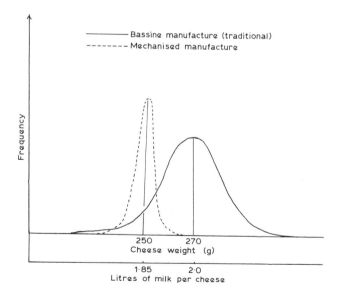

Fig. 10. Influence of mechanisation on cheese weight dispersion and volume of milk per cheese.

In some traditional factories still using small volume bassine manufacturing methods and hand moulding, a certain amount of mechanisation has been introduced in the form of bassine emptying, mould stacking, turning and conveying. Various manufacturers, such as Pierre Guerin, Alpma, Waldner, Burton Corblin and Cartier, produce such equipment which can considerably reduce the amount of labour required.

Several mechanised soft cheese manufacturing systems have evolved over the years and the various lines of development will now be considered.

Several systems, aimed at eliminating cheese weight dispersion due to irregular filling of small vats or bassines, utilise large capacity vats from 1000 to 5000 litres. The vats, usually of semi-circular section with a curd discharge opening at one end, are equipped with stainless steel partitions which allow the vat contents to be divided into sections for subsequent transfer to multi-mould systems. Pierre Guerin, Waldner, Burton Corblin, Alpma and Steinkecker manufacture soft cheese vats of this design.

Cheeses manufactured using this method are still subject to some weight dispersion due to variations in distributing the curd into the multi-mould systems. This problem has been partially resolved by the use of mechanical curd draining equipment prior to mould filling. Two such systems are manufactured by Alpma and Waldner where curd, after production in vats, is fed onto a mechanical whey drainage belt and finally into vertical moulding tubes, where portions of cheese are cut-off at the bottom by pneumatically operated knives. Each portion is automatically transferred to a mould for further drainage and handling.

The second line of development, also aimed at eliminating cheese weight dispersion, involves coagulating milk in small capacity 'vats'. A quantity of milk, corresponding to the volume required for a single cheese, is processed in a 'micro-vat' or 'micro-bassine' which is part of a mechanised handling system. Coagulation and whey drainage operations are carried out in the micro-bassines, and very good cheese weight dispersion can be achieved.

Two systems have evolved in France — the Rematom and the Hugonnet processes, but they have not gained much popularity due to processing constraints. Problems are experienced in ensuring good distribution of manufacturing ingredients such as starter, mould culture and rennet, in controlling uniformly the temperature, in cleaning the micro-bassine system, and in general mechanical control.

A third line of development has proceeded with milk coagulation being carried out in medium capacity units with uniform distribution of the curd at the moulding stage. Both Cartier and Pierre Guerin have developed systems based on shallow rectangular bassines of between 30 and 80 litre milk capacity (see Fig. 11). Milk, to which rennet, starter culture and mould culture are metered, is filled into the small rectangular bassine and coagulation occurs while moving on a conveyor. After curd cutting, a divider is placed into the bassine which serves to segregate a quantity of curd sufficient to produce one unit of cheese. The curd is effectively transferred to a block mould system by placing a block mould with corresponding mat and whey drainage tray on top of the bassine and divider. Bassine and moulding system are turned mechanically and the curd transferred to moulds. The bassine and divider are then cleaned and re-used, while the block moulds and drainage trays are stacked, regularly turned to allow good whey drainage, and conveyed mechanically to the store (Fig. 12).

A system currently gaining favour in France and Germany represents a fourth line of development. This is the Alpma coagulator (Fig. 13) which consists of a sectioned, slow moving belt of semi-circular cross-section on which all the different sequential manufacturing steps

Fig. 11. Cartier system for the manufacture of soft mould-ripened cheese using small bassines or vats (30 litre). (Courtesy of Pierre Guerin.)

involved in traditional manufacture are carried out (see later for further details).

A fifth line of development involved a method of continuous milk coagulation and curd draining. The Stenne-Hutin process, for example, consists of concentrating milk to a total solids content of about 36 per cent by evaporation, and then cooling, ripening with bacterial starters, and renneting. The cold, renneted, concentrated milk is then dosed with hot water which has the dual purpose of reconstituting the milk to the original total solids level and heating the milk to the coagulation temperature of 32 °C. Coagulation follows in approximately 45–60 s, when the curd is cut and tipped conventionally into block moulds. The APV Co. Ltd, in conjunction with Stenne-Hutin, developed a piece of equipment to allow continuous coagulation of the cold renneted milk (the Paracurd machine). Other processes, for example, Berridge, Nicoma and Multitube Schulz, all make use of this cold hydrolysis of casein, but none have passed the stage of experimental use as they rely greatly on homogeneity of raw materials and very precise control of the manufacturing parameters.

Finally, a sixth line of development is being actively pursued at the present time, and this is the application of ultrafiltration (UF) to milk for cheesemaking. Using the process of membrane ultrafiltration, it is

Fig. 12. Mechanised handling of the block moulds used for the production of soft, mould-ripened cheese — the turning facilitates drainage of the whey. (Courtesy of Alpma, UK.)

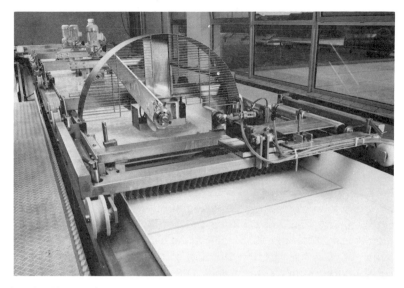

Fig. 13. The continuous Alpma coagulator showing curd cutting and stirring stages. (Courtesy of Alpma, UK.)

possible to remove the necessary amount of water, lactose and minerals before coagulation and acidification of the cheese milk. A retentate or 'pre-cheese' can be produced having a similar chemical composition to a drained cheese. Good control of the cheesemaking process can be achieved, and the procedure has the additional advantage of increasing cheese yield as the soluble whey proteins, normally lost in the drained whey, are retained in the concentrated 'pre-cheese'.

Alongside these equipment developments, work has been proceeding with a technique known as 'Curd Stabilisation', which utilises a particular mix of starter cultures. The development has arisen due to a consumer preference for more mild soft cheeses with a longer shelf-life. The products generally have the same characteristics in terms of dry matter, size and surface appearance, but their pH after salting is higher, the fat content may be higher (e.g. 60–70 per cent FDM), and the shelf-life is longer.

The technique involves using partial or total replacement of the mesophilic starter strains with a thermophilic species such as *Str. thermophilus*. Modifications to the cheesemaking procedure are necessary, in that the set temperature is increased from 30–32°C to 37–40°C. Thus, acidification proceeds at an elevated temperature, but after moulding and cooling, the acidification rate is decreased and the curd pH stabilises in the region of 4·9–5·1 after salting. This results in a relatively stable product whose textural characteristics remain fairly constant throughout a shelf-life of 12–14 weeks.

Surface/internal mould-ripened soft cheese
A number of cheeses in this category have been developed over the last five years in Europe, and these include Lymeswold, Blue Brie, Cambazola, Saga, Bavarian Blue and Opus 84. Production methods are generally as outlined in the previous section for soft mould-ripened cheese, and the products are further characterised by the presence of blue veining or pockets throughout the body of the cheese, as well as a white mould surface coat. The products are generally of comparatively high fat content, (60–70 per cent FDM), and possess a mild flavour due to the use of *Penicillium roqueforti* (blue mould) strains with low proteolytic and lipolytic activity. An extra processing stage is necessary after cheese brining, where piercing with needles is carried out to allow growth of the aerobic blue mould spores added to the vat milk. Some manufacturers use curd stabilisation techniques to prolong product shelf-life, particularly for export.

Surface bacterial-ripened soft cheese
Some surface mould-ripened cheeses, such as Pont l'Evêque and Livarot, are known as 'washed rind' cheeses, and these are produced by removing any white mould growth from the cheese surface and encouraging the growth of the red-pigmented bacterium, *Brevibacterium linens*. Other bacterial-ripened cheeses include Limburger, Liederkranz, and the semi-soft varieties Saint Paulin, Brick, Muenster and Port Salut.

Generally, the manufacturing steps for the different soft varieties are similar to those used for mould ripened varieties, but the flavour intensities can vary from weak to strong. Strong flavours result from increasing moisture levels, increasing the period and temperature of maturation, not removing the surface growth, and/or increasing the surface area of the cheese in relation to its volume.

The flavour-producing bacterium *Brevibacterium linens* produces a characteristic reddish-brown surface to the cheese, and the essential process step is known as 'smearing'. This involves rubbing the surface with warm salt water either by hand or using a mechanical brushing device. The 'smearing' does not introduce the *B. linens* onto the cheese surface, as the bacterium is generally engrained into the wooden shelves in the ripening room. At the commencement of manufacture of these varieties, it is customary to apply a culture of *B. linens* to the shelves. It is essential to maintain a high relative humidity environment for the growth of *B. linens*, and growth is limited to the cheese surfaces due to the highly aerobic nature of the organism.

In the initial stages of ripening, lactose fermenting yeasts establish themselves, growing well in the high salt, low pH (*ca* 5·2) environment. As the yeasts increase the pH of the cheese surface to the order of 5·9, the *B. linens* begins to grow, and further control of the growth is achieved by washing and brushing techniques.

Internal bacterial-ripened soft cheese

The cheeses in this category may be considered as soft mould-ripened cheeses without the white mould coat. Some Coulommier is manufactured by this route, as are the traditional English varieties — York and Cambridge. Because these varieties lack the stabilising influence of the mould coat and the moisture content is high, the shelf-life is limited to 2 weeks at refrigeration temperatures. Its manufacture is thus restricted to the small producer or farmhouse situation, where cheese can be marketed locally.

Internal bacterial-ripened and brine preserved cheeses

Manufacturing process

Cheeses included in this category include Feta, Domiati and Halloumi. This group is often referred to as pickled cheese, as they are matured in brine. Feta, traditionally made from sheep's milk or a mixture of sheep and goat's milk, is now more commonly made from bovine milk with the addition of a bleaching agent, and lipase enzyme to enhance flavour production. A wide range of Feta types are on the market today with differences primarily in texture and flavour. It is typically of soft crumbly texture, with a subtle aromatic flavour, and a very white colour. Figure 14 shows a typical manufacturing procedure for a traditional Feta.

Milk is pasteurised, standardised and, if using cow's milk, bleached by addition of a chlorophyll decoloriser; a starter culture (2 per cent) is added and milk is ripened. Rennet, and optionally, calcium chloride and/or kid or lamb lipase are added, and once set, the curd is cut. Curd is ladled into moulds where whey drainage proceeds. These are turned and sometimes subjected to slight pressure for an overnight period. The

Time/temperature *Operation*

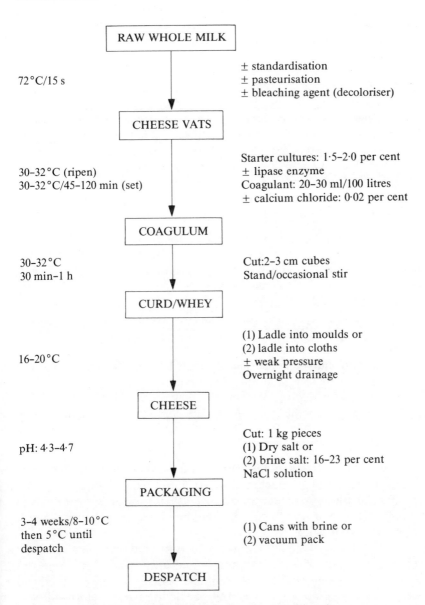

RAW WHOLE MILK

± standardisation
72 °C/15 s ± pasteurisation
± bleaching agent (decoloriser)

CHEESE VATS

Starter cultures: 1·5–2·0 per cent
30–32 °C (ripen) ± lipase enzyme
30–32 °C/45–120 min (set) Coagulant: 20–30 ml/100 litres
± calcium chloride: 0·02 per cent

COAGULUM

30–32 °C Cut:2–3 cm cubes
30 min–1 h Stand/occasional stir

CURD/WHEY

(1) Ladle into moulds or
(2) ladle into cloths
16–20 °C ± weak pressure
Overnight drainage

CHEESE

Cut: 1 kg pieces
pH: 4·3–4·7 (1) Dry salt or
(2) brine salt: 16–23 per cent
NaCl solution

PACKAGING

3–4 weeks/8–10 °C (1) Cans with brine or
then 5 °C until (2) vacuum pack
despatch

DESPATCH

Fig. 14. Steps in Feta manufacture (traditional).

following day, the cheese is removed from the moulds and cut into blocks, followed by dry salting and/or immersion in brine solution. The cheese is then usually transferred to sealed tins containing brine, for storage. Cheese may be consumed fresh, or held for up to 6 months.

Equipment and process developments

Although Feta cheese may be produced on a wide range of traditional and modern equipment, the majority of cheese is now made from ultrafiltered milk, and two types known as 'unstructured' and 'structured' are produced. The difference between these is that the unstructured type has a dense smooth texture, while the structured tends to resemble the traditional, crumbly product. UF based systems are available from a number of equipment suppliers, and these include the Pasilac 5600 coagulator, and the Alfa-Laval Alcurd continuous coagulator.

The UF-based process involves the following steps:

(a) Milk standardisation, pasteurisation, homogenisation.
(b) Milk concentration (usually 5:1) by ultrafiltration at 50 °C.
(c) Pasteurisation and homogenisation of the UF retentate.
(d) Addition of starter and lipase.
(e) Addition of rennet to the ripened retentate and transfer to the coagulator system — tubular in Alcurd and small batch tanks in the Pasilac 5600.
(f) The coagulum may be cut and subjected to a holding/heating stage usually in the permeate stream from the UF plant, followed by dosing into moulds. This effectively texturises the final product which is then known as 'structured' Feta.
(g) Alternatively the retentate may be dosed into moulds or directly into storage tins, to which brine is added. This is 'unstructured' Feta.

Cheese yield, as a result of UF manufacture, is in the order of 5500 litres milk per tonne of product, compared to 7300 litres milk per tonne product of traditional manufacture. This increase in yield is due to incorporation of whey proteins into the cheese, and also a more efficient recovery of milkfat.

The technique of direct acidification may be used for the manufacture of Feta instead of starter cultures, and this has led to a recent development whereby cooled, ultrafiltered milk is directly acidified and then dosed as a liquid pre-cheese into cartons on a vertical form-fill-seal packaging machine. The pre-cheese then coagulates in the carton to produce an unstructured type Feta.

SEMI-SOFT CHEESE MANUFACTURE

Figure 2 illustrates a functional classification of semi-soft cheeses into 3 categories, viz. surface/internal bacterial-ripened, internal mould-ripened and 'stretched curd' cheeses.

Surface/internal bacterial-ripened semi-soft cheese

Cheeses in this category include Saint Paulin, Muenster, Brick and Port Salut, and the surface flora contribute to the flavour development, although recently varieties, such as Saint Paulin, have been manufactured without the surface organisms, which are replaced by a dye, to satisfy consumer demand for a smooth, clean flavoured product. The establishment of the reddish-brown coloured, surface bacterium, *Brevibacterium linens*, has been outlined above.

St Paulin is a French cheese derived from Port du Salut made in a monastery near Laval. It has a creamy white, soft but sliceable texture, with a mild, slightly acid flavour and delicate, aromatic aroma. The rind is yellow to orange in colour, and this is usually achieved nowadays with the use of coloured plastic coatings, although some product may be found with the natural rind dyed with annatto, or even with a surface growth of the reddish-brown bacterium *B. linens*. Figure 15 illustrates the process steps involved in Saint Paulin manufacture.

The manufacturing process is characterised by the removal of 50 per cent of the whey and replacement with water (or brine solution in some cases), this effectively reduces the acidification rate and results in mild flavour development. The curd is subjected to a cooking temperature of between 34 and 36°C, followed by a slight pressing, and the cheeses are salted by immersion in a saturated brine solution.

Equipment and process developments

Curd manufacture is carried out in a range of traditional and modern vats with the appropriate development of systems for carrying out the curd washing stage, i.e. removal of a volume of whey and replacement with water. This is generally achieved with the use of a perforated boom, which can be lowered into the whey during the settling stage. Whey is then removed by vacuum and water introduced, either through the boom arrangement or from jets installed in a pipe running along the internal surfaces of the vat. The most significant development has been the introduction of curd drainage and moulding machines, available from Stork-Friesland, Koopmans, Alfa-Laval and Pasilac, for example.

Time/temperature   *Operation*

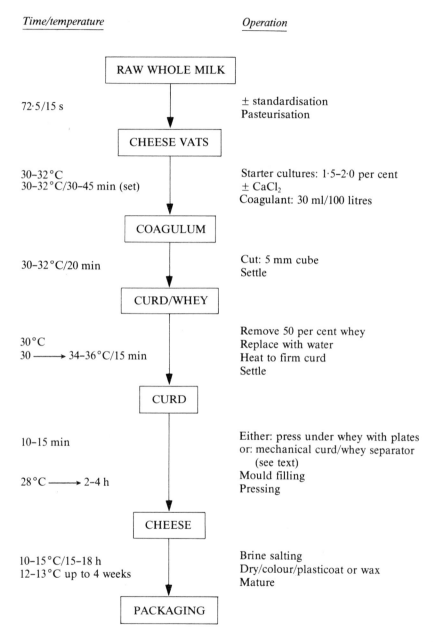

Fig. 15. Steps in Saint Paulin manufacture.

These consist basically of tubular drainage cylinders where curd is separated from the whey, and the curd is subsequently dosed into cheese moulds (see Chapter 2 for further details).

Other developments include single cheese, multi-row, automated tunnel press systems, automatic deep brining systems, mechanised equipment for plasticoating cheese, and automated cheese maturation stores. More recently work has been carried out on the application of ultrafiltration to St Paulin manufacture. This approach has resulted from the development of mineral membranes, which have allowed milk to be concentrated to the higher solids levels necessary for semi-soft cheese manufacture.

Internal Mould-ripened Semi-soft cheese

Cheeses in this category include Stilton, Roquefort, Gorgonzola and Danish Blue. Stilton manufacture is officially limited to the three English counties: Derbyshire, Leicestershire and Nottinghamshire, while Roquefort must be made from sheep's milk in the Roqueforte region of Southern France.

The blue cheeses are characterised by the internal veining resulting from the growth of the aerobic mould species, *Penicillium roqueforti*, which is available exhibiting a range of colours from green to deep blue, and varying proteolytic and lipolytic activities. The blue cheeses are generally produced in round-wheel forms, containing 3–5 per cent salt and possessing a spicy, piquant flavour associated with certain free fatty acids and ketones resulting from the mould's lipolytic activity. The textures are generally moist, with a slight stickiness and crumbliness.

Manufacturing process
The steps involved in manufacturing a semi-soft blue cheese are outlined in Fig. 16. The process is characterised by the homogenisation step which increases the susceptibility of the fat to lipolysis, and consequent flavour development, while also assisting in the production of a porous curd and open-textured cheese required for good blue mould growth. The *P. roqueforti* is either added to the milk, or at a later stage, such as onto the curd at salting. The curd is not pressed, but knits together under its own weight in open-ended moulds which are turned frequently.

During ripening, the cheese must be pierced to allow the escape of carbon dioxide and the entry of oxygen which promotes mould growth.

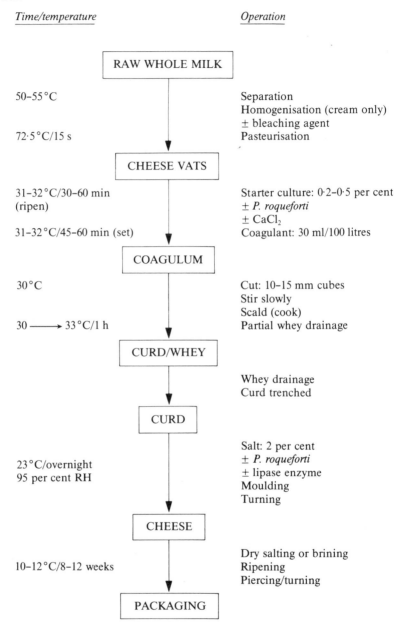

Fig. 16. Steps in semi-soft blue cheese manufacture.

Stilton manufacture differs from the typical blue cheese process, as illustrated, in that:

(a) there is no homogenisation;
(b) starter inoculation is very low (0·01 per cent);
(c) there is no cooking stage;
(d) the curd is left overnight in coolers/drainers;
(e) the curd is cut into chunks, salted and milled twice before filling into moulds;
(f) cheese drainage occurs in moulds for 4–5 days;
(g) the cheese is pierced fairly late in ripening, at 6 and 8 weeks.

Equipment and process developments
The manufacture of Stilton and other blue cheeses is considered very much a traditional process, and consequently much production is still carried out as a labour intensive operation using basic cheese vats and curd drainers. However, some cheese is now manufactured using modern, mechanised vats, such as the Damrow 'Double O', curd drainage belts, and mechanised mould-filling equipment. Some curd has been manufactured using the Alpma continuous coagulator and, as with other categories of cheese, the use of ultrafiltration has been investigated. Significant advances have been made in the storage of blue cheeses with environmentally controlled, fully automated stores being available from manufacturers, such as Elten and Koopmans, and automated piercing, cheese washing and packaging operations are being introduced. Many manufacturers have gone away from the long maturing, less salty types of blues, in favour of high yielding, quick maturing types capable of being manufactured on modern, mechanised equipment with reduced labour involvement.

Stretched Curd Cheese

The major cheese types included in this category are Mozzarella and Provolone, which are also referred to as 'Pasta filata' or 'plastic curd' types. Mozzarella was originally manufactured from high fat buffalo milk, but it is now made all over Italy, in other European countries and the USA from cow's milk. Some cheese is made from partially skimmed milk, especially for use in catering as a pizza cheese. It is an unripened semi-soft cheese which may be consumed shortly after manufacture. It has a soft, waxy body with a bland, but mildly acidic, flavour.

Manufacturing process

The steps involved in the manufacture of Mozzarella are illustrated in Fig. 17. Although traditionally manufactured from raw milk, it is more common to pasteurise nowadays. Often a bleaching agent, such as titanium dioxide, is added to mask the natural β-carotene yellow colour, and high temperature, or thermophilic starter cultures, such as *Lactobacillus bulgaricus* are used together with *Streptococcus durans* for flavour production. The use of high temperature starters is necessary because of the high cook temperature used during processing, and the need to produce curd of optimal pH for the stretching operation. Lipases from kid, lamb, or calf are often added in order to stimulate the production of piquant flavours resulting from free fatty acid production. At a critical pH of 5·1, the curd is heated under hot water, stretched and moulded into shape, and finally filled into moulds.

The chemistry of the stretching operation involves conversion, by lactic acid, of the dicalcium paracaseinate, produced as a result of rennet action, to monocalcium paracaseinate, which, when heated to 54°C or higher, becomes smooth, pliable, stringy and retains fat:

$$\text{calcium caseinate} + \text{rennin} \xrightarrow{\text{pH } 6\cdot2} \text{dicalcium paracaseinate}$$

$$\text{dicalcium paracaseinate} + \text{lactic acid} \longrightarrow \text{monocalcium paracaseinate}$$

Equipment and process development

Traditionally Mozzarella manufacture was carried out in wooden vats with the curd mass being manually manipulated under hot water. Now the curd is produced in modern automated vats and curd drainage equipment (see Chapter 2), and the downstream operations of milling, heating, stretching, moulding, cooling and brining have all been mechanised and automated by various equipment manufacturers, such as Modena and Alfa-Laval. The ripened curd blocks are reduced to smaller size particles using a mill which may be separate or part of a complete line. The milled curd is then heated with hot water or steam, and subjected to a gentle pulling or stretching into a smooth plastic mass, which is then transferred to a moulding module which forms specific weights into desired shapes and units by weight. Finally the moulded cheese is passed through a cooler into an automatic brining section.

Other developments include the use of direct acidification techniques

Time/temperature   *Operation*

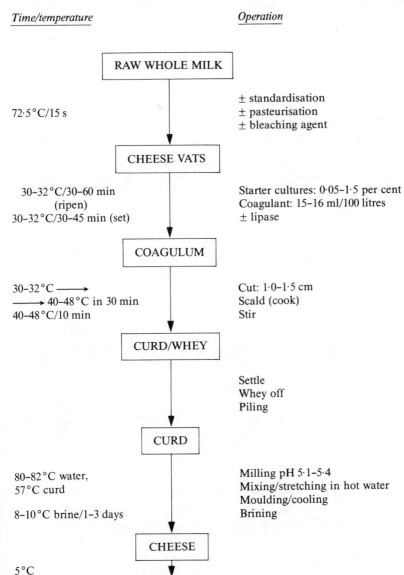

Fig. 17. Steps in Mozzarella manufacture.

instead of, or as well as, the use of starter cultures, and the application of ultrafiltration technology.

Other Soft/Semi-soft Cheeses

Figure 2 refers to three other categories of cheese which may contain soft and semi-soft types: Whey cheese, processed cheese and cheese substitutes.

Whey cheese

Soft cheeses made from whey include Ricotta, originating from Italy, and Mysost and Gjetost made from cow's milk and goat's milk whey respectively and originating from Norway. The Ricotta, although originally made from the whey from Provolone and other cheese manufacture, is now made from mixtures of whey and milk, or from whole milk alone. The cheese resembles cottage cheese curd, is soft and creamy, with a light delicate texture and slight caramel flavour. The Mysost and Gjetost are light tan in colour, with a smooth creamy body and a caramelised flavour.

Manufacturing process

Ricotta is made by heating whey to 85 °C by direct steam injection followed by the addition of citric or acetic acid. The whey proteins will precipitate out, and the curd can be transferred into drainage moulds for 4–6 h. Following drainage, the cheese is ready for packaging and consumption.

Mysost and Gjetost are manufactured by evaporating whey to 60 per cent total solids. The cheese is a condensed mass of caramelised lactose, protein, fat and minerals. The mass (60 per cent T.S.) is transferred to a vacuum cooker and further concentrated to 80–85 per cent total solids. The brown concentrate is then further heated until the required flavour and brown colour is reached. The product is kneaded to produce a butter-like texture, and is then moulded into blocks and packaged. It has a good shelf-life due to the low moisture content (15–20 per cent), and it can be stored for up to 6 months at 5 °C. The product does not undergo ripening.

Processed cheese

A number of soft/semi-soft processed cheeses may be found in the market place in the form of spreads, dips and dressings.

Processed cheeses are manufactured by blending and heating one or more base cheeses with a suitable emulsifying salt until a homogeneous mass results. Processed cheese spreads may also contain other dairy ingredients, such as skim-milk, cream, butter and whey powder. Many 'value added' spreads are available containing such additives as shrimp, crab, pepper, horseradish, nuts, mushrooms, garlic, herbs, port or kirsch.

Manufacturing process

Processed cheese spreads are made by selecting suitable cheeses according to age, flavour, body and texture. The cheese is then ground and mixed with emulsifying salts and water, followed by heat treatment in a kettle at 70–85 °C for 5–15 min, dependent on the product being made. Heating can be indirect, or direct by steam injection. The hot, plastic cheese mass, to which additives such as herbs, nuts or fruit may be added, is transferred to filling machines, where it is dosed into the final package. The emulsifying salts used are citrates, monophosphates and polyphosphates, and these cause an increase in pH which, in turn, solubilises the protein and results in a smooth, homogeneous mass.

Equipment and process developments

Many large production facilities now exist in the USA and Europe for processed cheese manufacture. There have been many types of batch and continuous cookers developed. In Europe, round-bottomed, stainless steel vessels equipped with direct or indirect heating and vacuum operation are widespread. In the USA, stainless steel, horizontal cookers with direct steam injection are popular, and more recently, continuous cookers, such as the swept-surface, heat exchanger (e.g. the Votator or Kombinator), have been introduced.

The use of enzyme modified cheeses (EMC) for processed cheese manufacture has recently become widespread. EMC's are made by treating cheese with a mixture of proteolytic and lipolytic enzymes, which results in the production of an intensely flavoured material which can substitute in a processed cheese formulation for a proportion of the mature cheese; it thus introduces economies to the processed cheese operation. Many EMC's are available commercially from flavour houses in the form of pastes or powders.

Cheese substitutes

The cost of producing dairy products has risen considerably over the years, and this has given an impetus to the development of a range of dairy product substitutes, including cheese in the form of blocks, slices, spreads and dips. The major application for cheese substitutes (or imitation, analogues, artificial) is in formulated foods manufactured by catering or industrial establishments, e.g. a Mozzarella substitute for use in pizza manufacture.

Manufacturing process

Substitutes may be manufactured by two routes, and these involve the use of fat and/or protein sources other than those native to milk, together with a suitable flavour system. The one route uses a liquid 'milk', whether skimmed milk plus vegetable oil, or totally synthetic, and involves conventional in-vat cheesemaking methods, the products often being referred to as 'filled' cheeses. The other route involves blending various raw materials together using techniques similar to those for processed cheese manufacture. The raw materials used can be various vegetable proteins and oils, or alternatively partial-dairy ingredients, such as casein and caseinates, together with a hydrogenated vegetable oil, such as soya bean, palm, cotton seed, coconut or corn.

The flavour system might be artificial, or of natural origin using enzyme modified cheese for example. As well as savings in the manufacturing process, raw materials are considerably cheaper per tonne of product than milk, with vegetable oil some 75 per cent cheaper than butterfat, and casein up to 50 per cent cheaper than skim-milk powder, if calculated on an equivalent protein content basis.

Methods of manufacture vary but, generally, processed cheese equipment can be utilised, for example batch cooker/mixers or continuous scraped-surface cookers.

The future of cheese substitutes depends on the relative prices of vegetable fats and proteins compared with milk ingredients, and perhaps more importantly on a greater consumer acceptance of the products.

RECENT DEVELOPMENTS

Over the last ten years a wide range of cheese products have been finding their way onto the supermarket shelves. In the UK for instance, while there has been a static Cheddar and territorial market, there has been considerable growth in the soft and semi-soft cheese sector. Imported products, such as Brie and Camembert, have proved popular and this has led to the development of home produced cheeses such as Melbury, Lymeswold and Somerset Brie. Changes in dietary fashion have also resulted in diversity of product development, with a number of low or reduced fat cheeses now being produced, as well as higher fat 'luxury' products for special occasions. Export opportunities for products, such as Feta and mould-ripened cheeses, have also stimulated production demand. Consideration of the high labour and raw material cost have led to a need to increase process efficiency, and this in turn has stimulated equipment manufacturers to develop mechanised, automated lines capable of producing good quality, high yielding cheeses at minimal unit cost. Many of the equipment and process developments have been reviewed in the sections relating to each category of cheese, but the following are worthy of further consideration.

Ultrafiltration

The technique of ultrafiltration has been applied to the manufacture of many soft cheeses, including Camembert, quarg, cottage cheese, Feta, Mozzarella and Danish Blue.

Ultrafiltration (UF) is a membrane separation process operating at a molecular level. The membrane has extremely fine pores (about 1–20 nm in diameter), and small molecules such as water, lactose and dissolved salts can be forced through the pores under pressure, while larger molecules, such as proteins and fats cannot go through and can, therefore, be progressively concentrated. UF membranes are based on cellulose acetate (first generation), polycarbonates or polysulphonates (second generation), and zirconium oxide (third generation), and UF systems are available in various forms: hollow fibre, tubular or plate, supplied by, for example, Alfa-Laval (Romicon), Abcor, and Pasilac (DDS), respectively.

The advantages of using ultrafiltration for cheesemaking are:

1. Higher protein recovery due to incorporation of whey protein, (over 20 per cent of total protein in milk);
2. higher fat recovery due to reduced loss into whey;
3. reduced volume of fluid to be processed;
4. less rennet and starter required;
5. smaller plants required;
6. continuous processing is made easier.

Ultrafiltration can be utilised in four ways:

1. Pre-concentration of milk up to 2 times (i.e. half volume of original milk). Concentrate is processed using conventional cheesemaking equipment, and the major advantage is increased plant utilisation. There may be a slight increase in yield due to increased recovery efficiency of fat and protein.
2. Production of a 'pre-cheese'. Here the milk is effectively concentrated to the final solids content of the cheese type with the concentrate or retentate being dosed, after addition of starter and rennet, into moulds. As there is only limited whey drainage, increase in cheese yield can be significant. Some Feta (unstructured), soft unripened cheese, and Camembert are produced by this method. The UF concentrate cannot be processed in conventional vats and, therefore, specialist equipment has been developed for UF cheesemaking.
3. Production of a concentrate, followed by cutting of the coagulum produced after addition of rennet and starter. This method allows some texturisation of the final product as in the case of structured Feta, Mozzarella and Danish Blue manufacture. Again yield increase can be significant, but specialist equipment is required for processing the concentrate into cheese.
4. Ultrafiltration of whey from cheesemaking to concentrate whey proteins which are subsequently returned to the cheese vat as a slurry. Only relatively small increases in yield can be achieved by this method without affecting product quality.

A number of continuous systems based on UF have been developed recently, and these include the Camatic system from Alfa-Laval for the manufacture of soft ripened cheeses, such as Camembert, the Alcurd system from Alfa-Laval for production of structured Feta and Mozzarella, and the CC2500 coagulator from Pasilac, again for the production of structured Feta and Mozzarella.

Direct Acidification

Acidification of milk during cheesemaking is generally as a result of lactic acid production by lactic bacteria (starter cultures) fermenting milk lactose.

However, direct acidification has been traditionally practised in the manufacture of a number of unripened cheeses, such as Ricotta and Queso Blanco, where lactic, acetic or citric acid is used to adjust the pH of the milk to 5·0.

More recently work has been carried out on the application of direct acidification techniques to the manufacture of cottage, Mozzarella and Feta cheese.

The advantages of direct acidification or the direct set method are:

1. elimination of problems associated with starter cultures;
2. decreased production times;
3. improved product consistency;
4. greater control of process.

The process involves acidification of cold (3–5 °C), milk with a mineral acid such as phosphoric, and this is usually achieved in-line using a helical mixer coupled to an acid dosing pump. The system is linked to a pH controller, which monitors and holds the pH in the range 4·90–4·95 (for cottage cheese manufacture). Following this operation, the acidified milk is heated in the vat to 30–32 °C and an acidogen, D-glucono-delta-lactone is added. This hydrolyses producing gluconic acid at a controlled rate (cf. the production of lactic acid by starter cultures), until pH 4·70–4·75 is attained. A small amount of coagulant may be added to assist coagulum formation, and cheesemaking proceeds as for production using starter cultures.

As no enzymes are formed with this method, compared to the use of starter bacteria, cheese flavour is bland, and often a starter distillate is added to the cottage cream dressing in order to give a 'creamy' note to the flavours. Presently, direct acidification has only found applications in the manufacture of soft unripened cheeses, where flavour development is somewhat irrelevant, but no doubt developments in the field of enzyme technology will result in a greater use of the technique in the future.

Direct set materials are available commercially from the Diamond-Shamrock Corporation who market two products: Vitex 750 and 850.

Automation, Mechanisation and Continuous Cheesemaking

Reference to the specific sections reviewing equipment developments illustrates how many technical solutions have been found for automating and mechanising the various stages of cheesemaking. Cost considerations have been the major impetus for these developments, and nowadays the labour requirement has been minimised with large throughput, process efficient plants controlled by centralised computer systems being available for the manufacture of many cheese varieties.

Modern cheese vat design achieving good milk solids conversion coupled with efficient curd/whey separation, and cheese handling equipment allows traditional cheesemaking procedures to be mechanised and automated, while the development of new equipment for handling UF concentrates, for example, has opened the door to a new era of cheesemaking technology. Recently, continuous curd production has become possible on equipment such as the Alpma and the Alfa-Laval Alcurd coagulators. The Alpma coagulator, which utilises either milk or partially UF concentrated milk, consists of a sectioned belt of semi-circular cross-section on which all different sequential manufacturing steps involved in traditional manufacture are carried out. Milk, to which rennet, starter and mould culture have been metered is filled into a section of rubberised flexible belt which is formed into a semi-circular cross-section by the stainless steel walls. Milk is compartmentalised by the lowering of a stainless steel spacing plate, and coagulation proceeds while the belt slowly moves forward. The spacing plates are removed prior to cutting of the coagulum and return, via an automatic cleaning system, to the beginning of the belt for further use. The belt advances the curd through automatic cutting devices which give uniform curd cubes, and these pass from the end of the belt system where some whey drainage occurs. The partially drained curd is filled through a tubular device into block moulds, which are then conveyed, automatically stacked, turned and forwarded for demoulding, salting and storage. Many products including Camembert, Feta, cream cheese and Lymeswold are, at present, manufactured using this system.

Recently, further downstream equipment has been developed by Alpma, which allows the manufacture of cottage cheese from curd produced on the coagulator.

The Alfa-Laval Alcurd continuous coagulator has been designed to operate using UF concentrates only. Concentrate, starter culture and rennet are mixed in-line, and coagulation takes place in an enclosed pipe-line system. On leaving the coagulator, the coagulum is cut into cubes and further curd treatment, such as whey drainage, varies according to the type of cheese being produced. Feta and Mozzarella are presently being manufactured on this system.

FUTURE DEVELOPMENTS

The market for soft and semi-soft cheese products has and will continue to be competitive. Cost considerations such as labour, raw materials and processing create an environment for further development on two fronts:

1. Further development of cost effective production techniques, both equipment and processes, particularly for existing products.
2. Development of new value added products.

The areas of process development will include:

1. Membrane based systems — ultrafiltration.
2. Tailored starter systems and/or direct acidification procedures.
3. Enzyme systems: new coagulants and flavour producers.
4. Continued mechanisation and automation of traditional cheesemaking operations.
5. Consideration of new/novel packaging systems.

The areas of new product development will include:

1. reduced/low fat cheese;
2. cheese substitutes;
3. compound products utilising cheese with other foods;
4. cheeses manufactured from UF concentrates;
5. extended shelf-life products, using techniques such as irradiation, UHT, and curd stabilisation;
6. range extension based on existing products, i.e. the creation of 'value added' products to cater for changing consumer needs.

REFERENCES

Davis, J. G. (1966). *Cheese, Vol. III: Manufacturing Methods,* Churchill Livingstone, Edinburgh.
Eck, A. (Ed.) (1985). *Le Fromage,* Lavoisier, Paris.
Guerault, A. M. (1966). *La Fromagerie devant les Techniques Nouvelles,* Editions Sep., Paris.
Kosikowski, F. (1977). *Cheese and Fermented Milk Foods,* 2nd edn, Edwards Brothers Inc., Ann Arbor, Michigan.

214   *M. B. Shaw*

Scott, R. (1981). *Cheesemaking Practice,* Elsevier Applied Science Publishers Ltd, London.

UK Cheese Regulations, S.I. 1970 No. 94 as ammended by S.I. 1974 No. 1122, S.I. 1975 No. 1486, S.I. 1976 No. 2086 and S.I. 1984 No. 649, HMSO, London.

Van Slyke, L. L. and Price, W. V. (1979). *Cheese,* Ridgeview Publishing Co., Reseda, California.

Chapter 4

Developments in Frozen-products Manufacture

H. L. Mitten*
Formerly APV Crepaco, Inc., Chicago, USA
and
J. M. Neirinckx
APV Crepaco International, Inc., Brussels, Belgium

Frozen dairy products are well known in most of the world, but there is a wide range of composition and physical characteristics from area to area. In those areas where frozen dairy products are made commercially, several types are defined either legally or by the industry.

CLASSIFICATIONS

There are two broad categories: hardened products and soft-serve products. Hardened products are packaged in the semi-frozen state as extruded from an ice cream freezer, then frozen in a hardening room, tunnel or other device, and stored prior to distribution. Soft-serve products are those which are served directly to the consumer in the semi-frozen state as extruded from the ice cream freezer.

Frozen dairy products may also be classified as ice cream, dairy ice cream, ice milk, sherbet, sorbet and water ice. All of these except sorbet and water ice contain some milk solids, and all of them may be aerated to some extent.

Frozen novelties are variations in the form in which the frozen products are marketed. They are usually stick and stickless lollies or bars, extruded bars, small cups and ice cream cakes or pies. Many of these have chocolate or other coatings, and some are elaborately decorated.

Ice cream usually contains 8–18 per cent fat, 9–12 per cent milk solids-

*Present address: 336 Kenilworth Avenue, Glen Ellyn, Illinois 60137, USA.

not-fat (MSNF), 13–16 per cent sugar other than the lactose of the MSNF, and small amounts of stabiliser and emulsifier. In the United Kingdom, ice cream may contain milk fat, vegetable oils or animal fats other than milk fat or combinations of these fats. If the fat is all milk fat, the product is known as dairy ice cream. This is the practice in much of the world, and because of the relative costs of the fats or oils, only about 20–25 per cent of the ice cream is dairy ice cream. This is not the case in the US where the legal standard of identity requires that the product contain at least 10 per cent milk fat if it is to be called ice cream. When vegetable oils or other animal fats are used, the product is called Mellorine or artificial ice cream.

Ice milk is similar to ice cream, but has a fat content of about half that of ice cream, usually 4–6 per cent. It also has more MSNF and sugar. Ice milk generally has a lower calorie content than ice cream, and this has made it attractive to diet-conscious consumers in recent years.

Ice cream and ice milk are aerated during freezing so that the overrun (percentage volume over volume of non-aerated mix) is from 25 per cent for the new super-premium products to 120 per cent or more for standard products.

Sherbets vary greatly in composition, taste, colour, and eating characteristics. Generally, they contain fruit or fruit juice, some added acid for greater tartness, about twice as much sugar as ice cream, 1–2 per cent milk fat, from 2 to 5 per cent by weight of total milk solids and from 0·2 to 0·5 per cent stabiliser, depending upon the type and strength. Sherbets are aerated to give overruns ranging from 20 to 35 per cent.

Water ices are similar to sherbets except they do not contain any milk solids.

Sorbets are often considered to be water ices, but in some areas small amounts of milk solids may be added and whipping agents may be used to permit somewhat higher overrun (30–35 per cent) and to control brittleness and smoothness of body.

These definitions or descriptions are very general. The manufacturer of frozen dairy products should familiarise himself with the legal regulations in his marketing areas.

HISTORY OF FROZEN DESSERTS

The history of frozen dairy products is given briefly by many writers, but a publication of the International Association of Ice Cream Manufacturers

(1966) is interesting and comprehensive. It traces the beginnings of ice cream and frozen desserts to iced drinks enjoyed by Alexander the Great in the fourth century before Christ down through the years of Nero in the first century AD, Marco Polo in the 13th century, through Italy and France to England in 1643–1649, and to America around 1700.

In 1846, an American woman, Nancy Johnson, invented the hand turned ice cream freezer which was widely copied, leading the way to a new industry.

The first wholesale ice cream business in the US was started in Baltimore in 1851 by Jacob Fussell. After the introduction of mechanical refrigeration around 1890–1900, the frozen dairy products industry grew rapidly throughout America and Europe.

The first continuous ice cream freezer was developed by The Creamery Package Mfg Company of Chicago about 1908. It was a crude machine with refrigerated discs rotating in an ice cream mix flowing through a vat-like trough so that the discs were partially submerged in the mix. As the mix froze to the discs, it was scraped off and blended into the liquid. Very softly frozen product flowed out one end and into containers.

This was followed shortly by a batch-type freezer invented by Miller. It had an improved, brine refrigerated, freezing cylinder and a rotating dasher with a beater and scraper blades. Batch freezers were improved by direct expansion and full-flooded refrigerating jackets, and were soon manufactured by several equipment makers in America and Europe. This type of freezer really started the era of industrially-made ice cream, and was the predominate type of freezer until around 1938.

In 1926, Clarence Vogt invented the first truly successful continuous ice cream freezer. It aerated and froze the product in a closed cylinder under pressure with the product in continuous flow. Within the next decade, others developed similar machines, and by 1940, about half the ice cream was made with continuous freezers.

High speed packaging equipment and rapid hardening equipment developed along with ice cream freezers led to the mass production of frozen dairy foods, taking them from a high cost luxury dessert to a readily available, nutritious, and nearly universally desired food.

The manufacture of frozen dairy products is an economically important segment of the dairy products industry. Information published by the International Association of Ice Cream Manufacturers (1980) indicates that the total production of frozen desserts in the 58 countries listed, not including the United States was 4 779 307 300 litres in 1979; Of this, 206 570 000 litres were made in the UK. In addition, 4 583 628 000 litres were produced in the US for a world total of 9 362 935 300 litres; of this world total, 8 033 275 000 litres, or about 86 per cent, was ice cream and ice milk. The trend is growth world-wide, but there is a slight decline or static production in the US and the UK where the aging population is increasingly diet conscious.

MANUFACTURING METHODS

Ice cream and several of the other frozen dairy products are very complex in their chemical and physical properties. Readers are referred to books or papers on the more scientific aspects of the products, such as those of Arbuckle (1977), Berger (1976) and Keeney (1972).

Raw Ingredients

The raw ingredients necessary to provide the components of ice cream or ice milk must contain sufficient fat and milk solids in proportions that can be combined to make a mix of the desired composition. The usual sources of milk fat are milk, cream of 30–35 per cent fat, butter and sometimes concentrated whole milk. The milk solids-not-fat of the milk and cream alone are seldom sufficient for the mix. Therefore, concentrated whole or skim-milk or dried skim-milk solids are included in the necessary raw ingredients. Sugar is required for sweetness, and for additional solids desired for the characteristic body and texture of the finished ice cream. The amount of sugar depends upon market preferences to a large extent. Corn syrup of moderate dextrose equivalent (DE) can provide solids to build body and give chewiness to the ice cream without making it overly sweet. Both sucrose and corn syrup solids are available as syrups or in granulated or crystalline form. Other ingredients, such as the stabilisers and emulsifiers, are usually in dry form and are used in small amounts compared to the others. Water may also be considered as one of the ingredients, especially if dried milk and granulated sugar are used.

The choice of raw ingredients depends upon such influencing factors as quality of ice cream to be made, availability of the ingredient, relative costs, storage equipment and space available, the blending or batching methods to be used and the size of the operation. Nearly all modern ice cream makers having a yearly production of 2 000 000 litres mix or more, use liquid raw ingredients including sugar for most of the year. They may also use dry ingredients some of the time. Small ice cream makers may find that ease of storage and lower costs make dry ingredients more attractive, and will choose processing methods and equipment designed primarily for convenience in accommodating these ingredients. Occasionally, very large ice cream makers design their operations around dry ingredients, because of the location of the plant and availability of the ingredients. When there is no great economic advantage of one type of ingredient over another, most ice cream makers prefer liquid sources which can be pumped through closed systems with relatively lower labour costs for handling and for equipment cleaning.

Receiving

Receiving operations are similar to those of any other dairy processing plant. Milk, cream and concentrated milk or concentrated skim-milk are delivered by transport tanker, weighed, and pumped into storage tanks. The storage tanks may be of the horizontal type or the vertical or silo type. They should be insulated sufficiently to allow holding the cold ingredient for two days or more without a temperature rise of more than 4 °C, when the temperature difference from ambient to product is 27 °C or less. Air agitation may be used in milk storage tanks, but it may cause the incorporation of air and foaming when cream and concentrated milk are involved. Mechanical agitators of special design are required for these products.

Liquid sugar and syrups are also received from tankers and pumped to storage tanks designed especially for syrups. Because liquid sugars and syrups are very viscous at lower temperatures, they are usually transported and stored at temperatures of 35–45 °C, where viscosity is low enough for ease in pumping and for good drainage characteristics. Care must be exercised to prevent the temperature of these syrups from exceeding 45 °C where changes in flavour and colour may occur, and result in shortening the storage time and quality of the final product.

Storage tanks for liquid sugar and syrups should have stainless steel product contact surfaces or the equivalent in cleanability and corrosion resistance, should be insulated and should have means for warming the syrup or keeping it warm. As an alternative, single shell, uninsulated tanks located in an insulated, heated room are often used. The temperature of the room is easily maintained by ordinary steam-heated blower units with thermostatic controllers.

The concentration of liquid sugar and syrups is sufficient to prevent or minimise growth of micro-organisms within the liquid; however, there may be some growth of yeasts and moulds on the surface due to contamination of air in the head space of the tank. Such contamination can be prevented by an ultraviolet lamp in the air vent line combined with an absolute filter on the air entry to this line. Sometimes an ultraviolet lamp is installed within the headspace of the tank. While this is effective in destroying yeasts and moulds, it must be removed for cleaning the tank and is a hazard in case it breaks.

Granulated sugar when used in large quantities may be received in bulk and conveyed into bins or silos. More often, granulated sugar and corn sugar solids are received in bags and stored in a dry area near processing equipment.

The emulsifiers, stabilisers, cocoa and other ingredients such as fruits and nuts are usually received in drums and stored in appropriate areas near their point of use.

Making the Mix

Putting the ingredients together in the correct proportions to make a mix of the desired composition, and preparing it for further processing, is known as blending or batching the mix. The procedures for making the mix are influenced by the size of operation, and whether pasteurisation is to be by batch (vat or long-hold) method or by continuous high-temperature, short-time (HTST) pasteurisation.

In small volume operations, where less than 2000 litres of mix per hour for about 5 h per day is made, pasteurisation is often by the batch method. The mix ingredients are usually blended in the pasteurising vat, with the liquid ingredients added by pumping and dry ingredients dumped by hand directly into the liquid portion in the pasteuriser tanks. The amount of each ingredient required for the recipe is determined prior to the beginning of the blending operation. The amounts of liquid ingredients may be added by measuring the volume by calibrated dip-stick, pumping through a flow meter or by counting cans of a given volume and dumping the contents into the blending vessel. Dry ingredients are usually measured by the bag with part amounts weighed in a separate container. The ingredients which are used in small amounts, such as emulsifiers and stabilisers, are measured or weighed separately, and usually added after the other ingredients are in the blending tank.

Sugars, dried milk solids and stabilisers are much easier to blend when the liquid portion of the mix is warm. Some of the stabilisers disperse and hydrate best with the mix blend at 60°C or higher.

With batch pasteurisation, mixes are usually blended warm. Heating can start with the addition of the first liquid ingredient, and by the time the batch is completed, the temperature is at or near the pasteurising temperature. Little or no time is lost.

As homogenising and cooling following batch pasteurisation are continuous processes, two or three blending/pasteurisation tanks are often employed to provide an uninterrupted flow of mix.

Mix Blending Systems

Those ice cream makers who produce more than 25 000 litres of mix per day usually employ mechanical or automated mix blending systems and HTST continuous pasteurisers. The larger the operation, the more sophisticated the mix making system, and when more than 8000 litres per hour of mix is made, multiple lines are often employed. For example, in a factory which makes 20 000 000 litres or more of mix in a year, 120 000 litres per day might be made during the four months of peak production and 60 000 litres per day during the other eight months. With a 12-hour mix making day during the peak season, the average rate is 10 000 litres per hour. Since two 5000 litre batches can be made each hour, a line of a 5000 litre batching system and a 10 000 litres per hour pasteuriser can handle the production requirements. Most ice cream makers find it necessary to make some of the mixes in smaller volumes, and therefore, in order to maintain a high average production rate, prefer to have two mix making lines. In the example here, two 4000 litre batches per hour feeding an 8000 litres per hour pasteuriser, plus another line capable of about half the capacity of the main system, would provide greater flexibility in production and keep the production day within the desired time restraints.

Blending of ingredients to a recipe for the proper mix composition can be done in many ways. If all the ingredients are liquids, volumetric or mass flow meters can be used. Metering systems with a meter on each ingredient line are the most rapid of batching systems, for all of the ingredients can be metered into the blending tank simultaneously. Weighing systems require adding ingredients individually, one after the other.

Meters can be set manually or electronically to activate valves or pumps or both for predetermined amounts, so that flow is stopped when the desired quantity of ingredient has passed through the meter. Metering systems can be fully automated using computers or microprocessors to compute the recipe, to set the quantity of each ingredient, to start and stop flow when the quantity has been satisfied, and to total and record data desired for inventory and production control.

Properly designed metering systems can be very accurate provided they are kept in good mechanical condition, and are recalibrated after a change in piping or a change in the ingredient to be metered. In smaller systems, one meter may be used for two or more ingredients. In this case, the density of the ingredients must be nearly the same if satisfactory accuracy is to result.

When it is necessary to add dry ingredients or if dry ingredients are used regularly, they can be weighed into hoppers above the blending tank and dropped into the liquid ingredients, or they may be fed by air or screw conveyors to a continuous weighing device and dropped into the blending tank. The blending tank should have a turbine agitator which will produce a deep vortex into which the dry ingredients can fall. This will assure rapid wetting and dispersion of the powders. The computerised control system can be designed to integrate the data from the meters with data from the dried solids weighing equipment.

Some mix makers prefer to make up a concentrate or slurry of the dry ingredients which can be stored and then metered to the blending system. When this is done, care must be exercised to assure that the composition of the slurry or concentrate is the same from batch to batch, and that the slurry or concentrate is cooled to 7 °C or lower if it is to be held longer than 4 h before being used.

There are some worthwhile advantages to pre-liquefying dry ingredients. Air or foam tends to come out of the mixture during holding, and holding allows time for the dried skim-milk solids, whey solids, etc., to hydrate for a better mix later on.

Dry ingredient incorporators, high shear mixers, funnels and similar equipment may be used in mixing the dry ingredients with water.

Weighing systems are generally preferred over metering systems since weighing is a direct or basic process without the need for converting volume to weight. Weights are not influenced by differences in density due to incorporated air or composition of the ingredient, and properly designed weighing systems are very accurate. The batching tank may be suspended from beam scales or from a straingauge-type load cell, or it may be mounted on multiple load cells. The suspended types are often located above blending tanks so that after the first liquid is in that tank, subsequent additions can be subjected to mixing for a considerable part of the batching time. Figure 1 shows a load cell batching system.

Load cell weighing systems are well suited to computer operation and control, and fully automated systems are possible. The fully automated systems may be equipped with a very small separate unit for the micro-ingredients, such as the stabilisers and emulsifiers. These in dry or liquid form are weighed by the micro-system for greater accuracy with very small increments, and dropped into the main weigh tank. The stabilisers and emulsifiers are more often weighed manually and added by hand to the main load cell tank where they are included in the total weight of the batch. This small manual operation requires little time and is nearly always preferred over the expensive and delicate automated micro-load systems. After batching the mix with either the metering or weighing system, it may be pumped directly to the surge or balance tank of the HTST pasteuriser or to a larger surge tank, depending upon the overall design of the plant.

Mix batching systems and their controls can be simple or very complex and expensive. With computer or micro-processor control, almost every operating desire can be engineered into the system if the budget permits. Much of the time, simplicity in both the ingredient handling and in the batching system is prudent, less costly and results in greater flexibility for future operations. The control room of a computer controlled batching and pasteurising system is shown in Fig. 2.

In recent years, micro-processors have replaced mini-computers as the choice for automated mix-making systems. Micro-processors are small, can be expanded when operations are changed, and can be located in the processing room near or on the equipment to be controlled. Individual micro-processors located adjacent to specific operations may be linked to a central control room, if desired, for distributive control. Distributive control systems minimise the amount of field wiring for the plant and are less subject to hazards such as fire, physical damage to a central system, etc., for the separate individual units can function without the others.

Aids for incorporating dry ingredients

When dry ingredients such as powdered milk, whey solids, and

Fig. 1. Automated mix batching system. The weigh tank with its load cell is suspended above two blending tanks. After each ingredient is weighed, the control system indicates or records the weight and opens a valve to allow the ingredient to flow into one or the other blending tank. After a batch is completed, the next is weighed and dropped into the other blending tank. (Courtesy of APV International, Ltd.)

Fig. 2. Computer control room with a view of the mix batching and pasteurising operations. (Courtesy of APV Crepaco, Inc.)

granulated and crystalline sugars are used, mix blending may be slowed, and incorporation and dispersion incomplete in the time available. There are items of equipment which can assist the mix maker in improving his operations.

The simplest mechanical aid to speed up the addition and dispersion of dry ingredients into liquid is a funnel and centrifugal pump combination. A simple funnel or hopper is attached to a tee in a pipeline just upstream from a centrifugal pump. Liquid from the pasteurising tank or blending vessel flows by the tee to the pump and is returned to the tank. Dry ingredients are drawn from the funnel through the tee into the liquid by the partial vacuum developed by the centrifugal pump, and are wetted and dispersed very quickly by the turbulent flow in the line and the pump. A valve may be installed between the funnel and the tee so that after the dry ingredients have been added, recirculation can be continued for additional mixing without drawing air into the product. A more sophisticated funnel and pump incorporator or mixer is available as a unit, complete with the funnel, valve, and special pump which has a mixing chamber. These aids are especially useful for small volume, manual operations. For greater volumes and more rapid mixing, two other aids are available, and are especially helpful with high volume mix blending systems when dry ingredients are used. One is a high shear mixer, and the other is a dry ingredient incorporator. In use, a portion of the liquid ingredients are put into one or the other of these mixers, the agitator turned on, and the pre-weighed dry ingredient added. Mixing is very rapid, sugars are dissolved quickly and dried milk solids dispersed almost instantaneously. The dry ingredient incorporator has a high speed turbine type agitator which forms a deep vortex in the liquid. Dried ingredients flowing into the vortex are wetted and dispersed quickly. Air is not intimately mixed, but forms large bubbles that break out at the top surface as the mix circulates outward along the bottom, away from the turbine and up the side walls of the tank. A turbine agitator with a closed bottom is especially effective in dispersing powders of lighter density than the liquid portion of the mix. A turbine with a partly open bottom lifts heavy ingredients off the bottom of the tank and disperses them effectively, and easily handles the low density ingredients as well. The open bottom turbine agitator is especially useful in rapidly liquefying granulated sugar and heavy syrups which tend to accumulate on the bottom of the tank.

Other considerations in mix blending

Some stabilisers and emulsifiers are not readily incorporated or dispersed with cold blending. The choice of stabiliser should be based upon the desired effect on the product, and then upon how it can be blended. As a compromise may satisfy both requirements, ice cream makers often experiment from time-to-time to optimise their recipes and to evaluate new commercial stabilisers. Generally, sodium alginate must be added when mix temperature is $70\,^{\circ}C$ or higher. Gelatin, propylene glycol alginate, carboxymethylcellulose and mixtures of carrageenan with locust bean gum or guar gum can be dispersed into a cold mix, and are suitable for pasteurising through an HTST system.

Emulsifiers are usually not a problem in blending either hot or cold. Sometimes emulsifiers are combined with stabilisers, but many ice cream makers prefer separate emulsifiers because they permit variations in the amount used for different products.

The choice between blending hot or cold is an important consideration when continuous pasteurisation is to follow blending. As seen earlier, when vat or batch pasteurisation is employed, heating during blending is an advantage and no choice is necessary.

When continuous high-temperature–short-time (HTST) or ultra-high-temperature (UHT) pasteurisation is used, much greater economies in energy utilisation are realised when cold rather than warm mix enters the pasteuriser, for it is possible for regenerative heating and cooling to act over a greater product temperature range. For example, in an HTST pasteuriser designed for 80 per cent regeneration, pasteurising at 80 °C, and final cooling to 4 °C, mix entering at 16 °C after blending is heated to 67·2 °C by regeneration, to 80 °C by steam, cooled to 28·8 °C by regeneration and to 4 °C by refrigeration. If the mix is blended warm and enters the HTST at 45 °C, it would be heated to 73 °C by regeneration, to 80 °C by steam, cooled to 52 °C by regeneration and to 4 °C by refrigeration. Summarising the net energy requirements, it is seen that with cold blending, the steam heating range is 12·8 ° compared to 7 ° with warm mix, and the refrigeration range is 24·8 ° with cold mix compared to 48 ° in the case of warm mix. At first it appears that the warm mix offers savings in the steam requirements, but heating of the raw ingredients for warm blending must be added to that of the HTST unit. This example assumes that raw dairy ingredients at 6 °C and liquid sugar at 43 °C are used in about a 74 per cent to 26 per cent proportion to blend at approximately 16 °C. For warm blending at 45 °C, the 29 ° difference requires additional steam and this added to the 7 ° in the HTST total 36 ° net steam heating range for the warm blended mix. As steam and refrigeration requirements are directly proportional to the temperature ranges involved, the energy savings for cold blending compared to warm blending is 35·6 per cent in steam and 51·7 per cent in refrigeration.

There are other factors to consider in deciding the temperature for blending. Sugars, especially heavy syrups, dissolve more slowly in cool liquids, and dried milk solids are difficult to wet, disperse and hydrate at temperatures below 35 °C. Air entering the mix with dried milk solids and with the cream and concentrated milk, if used, tends to remain in the mix until heated to 35–40 °C. When this occurs in a closed, continuous pasteuriser, the resulting mix and air may cause fouling of the heat exchange plates or surfaces, reducing the heat exchange rates and shortening the pasteurisation run. Hydration or holding tanks used between blending and the balance tank of the pasteuriser may permit enough of the incorporated air to escape to prevent problems downstream, but blending equipment which minimises the incorporation of air should be employed.

PASTEURISING, HOMOGENISING AND AGING

Pasteurisation

Both batch or long-hold pasteurisers and continuous HTST pasteurisers were discussed briefly, and are generally understood by processors of dairy foods, but some additional comments on continuous pasteurising equipment are in order.

While the great preponderance of continuous pasteurisers are of the plate type, other types are sometimes used. These are of the tube-within-a-tube, triple tube, and small diameter, high velocity tubular types. The reasons for selecting these in preference to plate-type units are various and usually not related to operating costs, for plate pasteurisers with a high degree of regenerative heating and cooling offer low operating costs seldom equalled by other types. Tubular units are sometimes chosen because of limits on floor space, for they can be mounted along a wall or suspended from the ceiling.

Small diameter, high velocity tubular types are usually limited to 6000 litres h^{-1} or less capacity with most makes having a maximum capacity of half that. These are generally available as a complete processing unit with controls, heater, pumps, surge tanks, holder, and homogeniser all mounted on a base. This type of pasteurising-homogenising system is especially attractive for ice cream makers in remote areas where installation skills are lacking, or when a smaller, complete line is required which can be installed with minimum disturbance to existing operations. These systems are also used for ultra-high-temperature (UHT) pasteurisation or sterilisation of mixes. The latter is occasionally used to give long shelf-life to mixes for soft-serve products, or for small volume freezing plants not making mix on site.

UHT pasteurisation or sterilisation of mixes is also done with direct contact steam heaters combined with a flash chamber to remove the added steam, and with plate heat exchangers for pre-heating, cooling and regenerative heat exchange. The direct steam heaters are of two types — injector and infusor. With injector heaters, steam is injected through a venturi into the flowing liquid mix; with infusion, the liquid mix flows into an atmosphere of steam. Both heat the product to almost the steam temperature nearly instantaneously, and by controlling pressure, very high temperatures — 150°C or higher — can be obtained. A short holding tube providing 0·2–6 s or more residence time, as desired, assures complete heating of all the mix.

The flash removal of the vapour added is usually done in an evaporation/vapour separation chamber with controls to assure that vapour removed is equal to that added, so that no change in the composition of the mix occurs. Of course, the controls can be adjusted to concentrate or dilute the product, if that is desired. Vapour removal also removes some volatile odours associated with weeds and feeds ingested by the cattle, and much of the air and oxygen is also removed.

Cooked or heated flavour associated with UHT treatment is minimised by the direct heating pasteurising and sterilising systems; Fig. 3 shows a typical system.

Homogenisation

Homogenisation of ice cream mix reduces fat globule size to prevent churning in the ice cream freezer, improve the smoothness of the ice cream, and permit more of the milk proteins to adsorb onto the fat globules, which increases the viscosity of the mix and produces a smoother body and texture in the frozen ice cream.

Homogenisers are specialised reciprocating pumps having from three to seven pistons and an homogenising valve or valves. Mix is pumped through the homogenising valve, usually with quite a high pressure which is then released suddenly, dissipating the pumping energy into the product as heat.

How the fat globule reduction is accomplished has been studied since homogenisation was introduced, and even today there is no general agreement. Probably hydraulic shear forces, developed as the mix flows through the homogenising valve, distorts the liquid fat globule beyond the limits where the surface forces can hold the fat in globule form. When these forces are exceeded, the fat is attenuated, and during turbulent flow, separated into several new, smaller globules. When the average diameter of the fat globules is halved, the number of globules is increased by eight times and the total surface is doubled, allowing more proteins to adhere to the surface.

Homogenisation occurs only when the fat is liquid. This means that

Fig. 3. Ice cream mix UHT pasteurising/sterilising system of the direct steam infusion heating type. The upright vessel in the left foreground is a jacketed steam infusion heater. The vessel in the centre is the steam flash (removal) chamber. The plate regenerator and homogeniser are beyond these two vessels. (Courtesy of APV Crepaco, Inc.)

the mix for frozen desserts must be at about 50°C or higher for homogenisation, and the higher the temperature, the easier it is to produce good homogenisation with smaller and more uniformly sized globules.

Fat globules in ice cream mix can be made so small that the amount of natural phospholipids available to form the globule membrane is insufficient to cover the increased surface. In this case, the globules may coalesce to form larger globules or clumps which have the effect of larger globules.

Over-homogenisation may also affect the fat globule surface phenomenon to allow clustering or the incomplete dispersion of individual globules; clusters act physically much the same as larger globules.

In practical terms, large fat globules, clumps and clusters indicate instability in the fat-in-water emulsion of the mix, which may lead to partial reversal of phase during freezing due to the severe agitation of the dasher or mutator. The partial reversal of phase shows up as churning or smearing of the fat with buttery deposits on the scraper blades, the inside surfaces of bends, the edges of extrusion nozzles, and as visible particles in the product. When this occurs, the body and texture of the finished product is inferior.

The purpose of homogenising mixes containing fat is to produce a good frozen product without evidence of churning or smearing. This does not necessarily mean that the smallest fat globule size is necessary. A practical and economical approach to accomplishing satisfactory homogenisation is to place the homogeniser in the HTST pasteurising system at a point where the temperature is highest. This allows excellent homogenisation at lowest homogenising pressures. Further, the pressure should be adjusted to the lowest value which prevents churning in the ice cream freezer. It is important to keep homogenising pressures low to conserve energy and reduce production costs.

Another factor affecting homogenisation is the fat content of the mix. In general, the greater the fat content, the lower the homogenising pressure required. It is easy to over-homogenise or use too high pressure for high fat mixes.

Aging of Mixes

Following pasteurisation and homogenisation, mixes are cooled to 4 °C or lower and held in a holding or storage tank until they are required at the freezer. The tanks may be insulated and, in addition, may be refrigerated to maintain the low mix temperature. This prevents the growth of micro-organisms, and promotes crystallisation of fat and other changes which improve freezing, air incorporation, smoothness in body and texture, and resistance to melting. Holding time combined with a low mix temperature is called aging, and may be deliberate or incidental to the production procedures.

The changes which occur during aging are more complete hydration of stabilisers, increase in adsorption of protein to fat globules and continuation of fat crystallisation. The effects of aging are most pronounced with gelatin stabilisers, and improvements in ice cream and freezing performance are more pronounced as the aging time increases from about 4 to 12 or more hours. Because gelatin is expensive and must be used in greater amounts than other stabilisers, it is no longer used extensively.

Mixes containing most of the other types of stabilisers require much less aging time, and improvements in the mix may be more the result of increased protein adsorption to fat globules and fat crystallisation than a change in the stabiliser. Those stabilisers which hydrate completely during cold blending, or as the mix is heated during HTST pasteurisation, are affected very little by aging. CMC (sodium carboxymethylcellulose) and guar gum are in this category, but a small amount of carrageenan (Irish Moss) must be blended with them to prevent serum separation (wheying-off) in the mix during holding, especially if the cold mix is held longer than 4 h.

Many progressive ice cream makers have observed that aging can be shortened by cooling the mix to temperatures of 0–2 °C. This probably increases the crystallisation of the fat and improves adsorption of milk proteins to the fat surface. A more certain result is an increase in the ice cream freezer capacity, for the freezing point in the freezer is reached much sooner when the additional cooling load is done outside the freezer. To cool the mix to such low temperatures efficiently and conveniently, a direct expansion chilled scraped-surface heat exchanger is most often used.

FREEZING ICE CREAM

Ice cream is a very complex food. The mix is nearly always more than 60 per cent water. The water dissolves the sugars, both natural lactose and the added sugars, and also dissolves a portion of the salts from the milk solids. Then there is a colloidal system suspended in the water. The milk proteins and proteins from stabilisers and the insoluble salts make up the colloids here. Another system which depends upon the water is the fat-in-water emulsion from milk fat sources or from vegetable oils.

These systems coexist. When this mix is passed through an ice cream freezer, it is chilled to the congealing point where air is incorporated, and then chilled to freeze out some of the water. All of this is accomplished under the rather severe agitation of the scraper blades, dashers and beaters, if any.

As ice is removed from the mix system, the sugar in water solution is concentrated, as are all the other systems, and the freezing point is depressed or lowered further by the increased sugar concentration in the remaining water. As ice is frozen, the ice crystals are suspended in the water, and very small air cells are incorporated into the mixture. When the drawing temperature is −5·6 °C about 50 per cent of the water, in most mixes, is frozen. This means that in a mix having 38 per cent total solids, the semi-frozen ice cream extruded will have 69 per cent of its contents as solids suspended or dissolved in the remaining water, which amounts to only 31 per cent of the whole. If that mix started with 12 per cent fat or 19·4 per cent fat in water, it would be extruded with 38·7 per cent fat in water. If freezing continued to a drawing temperature of −9·4 °C where 67 per cent of the water is frozen out, the fat in water content would be 58·7 per cent. It is not unreasonable to expect such a concentrated emulsion to break or reverse phase to result in partial churning or a greasy texture. This is why some mixes, regardless of how well they are homogenised, cannot be frozen into the 'low temperature' range in an ice cream freezer with its severe agitation.

In addition, as ice crystals are formed, they add to the solids already in the mix, and this increases viscosity and motor load requirements.

In continuous-flow, industrial ice cream freezers, the air for overrun has very little effect in the freezer cylinder because it is compressed. In a freezer operating with 5 atmospheres absolute cylinder pressure, the air required to give 100 per cent overrun occupies only one-sixth the volume of the total mix and air. Thus, the density of the mixture in the freezer is not affected enough by the air to interfere with rapid internal heat flow to the cylinder walls. When the semi-frozen ice cream is extruded, it expands as the pressure is lowered to atmospheric and only when this expansion is complete is overrun fully realised.

Most textbooks deal with air incorporation as it occurs in batch type ice cream freezers at atmospheric pressure. In this case, the mix takes on air when chilled to its 'congealing' point where it changes from a liquid to a more plastic mass. This, of course, starts at the initial freezing point. At atmospheric pressure, air is incorporated readily between the initial freezing point and −5 °C to −5·3 °C.

In continuous freezers, the cylinders are under pressure and very high overruns at very low drawing temperatures can be obtained. Generally, air for overruns up to 130 per cent at drawing temperatures to $-7\,^\circ$C is easily incorporated with cylinder pressures of 3·5–5·5 atmospheres, depending upon the dasher and blade design and the condition of the blades. For overruns in excess of 130 per cent, additional cylinder pressures may be required. When the drawing temperature is lower than $-7\,^\circ$C, cylinder pressures may have to be increased by two or three atmospheres.

The temperature of the mix as it enters the continuous ice cream freezer is very important in ice cream freezer performance. If the temperature of the mix is uniform throughout the run, the overrun control and freezing rate are predictable, provided that the refrigerant supply and suction conditions are also uniform. If such mix is also supplied to the freezer's mix pump at a constant pressure, operating controls can be adjusted at the beginning of a run to give the overrun and temperature desired with no significant changes required throughout the day.

A mix temperature of $0\,^\circ$ to $-1\,^\circ$C will optimise freezer performance. It will also assure that the fat or vegetable oils are crystalline, and will practically eliminate the necessity for aging where formulation has not already done so.

The consistency of ice cream as it is drawn from the freezer is often referred to as 'wet' or 'dry' or 'stiff'. This consistency is influenced more by formulation than by any other factor, and while related to the drawing temperature, it is not directly so related. Some quite cold ice creams can still have some wetness and be flowable, and some stiff, dry ice creams can be quite warm when drawn. The design of the freezer dasher and the volume of product in relation to throughput does have some effect on this quality of product, but mix that produces a characteristic wet ice cream can be reformulated to give a drier product at the same drawing temperature. On the other hand, if the freezer itself causes the dryness and stiffness, the drawing temperature must be raised or an extensive reformulation made in order to obtain a wetter, more flowable product.

Freezers with displacement dashers combined with relatively high rotative speeds tend to cause a shearing action that may produce a physical change in the structure of the protein-fat-sugar complex. This is more pronounced as the drawing temperature is lowered, but it is definite at $-5\,^\circ$C. The ice cream produced this way is stiffer and drier at a given drawing temperature than that made from the same mix but with a more open, slower dasher. This ice cream is slower melting after hardening, and is characterised as a 'warmer eating' product. This eating quality is a result of slower melting in the mouth, and some would say that this product is less refreshing than the wetter-drawn, faster-melting ice cream. Flavour is not released so readily with the warm eating product, and melt down may be flaky in appearance. The 'warmer eating' ice cream, when its characteristics are not carried to extremes, is preferred by about 25 per cent of the consumers and producers in some markets.

From an engineering viewpoint, 'stiff, dry' ice cream is less than ideal, for it has poorer flowability, causes more pressure drop in pipelines and has poorer container filling characteristics than a 'wetter appearing' ice cream at the same temperature; stiff, dry ice cream also requires more power input from the dasher motor and the pump drives. The term 'wetter appearing' ice cream as used here means one which has a sheen or gloss indicating some unbound moisture, but a product that is stiff enough to stay in the container during packaging operations and which, if drawn colder, is stiff enough for extrusions for some of the stickless products.

Freezing point curves for ice cream and ice milk of various compositions are very similar (Fig. 4). They differ according to the depressing effects of the dissolved sugars and salts, but a curve for one mix composition falls slightly above or slightly below that of another. A freezing point curve is a practical reference for any ice cream maker, because it shows important general facts. For example, from earlier discussions, it was seen that because of the nature of ice cream, the most practical extrusion temperature is between −5·5° and −6°C. The amount of ice formed is from 52 to 55 per cent of the water in the mix. If ice cream extruded at −6°C were held long enough between extrusion and hardening to permit its temperature to rise to −5°C throughout the mass, the amount of water frozen is reduced to 48·5 from 54·5 per cent. This means that 6 per cent of the water has been affected or 11 per cent of the ice formed has melted; when refrozen, larger ice crystals result. To prevent coarse, icy ice cream, the product temperature should not be permitted to rise, but should be lowered continuously until well hardened. When hardened to −15°C, about 80 per cent of the water is frozen. If the temperature goes up to −14°C, about 78·5 per cent of the water is still frozen, but as only about 1·9 per cent of the ice has melted, refreezing will not produce as much coarseness as in the case of the same 1°C change from extrusion temperature. The freezing point curve

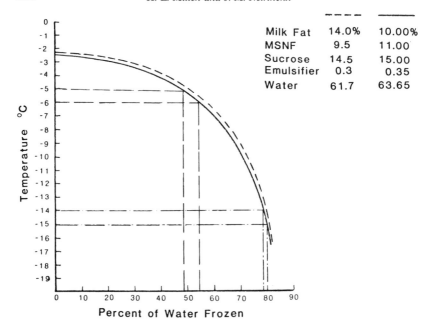

Fig. 4. Freezing point curves for two ice cream mixes. The horizontal and vertical lines indicate percentage of water frozen from the mix at a given temperature for the mix having 10 per cent milk fat. It is apparent that a temperature change of 1 °C at higher ranges affects more of the ice crystals than a change of 1 °C at lower ranges.

shows that temperatures should be lowered continuously, that fluctuations cause iciness, and that at lower temperatures, there is a diminished effect. It can be deduced that repeated cycles of temperature fluctuations will accelerate this problem. This information is important in the proper management of packaging, hardening and cold storage of the product.

ICE CREAM FREEZERS

There are two types of ice cream freezers: the batch type and the continuous flow type. Batch freezers are used in very small ice cream factories, or for small volume speciality lines and for pilot plant research in larger factories. They are also used rather extensively in ice cream stores which freeze and harden ice cream in-house.

Batch Freezers

Batch freezers made today are generally of the horizontal cylinder type with the freezing cylinder refrigerated with one of the halocarbon refrigerants, most often R22 or R502. There is a mix supply tank located above the freezing cylinder, so that mix will flow by gravity into the cylinder when a valve is opened. The usual procedure is to charge the freezer, turn on the dasher and start the refrigeration. When freezing has increased the power required to turn the dasher to a value which indicates sufficient freezing, the refrigeration is turned-off, and air is whipped in until the desired overrun is obtained; this is determined by drawing samples for overrun determination. At this point, any fruits, purees or flavours may be added directly into the open port of the front door and mixed into the ice cream. A slide or pivot valve on the bottom portion of the front door allows the ice cream to be drawn into containers or bulk cans. The dasher is designed to propel the product toward the discharge port. Subsequent batches are made in the same manner. A new charge is put into the mix supply tank as soon as the previous charge is dropped into the freezing cylinder. A batch of ice cream can be made every 6–7 min by experienced operators.

The sizes of batch freezers vary according to the manufacturers' designs, but 18–20 litres and 36–40 litres of 100 per cent overrun ice cream are the two sizes found most often. If greater than 100 per cent is desired, the mix charge must be reduced enough to prevent overflow of ice cream during whipping.

As freezing in a batch freezer is at atmospheric pressure, the whipping ability of the mix is important and must be considered in formulation. The dashers of the freezers are designed with beaters to promote whipping, but when freezing to temperatures below −5 °C, air is less readily incorporated, and some may even be expelled.

Generally, ice cream made with batch freezers has both larger ice crystals and bigger air cells than ice cream made with the same mix on continuous freezers. Overrun control to close tolerances is difficult with batch freezers, and it may vary by 10–15 per cent from the beginning of drawing the batch to completely emptying the barrel. This occurs because whipping continues all during the drawing time. Where most of the ice cream is hand packed and where it is sold by weight, close overrun control is regarded as unnecessary.

Modern batch freezers are not much different from the earlier models except in better hygienic design, more durable materials, and better refrigeration systems.

Continuous Ice Cream Freezers

At present, there are approximately ten manufacturers of continuous ice cream freezers in the world. These are in Europe, Japan and the United States. Three of these supply between 60 and 70 per cent of world requirements in numbers of units, and about 80 per cent in terms of ice cream capacity.

All, but a few, very small, continuous freezers, have horizontal freezing cylinders, and all freeze with the cylinder pressurised. What happens during freezing was discussed earlier, and applies regardless of make of freezer.

Some makes have one cylinder or freezer to a unit, others have one, two or three freezers in a single housing, and one make uses pairs of cylinders in series to provide sufficient heat exchange surface for its larger models.

The capacity of continuous ice cream freezers is difficult to rate, since products around the world have greatly different characteristics which affect refrigerating requirements. Then, the different refrigerants are not thermodynamically equivalent. While there is no generally adopted standard for capacity rating, most manufacturers list very similar conditions for their nominal maximum rated capacity. Some, but not all, have a margin of safety included in these ratings. The conditions for rating include the following:

1. machine is new or in excellent operating condition;
2. refrigerant is clean, free of oil (if R717) and free of non-condensible gases. Ammonia (R717) is the refrigerant used for rating;
3. mix is an ice cream mix of approximately 10 per cent fat, 15–16 per cent sugar and 37–38 per cent total solids;
4. mix enters freezing cylinder at 4·4–4·5 °C and ice cream is drawn at −5·6 °C. One major manufacturer rates at a drawing temperature of −5 °C;
5. refrigerant evaporating temperature is at −30·6 °C (saturated conditions);
6. rated capacity is either in terms of litres of 100 per cent overrun ice cream, or in litres of mix input.

The manufacturers' rating is a nominal rating which can reasonably be expected under the conditions stated for rating, but it does not assure that the ice cream maker will be able to operate at that rating. The types of sugars and some ingredients of specific formulations influence the viscosity of the freezing ice cream and its internal heat flow characteristics to such an extent that the maximum nominal capacity may not be attainable.

In 1984, the sizes of industrial continuous ice cream freezers available ranged from 300 to 3000 litres per hour per cylinder in terms of 100 per cent overrun ice cream. Small pilot-plant continuous freezers down to 100 litres per hour were also offered.

Pumps and Overrun Systems

Pumps on ice cream freezers are usually of the rotary type with the capability to pump against pressures of $7–14$ kg cm^{-2} (approximately $690–1380$ kPa) with reasonable volumetric efficiency.

There are two general pumping arrangements, both designed as a part of the overrun system. The first employs a pump (or a pair of pumps or a compound pump), to pump or meter the mix into the freezing cylinder, plus a hold-back valve at the ice cream discharge port. The hold-back valve may be spring loaded with manual adjustment, or it may have an air operator with adjustable air pressure supplying the operating power. The hold-back valve permits imposing a pressure on the cylinder during freezing which compresses the air admitted with the mix for overrun. Cylinder pressure of $3·5–4$ atmospheres keeps the volume of air in the freezing cylinder sufficiently small so that it does not significantly slow the internal heat transfer out from and through the mix, and that pressure is sufficient for good air dispersion and small air cell size. Higher pressures may be imposed on the cylinder, but in most cases, the improvement of heat transfer and air cell size is not great enough to offset the disadvantages of increased pumping costs.

The earlier continuous freezers using the pump and hold-back valve arrangement had two pumps in close proximity, both powered from a common drive, but with the second operating at about three times the volume of the first. As mix was pumped, a partial vacuum was produced between the pumps. Air for overrun was allowed to flow into the partial vacuum so that the difference in pumping volume between the pumps was made up with air. An adjustable snifter valve on the air intake allowed control of the amount of air to give the overrun desired.

Current freezers using this system make use of a compressed air source with an air regulator, sometimes combined with an air flow meter, to control air from a slight vacuum to moderate pressure at entrance to the mix-air pump. This allows better control as the pumps wear and permits greater versatility in overrun control. One of the current freezers having this pumping system has a combination pump using two metal, gear-type rotors and separate air and mix inlets, air entering the rotor cavities on one side of the pump, mix on the other. This combines the mix and air at the discharge of the pump in the line to the freezer cylinder.

The other general pumping arrangement has a mix pump or mix-air pumps metering the mix into the freezer cylinder, and another rotary pump at the discharge of the cylinder. The mix-air pumps operate in the same way as in the pump and hold-back valve arrangement, but when a single mix pump is employed, compressed air is admitted directly into the freezer cylinder or to the mix line between the mix pump and cylinder. The ice cream discharge pump operates in a ratio to the volume of the mix pump that will develop cylinder pressure. For example, if the ice cream pump is operated at 1·23 times the volume of the mix pump, it will cause a cylinder pressure of 4·4 atmospheres with 100 per cent overrun ice cream, and 5·2 atmospheres with 120 per cent overrun. Low cylinder pressures result with low overrun products, and if the pumping ratio is not changed or a restriction is not placed at the inlet to the discharge pump, the cylinder may be emptied when overruns are less than 25 per cent. One current model using this two-pump system has a hydraulic pump drive which, along with a cylinder pressure sensor and speed controller, permits a continuously variable ratio of pumping volumes between the mix and ice cream pump to maintain any preset cylinder pressure from 1 atmosphere for products without overrun, to in excess of 12 atmospheres for very high overrun products drawn at cold temperatures.

With the two-pump arrangements, the mix or mix-air pump works against only the cylinder pressure, while the discharge or ice cream pump works against the difference between the downstream line pressures and the cylinder pressure. With the mix or mix-air pump and hold-back valve arrangement, the mix-air pump has to work against the imposed cylinder pressure plus the downstream line pressures. This may total 10-13 atmospheres, near or exceeding the design limits of some of the rotary pumps found on freezers. In the case of such high pressures, pump and pump rotor life is short and pump slip is relatively great.

Both systems are capable of the uniformity of pumping necessary for excellent overrun control, when the pressure on the pump inlet and downstream line pressure is uniform throughout the run. This is an ideal which seldom prevails. Levels in the mix supply tanks affect the pressure at the inlet even when the mix is pumped to the freezer, and there are variations in pressure drop through the lines downstream of the freezer. Such changes may be caused by slight fluctuations in product viscosity resulting from variations in extrusion temperature and changes in heat flow into the lines as frost thickness changes. Any changes in downstream pressures are transmitted hydraulically to the cylinder through a hold-back valve. On the other hand, the cylinder is effectively isolated from downstream pressure changes with a product discharge pump. Thus, when pressures outside the freezer are unsteady, the two-pump system yields more uniform overrun.

Ice cream freezer pumps are driven by various means, but all of these provide for varying the pump speed. Usually the set of pumps for each cylinder is powered by one drive. Drives are of three types:

1. Electric motor powering a mechanical variable speed drive.
2. Frequency inverters with electronic speed control for standard electric motors. A gear reducer is nearly always used between motor and pump.
3. Hydraulic pumping systems connected to hydraulic motors on the pumps. The hydraulic pumping units may be located within the freezer housing or remotely outside the production room.

One freezer model series has a hydraulic pumping system, pumps, and controls for variable speed all mounted as a unit in a drawer-like arrangement which can be pulled out from the freezer housing for maintenance and servicing, or easily removed to the shop as an alternative to servicing in the freezing room.

Automated Overrun Control

Automated overrun control systems which measure the density of the extruded ice cream and, by feedback, adjust the air supply to attain and maintain the desired overrun are not yet available. This type of system has been sought since about 1948, but the very nature of ice cream and the way it is produced introduces so many variables, that such feedback systems are not successful.

One of the problems is choosing the point at which to measure the density. Ice cream and related products containing air for overrun are compressible and full overrun is not attained until the product has expanded to atmospheric pressure. This requires some time and, in a continuous flow, is not realised until the product is in its package.

Measuring weight in the package rather than density is probably an easier and more accurate control element, but this is subject to the variations in filler performance and container size, as well as differences in the cut-off of product from the extrusion nozzle. The second major problem is the time lag between density or weight measurement and the change in air input. In a typical continuous freezer with a dasher displacement of 35–50 per cent, the flow through the cylinder at 80 per cent of rated maximum capacity varies from 0·9 to 1·2 min (as air is compressed within the cylinder, mix flow rates are the governing values). This is a considerable lag, and the tendency is for the instrumentation and controls to overcorrect resulting in a hunt and seek cycling of overrun which may be greater than with manual control.

Two major manufacturers of ice cream freezers offer automated overrun systems which use micro-processors to regulate air input in relation to mix input. These provide for presetting the desired overrun. After operation is underway, the overrun is carefully adjusted manually, and the set points noted. These can be used for subsequent runs of the same products with excellent reproduction. With both makes, once overrun has been adjusted, the micro-processor will maintain the flow rates, pressures and other conditions to maintain accurate overrun control.

How accurately do these systems control overrun? With 100 per cent overrun ice cream, and with a properly designed installation of equipment, standard deviations in the order of 1·5–3 per cent in terms of overrun, or 1–1·5 per cent in terms of weight can be expected. These standard deviations are about half those with standard freezers.

A word of caution is in order here. If the mix has an excess of air incorporated in it from the blending operations, from a leaky seal on the suction side of a pump, or from unmelted rerun (refreeze) in the mix, no amount of automation will control the overrun until these undesired air sources are eliminated. Automation is not a replacement for good management practices.

With good management practices and proper operator skills, manual overrun control can be within the standard deviations expected for automated systems. Good practices include proper maintenance of all equipment and pipelines in the system, proper blending of mixes with sufficient hydration and aging time, minimising air incorporation, air removal, complete reprocessing of refreeze, keeping mix temperature low and constant throughout the day, supply of mix under uniform pressure to the freezer mix pump, assuring that there is nearly constant pressure in the suction mains of the refrigeration system, and keeping frozen product lines between freezer and packaging point as short as practical. Good production management will also provide for long operating runs of one product any one day to avoid unnecessary changes in products where freezing must be interrupted and restarted.

Dasher Design

Dasher types and their influence on ice cream was discussed earlier in the section on freezing ice cream. Most manufacturers of industrial ice cream freezers offer at least two different types of dashers or mutators for their machines, in order to produce the wide variety of frozen products being made around the world. One set of specifications for a freezer cannot be expected to produce optimum results for all of the products.

Dashers were originally designed for batch freezers as rotating carriers for the scraper blades, and for beating or whipping air into the congealing, partially frozen product. The first continuous freezers had a small diameter cylinder and a solid mutator to which scraper blades were attached. This solid dasher displaced almost 80 per cent of the volume of the cylinder, and having a small annular space for the product and a high rotative speed, the mix was subjected to rather severe shear. These conditions, combined with rapid freezing, produced a very smooth textured ice cream which, with many mixes, tended to be greasy, warm eating and slow melting. In operation, this type of freezer was sensitive to changes in refrigerant temperature and froze-up quickly.

The next continuous freeezer to be introduced had a much larger diameter cylinder and an open-type dasher with a beater similar in design to that of batch freezers. When ice creams from this freezer were compared to those from the other with the solid dasher, when drawn at the same temperature, the ice cream appeared wetter, melted faster and seemed to be more refreshing to the taste. It was also somewhat coarser and more nearly like that made with batch freezers.

During the years following the introduction of these first continuous freezers, designs of the cylinders and dashers of both makes changed, becoming similar in several ways. The first make increased its cylinder diameter and reduced the displacement of its dasher, making the freezer less sensitive to refrigerant changes and producing a very smooth ice cream with less tendency to be greasy. Dasher speeds were also reduced which produced better melt-down in the ice cream. Later on, other dashers were designed for different products. Open dashers with beaters operating with greater cylinder volume are the least sensitive to operator inattention, as they respond slowly to both intentional and unintentional changes in air, refrigeration or flow rates. Dashers with 30–40 per cent displacement, with or without beaters, are nearly universal in respect to the types of products handled, and are excellent for the currently popular low overrun, premium ice creams, and for mixes having a relatively high content of whey solids. Solid dashers with 65–80 per cent displacement are preferred for cold drawn extrusions. They require less power than the more open dashers and enhance quick freezing. They are very sensitive to changes in refrigeration, freezing-up rather easily unless monitored closely by the operator or by automatic controls.

Dasher types and rotative speeds may be matched to characteristics desired for special products. For example, ice crystal size is especially important in certain types of delicate water ices, and the proper choice of dasher and its speed can produce just the crystal structure desired.

Operating Controls and Automation

All continuous ice cream freezers have controls for operation which include on-off switches for power to pump and dasher motors, and for air compressor motors (when these are part of the freezer), for solenoid valves on hot gas defrost lines, air lines, and refrigerant supply lines, speed regulation of pumps, refrigeration supply and back pressure, pressure gauges for the refrigeration system and cylinder or air pressure, and dasher motor ammeter, wattmeter or motor load indicator.

In addition, the newer, more sophisticated machines may have a viscosity meter and controller, and a programmable controller or microprocesor to operate and control most functions of the ice cream freezer.

The viscosity controller takes a power signal from the dasher motor power line and, through a transducer and controller, adjusts the refrigerant back pressure to raise or lower the evaporating temperature which lessens or increases the degree of freezing. As evaporator temperature is lowered, the extrusion temperature of the product is also lowered, and more ice is frozen from it. This increases the viscosity or stiffness of the product within the freezing cylinder, thus requiring more power to turn the dasher. The dasher power for the desired stiffness or viscosity of product can be preset and changed any time during operation. Typical of the state of the art is an ice cream freezer with a micro-processor programmed to control all the functions of operation including overrun, viscosity of product, cylinder pressure, all operating steps such as start up, routine or emergency shutdown, resumption of operation after an automatic shutdown when the cause for shutdown has been corrected, preparation for cleaning, and the valve and pump bypass required for automated cleaning (see Fig. 5). The micro-processor on one such freezer shows the time of day, rate of mix flow, percentage of overrun, rate of production, hours of operation, accumulated production in that time interval, the program step in operation, and various warnings. In case of an impending freeze-up, the warning is displayed and corrective action is taken. If a freeze-up should occur, the micro-processor automatically causes defrosting of the cylinder and operation to be resumed when conditions are satisfactory. The display can be in one or more of several common languages.

Micro-processor programmed operation assures that all functions are performed in the proper sequences, and under the conditions envisioned by the designer of the freezer. This is especially beneficial to the ice cream maker in preventing damage to the freezer in emergency situations, thus avoiding the incidental, unplanned down-time in production.

Adding Ingredients and Flavours

Flavouring materials, other than chocolate, are nearly always added after the mix has been made. These may be added at the aging or holding tanks, or in flavour tanks located just upstream of the ice cream freezer. Fruit juices, flavour extracts, colours and similar materials are added at these points. Pieces of fruit and purees should not be added to the mix prior to freezing in continuous ice cream freezers, for they tend to settle out in the tank with subsequent poor distribution in the frozen ice cream. Further, seeds in fruit or other gritty content harms the close-fitting pumps, the dasher bearings and seals and dulls the scraper blades.

Ingredients such as pieces of fruit, purees, nuts and candies, are inserted into the ice cream after extrusion from the freezer. When the

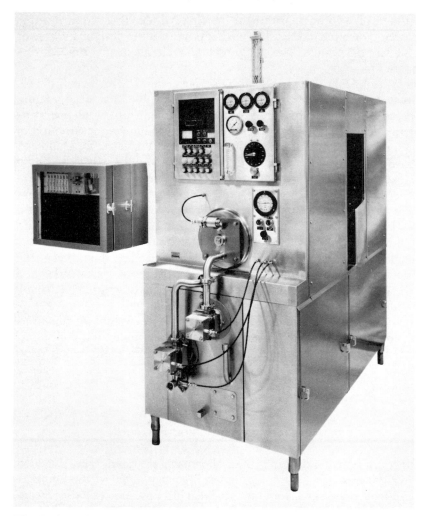

Fig. 5. A modern continuous ice cream freezer with micro-processor control. The specific machine shown here is a single cylinder model. The microprocessor, shown at the left of the freezer, can operate up to three cylinders and can be mounted on the side or rear of the freezer, or it may be located in a control cabinet at a distance from the freezer. (Courtesy of APV Crepaco, Inc.)

ingredients are purees, syrups or conserves, they may be pumped into the product carrying line by a small pump often called a ripple pump. A device within the line at the entry from the ripple pump may be used to

Fig. 6. Ice cream freezer installation. From the left is a three cylinder ice cream freezer, an ingredient feeder, and a vibratory dry ingredient feeder (not in its operating location). The long-radius, sweep bends in the lines which carry ice cream reduce pressure drops and minimise damage to the produce. (Courtesy of APV Crepaco, Inc.)

produce a variegated effect, or to distribute the ingredients more uniformly throughout the ice cream.

Ingredient feeders, often referred to as fruit feeders, have a hopper for the ingredient, an auger or other means for metering or proportioning the fruit, a rotor or plunger for inserting the ingredient, a mixing chamber or blender, and a variable speed drive. A typical ingredient feeder is shown in Fig. 6. An auxillary unit is a vibratory dry ingredient feeder used to feed nuts, marshmallows, or other dry ingredients into the enrobing throat of the ingredient feeder. Dry ingredients may be fed simultaneously with the fruits for a greater variety of flavoured ice creams. Ingredient feeders are capable of adding ingredients continuously at combined ice cream and ingredient rates up to 11 000 litres h⁻¹, and with ingredients constituting 10–12 per cent of the product. Fruits and other ingredients of similar composition, fresh, frozen or canned should be sugared to prevent iciness in the final product. The amount of sugar required varies with the fruit from none for banana and grape to as much as 50 per cent for raspberries. There is an excellent discussion of fruit preparation and recommended fruit to sugar ratios in Arbuckle (1977).

PACKAGING

Packaging should be done as close to the ice cream freezers as is practical, for the longer the pipelines, the greater the possible damage to the ice cream. In the design of the ice cream pipelines, every effort should be made to not only keep the lines short, but to eliminate as many bends and other fittings as possible. Long radius, sweep bends are preferred over standard elbows for all lines carrying semi-frozen products.

Packaging Equipment

Packaging equipment is available for nearly all types of containers. These machines have only a few features in common: an extrusion head, some with a spreader plate, and a means for holding the container, moving it to the point of filling, and moving it away from the filling point.

Packaging-filling machines designed to handle paperboard cartons may store the flat folded carton in a feed cartridge, set it up for filling, fill it and close it. Filling may be on a time-cycle basis or on a volume/weight basis. Closure depends upon the carton design which may be of the interlocking-flap type, or a type which is held closed by an adhesive applied upon closure.

Carton fillers are available for various sizes from 250 ml through to 4 litres. In the US, the pint and half-gallon sizes are most popular; the quart-size, rectangular carton is seldom used. In Canada, the litre and 2-litre sized rectangular cartons are popular, and in the UK and some areas of Europe, the 1-litre size is fairly common. A typical 1-litre, home-pack carton and its set-up and filling machine are shown in Fig. 7. Other containers are the cylindrical paperboard containers used

Fig. 7. Carton set-up and filling machine. The machine shown here is for 1-litre home pack cartons used primarily in the UK. (Courtesy of APV Anderson, Inc.)

extensively in the US in the pint, quart and half-gallon size for the expensive premium ice creams; plastic cups and tubs in various sizes; and form-seal plastic containers formed in a web, filled, sealed and cut into separate packs on the same machine.

There are also bulk containers of 2, 4 and 6 litre size in Europe, and 2, 2½, 3 and 5 gallon size in the US. These are often hand-filled, but filling and lidding machines are available. In addition, there are speciality machines which use extrusions of ice cream in various forms, such as slices and sandwiches, and packaged in cartons containing six, eight and twelve pieces.

In the filling of cartons or other containers, those machines which use weight of contents to trigger the cut-off of filling generally give greater accuracy of fill than the time-fill type.

After the containers are filled, they may be bundled into larger units by means of a shrink-wrap machine before being hardened, or they may go directly to the hardener.

HARDENING ICE CREAM

Ice cream as drawn from the freezer and packaged is only partially frozen. It was indicated earlier that the most practical temperature of ice cream as it is extruded from the ice cream freezer is between −5·5 and −6°C, where the amount of ice formed is 45–55 per cent of the water in the mix. The product is relatively soft as it is extruded, and can be pumped easily with moderate pressure drops. In filling, it flows into the corners of the containers and does not leave air voids as happens when extruded at lower temperatures. If the product is to be rigid enough to store properly, to transport and to maintain its overrun, body and texture, it must have additional freezing. This freezing is done quiescently after the product is in its final container or package. This further freezing is known as hardening.

Hardening should be done as quickly as possible from the drawing temperature to about −18°C core temperature. Generally, the faster this can be done the smaller the ice crystals and the better the quality of the ice cream. However, if the temperature progression is continuously downward, hardening over a period of 10–12 h will cause no significant increase in ice crystal size. Faster rates of hardening are desirable for better space utilisation in cold stores, and for faster turnover of inventory.

Hardening Rooms

Cold cells or hardening rooms within, or separate from, the cold stores are still common, especially for smaller volume operations. If the hardening cell is located within the cold store, it should be partitioned-off from the storage area, so that air can be circulated at greater velocity with better control of flow. The hardening room may be equipped with shelves or racks on which the product is placed in bundles of open-type baskets. For rapid hardening, the product must be arranged to allow air to pass on all sides. The air temperature should be $-25\,°C$ or lower, and air should be circulated with enough velocity to give good turbulence. Generally, there should be enough space in the hardening area for 24 h of peak production.

Hardening in a room of the type described will usually produce a core temperature of $-18\,°C$ in 10–12 h in bulk or bundled packages, and less time with 1 litre or smaller packages.

Hardening Tunnels

Hardening tunnels are used by most of the larger ice cream makers. Some are custom made, but most are standard models available from a number of manufacturers, and nearly all employ cold air at $-30\,°C$ to $-35\,°C$ circulated at a velocity in the order of 180 m min^{-1}. Hardening is accomplished in about 2 h for package sizes up to 2 litres, and around 5 h for 4 and 6 litre containers. Air cooler and blower units are arranged to circulate the air across the packages as they are carried by conveyors or trays through the length of the tunnel.

Spiral Tunnels

Spiral tunnels are one of the types of hardening tunnels used for ice cream. This type consists of a spiral conveyor, usually motivated by a centre drive, which carries the product up and around the tunnel for discharge to an exit conveyor. Large capacity units may have two spirals with one carrying the product to the top of the tower and passing it to the second spiral, which conveys it down for discharge to the palletising operations or to the storage area.

Spiral tunnels may have their own enclosures, or they may be located in a cold room with walls to separate the conveyors from the storage area, and to control the flow of air.

Recently, a new oval path, spiral tunnel has been introduced in the UK. It has a unique edge-driven belt, and its mechanical drive is located outside the tunnel. It offers advantages over other spiral tunnels in space savings, access to product and ease of maintenance.

Straight-through Tunnels

Straight-through tunnels are available with a variety of conveyor and moving tray or belt arrangements. While some of these are made up of standard modules, others are custom designed using some standard components. The three most efficient types available are of the hanging-tray type, with a number of shelves in each tray. The conveyor carries the trays vertically at the ends, and makes a double pass through the air stream. The trays move in short strokes, being indexed a shelf at a time to the loading-unloading position, where an automatic pusher moves fresh packages onto an empty shelf and simultaneously pushes hardened packages off the adjacent shelf onto an exit conveyor.

Contact Plate Freezers

Contact plate freezers (Fig. 8) with automatic loading and unloading for continuous operation are excellent hardening devices, freezing very rapidly, requiring a minimum of floor space, and operating at low maintenance costs. They are more efficient in the use of refrigeration than any other hardeners, for there are no fans with their mechanical heat added to the refrigeration load as in the case of hardening cells and tunnels. With plate hardeners, packages of product are moved into the space between two plates; the plates carry refrigerant through internal passages., When the entire space is filled with packages to be hardened, the plates close until they contact the packages on both top and bottom. Heat flows by conduction from the product through the container walls directly to the cold plates. Hardening is completed in 2 h or less at ratings up to 6800 litres of ice cream per hour.

Contact plate hardeners can be used for products in rectangular containers of one size, or of two sizes which have a common dimension. Where an ice cream maker produces a variety of packaged products but has a large volume in rectangular containers, the plate hardener is especially attractive because of savings in space, refrigeration, power, and ease of maintenance. Further, it is a small enough unit to allow easy relocation should that be necessary.

Contact plate hardeners are becoming quite popular in the US, where a considerable portion of the ice cream is packed in half-gallon rectangular cartons.

Special Tunnels

The special tunnels used for Eskimo Pie, Glacier and Polarmatic systems are familiar to most makers of stickless novelties. Product is extruded onto plates which are conveyed through the air blast tunnel or

Fig. 8. Contact plate hardener. This type of hardener is the most efficient of hardening systems for ice cream in rectangular cartons. It will accommodate round containers which have large area flat tops and bottoms, but efficiency is reduced because package configuration on the plate prevents contact with as much container surface as in the case of rectangular cartons which fully fill the plate area. (Courtesy of APV Crepaco, Inc.)

cabinet for hardening. After hardening, the product is carried to a discharge point where a mechanical hammer blow to the plate loosens the hard extrusion. It is then swept to another carrier for coating, drying and packaging.

Other special conveyors of various designs are used for other types of extruded products.

COLD STORES AND DISTRIBUTION

Cold stores design, management practice, and distribution methods are beyond the scope of this chapter; however several considerations are important here. Temperature changes in ice cream and similar products causes ice crystals to melt as the temperature rises, and to refreeze when the temperature is lowered again. The stabilisers in the mix act to reduce migration of the free water, but there is a tendency for a portion of this water to collect and refreeze into larger crystals, or to refreeze onto existing crystals when the temperature is lowered. When this melting and refreezing of some of the ice crystals is repeated a number of times, the product gets coarse and icy. Fortunately, frozen dairy products, especially those with overrun are poor conductors of heat, so there can be considerable change in the ambient temperature cycle before much of the product is affected. However, the product in contact with the container walls is much more subject to the adverse effects of temperature cycling, and such temperature fluctuations should, therefore, be avoided or kept as small as possible. These effects cause more damage when they occur at higher temperatures, but affect very little of the ice at low temperature.

A general rule is to keep cold store temperatures as low as practical and consistent with the expected storage time ($-18°$ to $-25°C$ is desirable), and keep temperature fluctuations at a minimum. This is a good rule to apply to conditions of distribution, as well, but since frozen product is held in retail cabinets for much shorter periods, one or two weeks instead of up to about four months, higher temperatures in the range of $-13°$ to $-18°C$ can be tolerated.

NOVELTY PRODUCTS

Novelty products include ice lollies, ice cream bars, stickless bars, sandwiches, small single servings in slices or small cups, sundaes, fancy moulded novelties, pudding and gelatin bars, fruit and juice bars, ice cream cakes and pies and others.

Novelties without Overrun

A major portion of the novelties are ice lollies and similar products, which are made from an appropriate mixture that is measured into moulds without incorporating air for overrun and frozen quiescently. Sticks are inserted when the product is frozen enough to hold the sticks in position. The frozen pieces are then removed from the moulds, packaged and transferred to cold stores.

Ice lolly mix is made up of water, sugar, acid (usually citric), flavourings, colours and stabilisers. Sugars may total 17–20 per cent, and stabilisers, 0·3–0·5 per cent. The amount and type of sugar is chosen to produce the degree of sweetness desired, and to yield a relatively high freezing point. A high freezing point reduces the refrigeration required in manufacture, and makes the product less subject to melt and refreeze cycles in distribution.

Stabilisers must be compatible with the acids of the mix and be of the type which produces the body structure and flavour release desired in the frozen product. Gelatin, pectin and xanthan gum are often used. Pectin produces a short structure and a very quick flavour release, while the other two produce similar results but with slightly diminished flavour release. CMC and guar gum are also used, but flavour release is still somewhat slower.

Although citric acid is the most commonly used acid in water ices, tartaric, lactic, malic, ascorbic and phosphoric acids are also satisfactory. The amount of acid to be used in water ices depends on the sugar content, amount and type of fruit juice, and the degree of tartness desired. Generally, acid content increases as sugar content increases; the range is usually from 0·3 to 0·4 per cent by weight.'

Other stick bar products with little or no overrun are pudding bars which may contain some starch and milk solids, fudge bars and fruit and fruit juice bars.

Novelties with Overrun

Novelties with overrun are numerous, but most of them have ice cream or ice milk as the base. Those made on stick bar machines are handled in the same manner as the ice lollies, except that the ice cream or ice milk mix is partially frozen and overrun produced on an ice cream freezer before being transferred to the moulds for quiescent freezing. After extraction from the moulds, the bars are usually coated with chocolate or other candy coating. Sometimes ground nuts, crumbs or small bits of special candies are added after the main coating.

Other novelties with overrun are the stickless bars, sandwiches, cones and cups enumerated earlier. The stickless bars, such as made with the Eskimo Pie, Polarmatic and Glacier systems, are most popular and have many shapes. Some are highly decorated and sell at quite high prices.

A comprehensive discussion of novelties, their manufacture and the special equipment available is not possible in this chapter, but a few general remarks about trends and developments follow.

Trends in Novelty Manufacture

Because of the great variety of novelties produced and included in the sales lists, an ice cream manufacturer can make only a few different items in sufficient volume to be profitable. The trend, especially in the US, is towards factories which specialise in the production of novelties and package them in private label for other ice cream makers and marketing firms. These specialists may also produce a brand which is advertised and distributed on a national basis. Such novelty specialists may make only one type of novelty or they may make several types. Small cups, decorated ice cream cakes or pies and similar products made of ice cream are usually made by the regular ice cream makers.

This trend is likely to continue, for such specialisation makes better use of high cost capital equipment.

New Novelties

During recent years, there has been a proliferation of new novelties probably reflecting the higher consumer expenditure of two-income families. In Europe the fancy decorated pieces made on the Glacier and similar machines have increased rapidly, and this trend is now seen in the US.

In the US, there have been a number of larger ice cream bars and sandwiches come on the market. These are in 90-120 ml sizes, coated with chocolate, or sandwiched between wafers or various types of cookies. As the intended consumers are young business men and women who make these a major part of their luncheon, these products are often referred to as 'Adult Novelties'.

For a number of years, orange juice, with or without added sugar, has been frozen on ice lolly machines and used in school lunch programmes in Southern California and a few other areas of the US. These orange lollies provide a vitamin-C rich product in a form that is especially attractive to children.

Very recently, juice bars and fruit bars have been introduced. These, too, are stick novelties, quiescently frozen. Various juices and mixtures of juice and pieces of fruit are the major ingredients. Strawberries and grapes are especially popular. These are nutritious alternatives to ice lollies for children, and are appealing to adults as snack foods.

Pudding bars and, more recently, gelatin bars have entered the novelty market. Both are stick bars, but gelatin bars are made by partially freezing and adding air for overrun with an ice cream freezer before freezing in the mould machine. Pudding bars and gelatin bars are not as cold to the mouth as ice lollies and have a smaller ice crystal structure which is slower to melt. These are characteristics favourable to young consumers.

The introduction of new types of flavours and combinations of products maintains consumer appeal and seems to increase sales without reducing the sales of ice cream.

Equipment for Novelties

There have been no outstanding innovations in equipment for making novelties for some time. The straight-line and rotary types of stick bar machines are improved from time to time making them more efficient, adding accessories for special products, and increasing their trouble-free operating life.

The extruded product and freezing tunnel type machines have been improved, and new extrusion nozzles have been introduced, but many consider them to be very costly for their production rates.

Other developments are in the form of improvements rather than innovations, and these trends are likely to predominate over the next decade.

RECENT TRENDS AND POSSIBLE FUTURE DEVELOPMENTS

Ice cream technology as it applies to processing methods and equipment for manufacturing frozen dairy products has been rather static for some time, and it appears that developments in the near future will involve improvements rather than innovative technological changes.

Equipment developments during the past ten years have increased capacity, improved efficiency, reduced maintenance, assured better and safer operation, and reduced labour input, but with the exception of micro-processor control, there have been few really new developments. Possible future developments in processing and processing equipment will likely deal more with energy conservation and recovery systems, improved packaging systems, and more attention to refreeze (rerun) recovery systems. Automation and micro-processor control of processing and individual machines is just now being accepted. Developments in this area will continue and bring about significant changes in processing.

During the period from 1973 through 1982, when the effects of the world oil crisis were at their peak, more attention was focussed on conservation of energy. There was a trend to use more efficient electric motors, to process at lower, but safe temperatures, to recover heat from refrigerating condensers for processing and cleaning needs, to conserve water and product, and to reduce transport distances or increase payloads of fewer transport vehicles. It is likely that the quest for greater efficiency in processing and for conservation and recovery of energy will continue into the near future.

There have been some developments in new products which are changing processing and marketing methods. The major development here is the introduction of the premium, all natural, low overrun ice creams. While this type of ice cream is not new, it did not have much success in the market place until about 1978, when a newcomer to the industry began operations in New York with an effective marketing programme, a very high quality ice cream, and the advantage of the attention gained by natural food activists. Now there are many ice cream makers in the US, Japan and Europe producing similar products, and selling them at about three to four times the price of ordinary, good quality ice cream.

The premium ice creams are made from natural dairy products, contain no stabilisers or emulsifiers other than egg yolk solids, and are flavoured with natural vanilla or juices and fruits. These ice creams have a milk fat content of 13–18 per cent, milk solids-not-fat of 9·5–10 per cent, sucrose of 15–16 per cent, and egg yolk solids of approximately 0·8–1·4 per cent. The most common overrun for these new premium ice creams is 25 per cent, although a few have up to 50 or 60 per cent. They are packed in bulk cans for dipping stores and restaurants, and in pint (or half-litre) sizes for retail.

Because of the high price, the very cold taste and the heavy body of the product, conventional ice cream makers at first considered these premium products to be a short-term fad, but now after more than six years of continuous and rapid market gains, it appears that the appeal to affluent consumers is real and enduring.

CONCLUSION

The frozen dairy products industry is a dynamic segment of the broader dairy products industry with its own special technology and marketing methods. The *per capita* consumption statistics for much of the world indicates opportunity for considerable growth in the use of the more nutritious of the frozen dairy products.

REFERENCES

Arbuckle. W. S. (1977). *Ice Cream*, 3rd edn, AVI Publishing Co., Westport, Conn.
Berger, K. G. (1976). *Food Emulsions* (Ed. S. Friberg), Chap. 4, Marcel Dekker, New York.
International Association of Ice Cream Manufacturers (1966). *The History of Ice Cream*, Washington, DC.
International Association of Ice Cream Manufacturers (1980). *The Latest Scoop*, Washington, DC.
Keeney, P. G. (1972). *Commercial Ice Cream and Other Frozen Desserts*, Circular 553, University Park, Pa., The Pennsylvania State University.

Chapter 5

Physical Properties of Dairy Products

M. J. Lewis

Department of Food Science, University of Reading, UK

The physical properties of milk and milk products are of great importance to the Dairy Technologist, as they will affect most of the unit operations used during their processing. These include fluid flow, mixing and churning, emulsification and homogenisation, as well as heat transfer processes such as pasteurisation, sterilisation, evaporation, dehydration, chilling and freezing. Some of the rheological properties are also used for assessing and monitoring the quality of products, such as yoghurt, cream, butter and cheese.

Raw milk is extremely variable in its composition, and most dairy products can be produced in a variety of ways from this milk. Therefore, it is not surprising to find significant variations in the literature values for the physical properties of these products.

There are two approaches to obtaining data for physical properties. The first is to use data available in the literature, the second is to determine the values experimentally. Although the emphasis in this review will be placed on published literature values, some attention is paid to the experimental methods available, and to the variations that might occur due to modifications in the processing conditions. Wherever literature values are used in process calculations, it is important to ensure that the physical properties were determined under conditions similar to which they are to be applied. Before discussing the properties in more detail, the system of units and dimensions will be briefly discussed.

UNITS AND DIMENSIONS

All physical properties can be measured in terms of the fundamental dimensions mass (M), length (L), time (T) and temperature (θ). Three others, added for completeness, are electric current, luminous intensity and the amount of substance (mole). The systems of units commonly encountered are the SI (Système International), the cgs (centimetre, gram, second) and the Imperial System. Scientists, world-wide, now prefer the SI system of units, although in many operations and textbooks, the other systems are encountered.

Properties such as area, velocity, pressure and specific heat are termed 'derived variables' as they can be expressed in terms of the fundamentals. Table I shows how some of these derived physical properties can be expressed in terms of the fundamental dimensions, together with the conversion factors from Imperial to SI units.

TEMPERATURE

Temperature is defined as the degree of hotness of a body. In a spontaneous change, heat is transferred from a high to a low temperature, until thermal equilibrium is achieved.

The two scales encountered in practical measurement are Celsius (°C) and Fahrenheit (°F) with the conversion being; °C = (°F − 32) × 5/9. Temperature driving forces or differential temperatures are measured in degree F or degree C, where 1 deg F = 5/9 deg C. The rate of heat transfer is proportional to the temperature driving force. The absolute scale of temperature (Kelvin K) is based on the work performed by an ideal heat engine working between two temperatures; and absolute zero is taken as the lowest temperature attainable, i.e. the temperature at which all molecular motion ceases. Although absolute temperatures are not used in practical measurement, many equations require substitution in terms of the absolute temperature, e.g. the ideal gas equation and Arrhenius equation. The conversion factor is: K = °C + 273·15. Absolute zero on the Fahrenheit scale is given the value 0° Rankine (°R) and the conversion is: °R = °F + 459·7.

The reference temperature used for comparing the sterilisation efficiency during thermal processing operations is 121·1 °C (250 °F). More recently 135 °C has been selected for UHT processes (Kessler, 1981).

TABLE I
Fundamental and derived units

		SI unit*	Imperial unit	Conversion factor
Mass	M	kilogram (kg)	pound (lb)	1 lb = 0·4536 kg
Length	L	metre (m)	foot (ft)	1 ft = 0·3048 m
Time	T	second (s)	second (s)	
Temperature	θ	Kelvin (K)	°Fahrenheit (F)	
Luminous Intensity	cd	Candela		
Electric Current	I	Ampère (A)		
Amount of substance		Mole (mol)		
Area	L^2	m^2	ft^2	$1\,ft^2 = 9·29 \times 10^{-2}\,m^2$
Volume	L^3	m^3	ft^3	$1\,ft^3 = 2·832 \times 10^{-2}\,m^3$
Velocity	$L\,T^{-1}$	$m\,s^{-1}$	$ft\,s^{-1}$	$1\,ft\,s^{-1} = 0·3048\,ms^{-1}$
Acceleration	$L\,T^{-2}$	$m\,s^{-2}$	$ft\,s^{-2}$	$1\,ft\,s^{-2} = 3·048 \times 10^{-1}\,m\,s^{-2}$
Force	$M\,L\,T^{-2}$	Newton (N)	lb (force)	$1\,lb\,(f) = 4·448\,N$
Pressure	$M\,L^{-1}\,T^{-2}$	Pascal (Pa)	lb (f) in^{-2}	$1\,lb\,(f)\,in^{-2} = 6·895 \times 10^3\,Pa$
Work, Energy	$M\,L^2\,T^{-2}$	Joule (J)	Btu†	$1\,Btu = 1·055 \times 10^3\,J$

(continued)

TABLE I—*contd.*

		SI unit*	Imperial unit	Conversion factor
Power	$M\,L^2\,T^{-3}$	Watt (W)	Horsepower (HP)	$1\ \text{HP} = 745\cdot7\ \text{W}$
Density	$M\,L^{-3}$	kg m^{-3}	lb ft^{-3}	$1\ \text{lb ft}^{-3} = 16\cdot02\ \text{kg m}^{-3}$
Dynamic viscosity	$M\,L^{-1}\,T^{-1}$	N s m^{-2} (Pa s)	$\text{lb ft}^{-1}\,\text{s}^{-1}$	$1\ \text{lb ft}^{-1}\,\text{s}^{-1} = 1\cdot488\ \text{Pa s}$
Kinematic viscosity	$L^2\,T^{-1}$	$\text{m}^2\,\text{s}^{-1}$	$\text{ft}^2\,\text{s}^{-1}$	$1\ \text{ft}^2\,\text{s}^{-1} = 92\cdot9 \times 10^{-3}\ \text{m}^2\,\text{s}^{-1}$
Surface tension	$M\,T^{-2}$	N m^{-1}	‡	$1\ \text{dyne cm}^{-1} = 10^{-3}\ \text{N m}^{-1}$
Specific heat	$L^2\,T^{-2}\,\theta^{-1}$	$\text{J kg}^{-1}\,\text{K}^{-1}$	$\text{Btu lb}^{-1}\,{}^\circ\text{F}^{-1}$	$1\ \text{Btu lb}^{-1}\,{}^\circ\text{F}^{-1} = 4\cdot187\ \text{kJ kg}^{-1}\,\text{K}^{-1}$
Thermal conductivity	$M\,L^{-1}\,T^{-3}\,\theta^{-1}$	$\text{W m}^{-1}\,\text{K}^{-1}$	$\text{Btu h}^{-1}\,\text{ft}^{-1}\,{}^\circ\text{F}^{-1}$	$1\ \text{Btu h}^{-1}\,\text{ft}^{-1}\,{}^\circ\text{F}^{-1} = 1\cdot731\ \text{W m}^{-1}\,\text{K}^{-1}$
Thermal diffusivity	$L^2\,T^{-1}$	$\text{m}^2\,\text{s}^{-1}$	$\text{ft}^2\,\text{s}^{-1}$	$1\ \text{ft}^2\,\text{s}^{-1} = 92\cdot9 \times 10^{-3}\ \text{m}^2\,\text{s}^{-1}$
Latent heat	$L^2\,T^{-2}$	J kg^{-1}	Btu lb^{-1}	$1\ \text{Btu lb}^{-1} = 2\cdot326\ \text{kJ kg}^{-1}$
Heat film coefficient	$M\,T^{-2}\,\theta^{-1}$	$\text{W m}^{-2}\,\text{K}^{-1}$	$\text{Btu h}^{-1}\,\text{ft}^{-2}\,{}^\circ\text{F}^{-1}$	$1\ \text{Btu h}^{-1}\,\text{ft}^{-2}\,{}^\circ\text{F}^{-1} = 5\cdot678\ \text{W m}^{-2}\,\text{K}^{-1}$
Overall heat transfer coefficient	$M\,T^{-2}\,\theta^{-1}$	$\text{W m}^{-2}\,\text{K}^{-1}$	$\text{Btu h}^{-1}\,\text{ft}^{-2}\,{}^\circ\text{F}^{-1}$	$1\ \text{Btu h}^{-1}\,\text{ft}^{-2}\,{}^\circ\text{F}^{-1} = 5\cdot678\ \text{W m}^{-2}\,\text{K}^{-1}$

*SI units: reciprocal form is used here.

†British Thermal Unit.

‡More often measured in dyne cm⁻¹; very rarely seen in Imperial Units.

It is important to be able to measure, record and control processing temperatures accurately. Mercury-in-steel thermometers are useful for indication, whereas resistance thermometers and thermocouples are used most for recording and control applications.

PRESSURE

Absolute pressure is defined as the force divided by the area. SI units are expressed in $N\ m^{-2}$ or Pascals (Pa). This is a small unit of pressure, so for many operations, the bar or MPa is used, where:

$$1\ bar\ =\ 10^5\ Pa\ =\ 0 \cdot 1\ MPa$$

Other pressure units commonly encountered are:

$$\frac{lb\ (f)}{in^2}\ (psi)\ and\ \frac{kg\ (f)}{cm^2}$$

The absolute pressure is used in pressure, volume, temperature (P, V, T) relationships for gases and vapours.

It is well known that there is a relationship between the absolute pressure and the height or head of fluid, supported by that pressure. This is given by:

$$P\ =\ pgh$$

where P is absolute pressure ($N\ m^{-2}$), g is acceleration due to gravity ($= 9 \cdot 81\ m\ s^{-2}$), p is fluid density ($kg\ m^{-3}$) and h is head of fluid (m).

Differential pressures can also be expressed in similar terms, and are commonly used for pressure losses in pipes and fittings, and for pressures (heads) developed in pumping applications.

One standard atmosphere will support a column of mercury, $0 \cdot 76\ m$ high. In other units, one standard atmosphere is given by:

$$1\ atmosphere\ =\ 14 \cdot 69\ psi\ =\ 1 \cdot 013\ bar\ =\ 1 \cdot 033\ kg\ (f) cm^{-2}$$

Most pressure gauges measure pressure, above or below atmospheric, the reading being correspondingly referred to as gauge pressure (Fig. 1). A correction may be necessary if the absolute pressure is required:

$$absolute\ pressure\ =\ gauge\ pressure\ +\ atmospheric$$

If the pressure is below atmospheric, a vacuum gauge is required. On the Imperial System, vacuum is expressed in terms of inches of mercury

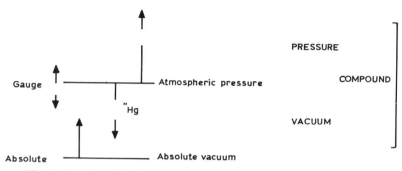

Fig. 1. Relationship between gauge pressure and absolute pressure.

(″Hg); this is measured *below* atmospheric pressure. Such units are still commonly used when expressing the vacuum in a can, or during evaporation. Using the metric system, vacuum is expressed as an absolute pressure, measured above zero absolute pressure. For applications, requiring very low pressures, two further units are used, namely the Torr (τ), and micrometre (μm)

$$1 \text{ mm Hg } = 1 \text{ Torr } (\tau) = 10^3 \text{ micrometres}$$

The most common type of pressure gauge is the Bourdon gauge. This is suitable for steam, hot and chilled water and compressed air, but not for foods, as the material enters the Bourdon tube and is difficult to remove. For hygienic operations, involving foods, diaphragm gauges, which are more expensive, should be used.

DIMENSIONAL ANALYSIS

The technique of dimensional analysis can be used to investigate fluid flow, heat transfer and mass transfer problems. The Buckingham pie (π) theory is used to develop relationships between the variables, in terms of dimensionless groups.

For example, a relationship between the heat film coefficient (h) and other physical properties can be derived, in terms of dimensionless groups, for a liquid flowing along a tube, as follows:

$$\frac{hD}{k} = \phi \left(\frac{vD\rho}{\mu}\right)^a \quad \left(\frac{c\mu}{k}\right)^b$$

(Nusselt (Reynolds (Prandtl
number) number) number)

This relationship can be investigated experimentally to evaluate the constant (ϕ), and the exponents a and b. Heat film coefficients are given for a wide variety of flow situations by ASHRAE (1981), Kessler (1981) and Loncin and Merson (1979). More information regarding dimensional analysis and dimensionless groups is given by Weast (1982).

DENSITY AND SPECIFIC GRAVITY

Density is defined as the mass of substance divided by the volume occupied; its dimensions are (ML^{-3}) and the SI unit is the kilogram per cubic metre ($kg\ m^{-3}$).

$$\text{At } 5\,^\circ C \text{ water has a density of } 1\cdot00\ \frac{g}{ml} \text{ or } \frac{10^{-3}\ kg}{10^{-6}\ m^3} = 10^3\ kg\ m^{-3}$$

The addition of most solids, e.g. minerals, sugars, proteins, will increase the density, whereas oils and fats will decrease the density. The density of fluids is usually measured with a hydrometer. The density is temperature dependent, so temperatures should always be recorded.

The density of bovine milk usually falls within the range 1025–1035 $kg\ m^{-3}$. It is generally measured with a special hydrometer, known as a lactometer, and the result can be used to estimate total solids.

The densities of the respectively solid constituents are regarded as fat (930), water (1000) and milk solids-not-fat (1614 $kg\ m^{-3}$). BS 734 (1937 and 1959) gives information on density hydrometers for use in milk, and tables are supplied for determining the total solids of milk, knowing the density and fat content; temperature correction tables are also presented. Fat contents range between 1 and 10 per cent, and the total solids determination is based on the following equation. Fat is determined separately, usually by the Gerber method.

$$T = 0\cdot25D + 1\cdot21F + 0\cdot66 \qquad \text{BS 734 (1937)}$$
$$T = 0\cdot25D + 1\cdot22F + 0\cdot72 \qquad \text{BS 734 (1959)}$$

where T = total solids (w/w), D = 1000 (density − 1) (density units are ($g\ ml^{-1}$)), F = fat percentage.

Thus, milk at 26 °C with a fat content of 3·5 per cent and a density of 1·0320 would be corrected to a value of 1·0322 at 20 °C, and have a total solids of 12·95 (1937) or 13·05 (1959). Total solids are normally expressed to the nearest 0·05 per cent. Obviously the 1959 tables are preferred, but several people are still of the opinion that they tend to overestimate total solids, and that the 1937 formula is more accurate (Egan *et al.*, 1981).

Kessler (1981) presents relationships for the density of whole milk and cream (20 per cent fat), over the temperature range 0–150 °C.

Whole milk: $p = 1033 \cdot 7 - 0 \cdot 2308\theta - 2 \cdot 46 \times 10^{-3}\,\theta^2$
Cream (20 per cent fat): $p = 1013 \cdot 8 - 0 \cdot 3179\theta - 1 \cdot 95 \times 10^{-3}\,\theta^2$
θ = temperature (°C)

Bertsch et al. (1982) present information on the density of milk and cream over the temperature range 65–140 °C.

Specific gravity is defined as the ratio of the mass of a fixed volume of liquid by the mass of an equal volume of water. Specific gravity is a dimensionless quantity. It is most conveniently measured using a specific gravity bottle, which can be used both for liquids and particulate solids; in the latter case, a solvent is selected which will not dissolve the solid, e.g. toluene.

Specific gravity is less susceptible to changes in temperature, compared to density. The relationship between the specific gravity and density of a material is given by:

$$SG_L = (p_L/p_W)$$

Therefore, if the specific gravity of a material is known at a temperature $T°C$, its density at $T°C$ will be given by

$$p_L = (SG)_T \times p_W$$

where p_L is density of liquid at $T°C$, $(SG)_T$ is specific gravity at $T°C$, p_W = density of water at $T°C$ (obtained from tables).

Density and specific gravity are useful for monitoring changes occurring during processing, e.g. evaporation, or for checking whether extraneous water has entered the product.

Figure 2 shows how the density of two samples of canned evaporated milk change with temperature. The density increases from 1053 to 1078 over the temperature range 75–25 °C. Often total solids is monitored during evaporation by measuring the density and, for the production of evaporated milk, a batch is struck when the desired final gravity is reached. Considerable errors may result if the temperature is not accounted for.

The density of most ice cream mixes falls within the range 1080–1100. Density measurement has been found to be an extremely simple means of quickly assessing whether an ice cream mix has been watered down during processing; this is important in UHT direct steam injection, where the condensed steam is subsequently removed during the flash

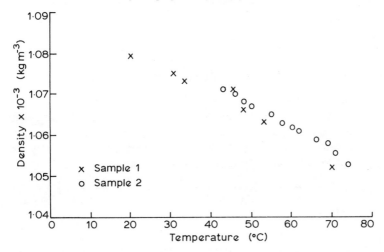

Fig. 2. Graph of density of UK evaporated milk against temperature.

cooling process. Dilution of the mix will result in ice cream with an icy texture. Such dilution could also occur during continuous HTST pasteurisation, and Table II shows how the specific gravity of an ice cream mix, originally at 33·7 per cent total solids changes, as it is diluted with water.

When dealing with solids, it is necessary to differentiate between solid density and bulk density, particularly with particulate matter and powders.

TABLE II
Variation in the specific gravity of ice cream with different total solids

Total solids (%) (w/w)	Specific gravity
33·7	1·0968
32·1	1·0918
30·8	1·0878
29·3	1·0826
28·0	1·0796

Author's unpublished data.

The solid or particle density refers to the density of the solid or an individual particle. It is defined, in the normal fashion, as

$$\frac{\text{mass of solid}}{\text{volume of solid}}$$

and it will take into account the presence of air within the solid. For particulate matter, it can be determined either by flotation using liquids of known specific gravity, or by using a density bottle.

The density of solid constituents have been summarised by Peleg (1983) and Walstra and Jenness (1984) as follows:

	(P)	(W and J)		(P)	(W and J)
Sucrose	1590		Fat	900–950	918
Glucose	1560		Salt	2160	1850[r]
Protein	1400	1400	Water	1000	998·2
Lactose		1780			

[r]residual milk solids.

Therefore, the density of the food can be estimated from a knowledge of the food composition, using the equation

$$p = \frac{1}{m_1/p_1 + m_2/p_2 + m_n/p_n}$$

where n = number of components, m_1 to m_n = mass fraction of components 1 to n, and p_1 to p_n = density of components 1 to n.

This equation may be useful for estimating the density of solids, such as butter, spreads, cheese, yoghurt and cream, as well as protein concentrates, sweetened condensed milks and ice cream mixes. This approach should be treated with caution for dehydrated dairy products, which may contain air spaces within the particles. This will substantially lower the particle density, and cannot be easily accounted for in this equation because the mass fraction of air is not known. However, an estimate of the volume fraction of air trapped within the particles can be made by calculating the solid density (assuming no air) p_c, and measuring the actual solid density p_s. The volume fraction of air trapped within the solid V_a is given by

$$V_a = \frac{p_c - p_s}{p_c}$$

The solid density is important in separation processes, e.g. centrifugation of cheese-fines, cyclone operation, and the pneumatic or hydraulic transport of powders and particulate matter.

Bulk density is an important property, particularly for the transportation and storage of bulk particulate material, e.g. fruit, grain, powders. It is defined as the mass of material divided by the total volume occupied. In most cases, it is important to have a high bulk density, and some values for full cream milk powder are recorded in Table III.

The method used for bulk density measurement has a marked effect on the value obtained; these methods are described for milk powders by Lovell (1980). Peleg (1983) discusses the compressibility of powders in more detail.

The bulk density of milk powders is affected by processing conditions, in particular, total solids, with bulk density increasing as the total solids increase. It should also be noted that particle density increases as the total solids increase, suggesting that less air is incorporated into the particles at higher total solids. Injection of air or nitrogen into the product immediately before atomisation may reduce the bulk density, and agglomeration, achieved by a re-wetting process, substantially decreases the bulk density.

In addition, the method of atomisation will affect the bulk density. Early designs of atomiser wheel produced powders of 0.45–0.55 g ml^{-1}, whereas later designs, typified by the vaned wheel, produced bulk densities of 0.55–0.65 g ml^{-1}; later designs have used steam to occlude air from the fluid. Jet nozzles can produce powders with bulk densities as high as 0.83 g ml^{-1}.

TABLE III
Bulk density, particle density and porosity of some whole milk powders

	Bulk density of powder (g ml^{-1})	Particle density (g ml^{-1})	Porosity (%)
11 per cent solids: 90°C outlet	0.39	0.802	51
34 per cent solids: 90°C outlet	0.55	1.017	46
47 per cent solids: 90°C outlet	0.66	1.150	42
47 per cent solids: 113°C outlet	0.46	0.913	50
Straight through, instantised	0.63	1.100	43
Re-wet instantised	0.33	1.100	70

All figures obtained by drop packing 50 g of powder under standardised conditions: Mettler (1980).

The characteristics of sprays produced by pressure nozzles, two-fluid nozzles and centrifugal atomisers are described in more detail by Masters (1972). The porosity of a material is defined as that fraction of the total volume which is occupied by air, between the particles. It does not take into account air within the particle:

porosity $= (p_s - p_B)/p_s$ (p_s is solid density, p_B is bulk density)

Mettler (1980) measured the porosity of 49 samples of full cream milk powder, and found porosity values ranging from 42·9 to 50·9 per cent. Kjaergaard-Jensen and Neilsen (1982) have reviewed some of the properties of milk powders necessary for recombination technology, and Peleg (1983) has described some other physical properties of powders, such as compaction, flowability and caking.

Over-run

When air is incorporated into a whipped or frozen product, the density decreases. The amount of air incorporated is measured by the over-run, where

$$\text{over-run} = \frac{\text{increase in volume}}{\text{original volume}} \times 100$$

In practical terms, it is most easily measured by comparing the weights of equal volumes of the original liquid and the final aerated product.

$$\text{over-run} = \frac{\begin{array}{c}\text{weight of original liquid}\\ -\text{ weight of same volume of aerated product}\end{array}}{\text{weight of same volume of aerated product}} \times 100$$

The major factors affecting over-run of ice cream and frozen desserts are the total solids content of the mix and the type of freezer used. Over-runs between 2 and 3 times the total solids content are recommended by some authorities; if the over-run is too low, the product becomes heavy, whereas if it is too high it becomes too light and fluffy. Other factors which affect over-run are reviewed by Arbuckle (1977): they are broken down into those which depress it, and those which enhance it.

For whipping cream, over-runs of 100–120 per cent would be expected, although the modern aerosol creams give much higher values. As well as total over-run, it is important to measure the stability of the whipped foam over a period of time (Society of Dairy Technology, 1975).

Specific Volume

An alternative way of quoting the density of materials, particularly gases and vapours, is in terms of specific volume (volume/mass), which has units of $m^3 kg^{-1}$. Gases and vapours are compressible, that is their density is affected by changes in temperature and pressure. The density of an ideal gas at moderate temperatures and pressures can be estimated as follows.

The molecular weight of any gas (expressed in kilograms) occupies $22.4 m^3$ at NTP. The volume occupied by the same mass at the experimental temperature and pressure can be calculated from the ideal gas equation, and is then used to determine the density of the gas.

The specific volume of wet, saturated and superheated vapours are summarised in thermodynamic charts and tables (see Thermal Properties).

RHEOLOGICAL PROPERTIES

Rheology is the study of the deformation of materials, subjected to applied forces. A distinction is usually made between fluids and solids. Fluids will flow under the influence of forces, whereas solids will stretch, buckle or break.

An ideal solid is represented by the Hooke solid, and the ideal liquid by the Newtonian liquid (Muller, 1973). Both are structureless (there are no atoms), both are isotropic (they have the same properties in all directions), and both follow their respective laws exactly.

Many materials can exert both types of properties, depending upon the environmental conditions and stresses they are subjected to. For example, butter at 20 °C is regarded as a solid, although if the shearing force is sufficiently high, it can be made to flow or if its temperature is raised to above 50 °C, it will melt and behave like a fluid. Some of these aspects of fluid and solid rheology will now be investigated.

Viscosity

The viscosity of a fluid is defined as the internal friction within the fluid. When a fluid is subjected to a shearing force (F) over a surface area (A), it will undergo a deformation, known as flow.

274   M. J. Lewis

The shear stress is defined as force/area and the rate of deformation, termed the shear rate, is determined by the velocity gradient (see Fig. 3(a)). For Newtonian fluids, there is a direct relationship between the shear stress (τ) and the rate of shear (dv/dy). The ratio of the shear stress to shear rate is known as the dynamic viscosity or coefficient of viscosity (μ):

$$\mu = \frac{\text{shear stress}}{\text{shear rate}} \quad \frac{\text{N m}^{-2}}{\text{s}^{-1}} \quad (\text{N s m}^{-2})$$

Viscosity data are often plotted as shear stress against shear rate, either in ordinary or logarithmic co-ordinates (Fig. 3). Such plots are known as rheograms.

The two main units used for viscosity measurement are the poise (p) (cgs) and the Poiseuille (Pl) (SI):

Shear stress	Shear rate	Dynamic viscosity
dyne cm^{-2}	s^{-1}	Poise (dyne s cm^{-2})
N m^{-2}	s^{-1}	Poiseuille (N s m^{-2}) (kg m^{-1} s^{-1}) (Pa s)

One Poiseuille is the dynamic viscosity of a fluid which, when subjected to a shear stress of 1 N m^{-2}, gives a shear rate of 1 s^{-1}. The conversion factor is 1 poise = 10^{-1} N s m^{-2}. The viscosity of water at 20 °C is 1.002×10^{-3} N s m^{-2} or 1.002 cp (note that the centipoise is still in common use).

$$1 \text{ cp} = 1 \text{ mPa s}$$

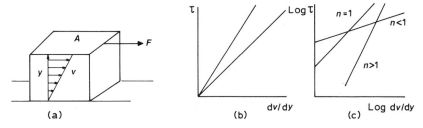

Fig. 3. (a) Deformation of a fluid: (b) and (c) rheograms on ordinary and logarithmic co-ordinates respectively.

Milk, skimmed milk, cheese whey and whey permeate can all be regarded as dilute solutions, and are usually considered to be Newtonian fluids.

The viscosity of all fluids is temperature dependent, the viscosity of liquids, pastes, suspensions and emulsions decreasing with increasing temperature, between 2 and 10 per cent for each degree Celsius; gases increase in viscosity as the temperature rises. Therefore, it is very important to control the temperature accurately when measuring the viscosity, and the temperatures should always be quoted with the results.

Occasionally it is more appropriate to use the term 'kinematic viscosity', which is defined as the dynamic viscosity divided by the density. The units of kinematic viscosity are as follows:

	Dynamic viscosity $(ML^{-1}T^{-1})$	*Density* (ML^{-3})	*Kinematic viscosity* $(L^2 T^{-1})$
cgs	poise	$g\ ml^{-1}$	$cm^2\ s^{-1}$ (Stoke)
SI	$N\ s\ m^{-2}$	$kg\ m^{-3}$	$m^2\ s^{-1}$

Kinematic viscosity is measured directly by the Ostwald capillary flow type viscometer.

The dynamic viscosity of Newtonian fluids can be measured by capillary flow viscometers (density is also required), falling sphere techniques, or by measuring the flow rate in a horizontal capillary tube, under streamline flow conditions. Probably the most sensitive are the capillary flow viscometers, particularly those with narrow capillaries, giving long efflux times. Methods for measuring viscosity are summarised by Lewis (1986).

Concentration Dependence

The viscosity of solutions increases as the concentration increases in a non-linear fashion. At high concentrations, small additional changes in the concentration will lead to rapid changes in the viscosity. This could result in reduced flow rates, higher pressure drops, decreased turbulence and, in heating operations, severe fouling. In concentration processes, such as evaporation, reverse osmosis and ultrafiltration, the extent of concentration may well be limited by viscosity considerations. There is often a transition from Newtonian to non-Newtonian conditions as concentration proceeds.

Non-Newtonian Fluids

As the complexity of fluids increases, there are considerable interactions that result in non-linear relationships between shear stress and shear rate. Such fluids are termed non-Newtonian. Thus in solutions containing macro-molecules, either dissolved or in the colloidal form, in suspensions, pastes or emulsions, there may be complex interactions between various components. This effect increases as the concentration increases. Various types of non-Newtonian behaviour are recognisable. To detect non-Newtonian behaviour requires the use of variable speed rotational viscometers. Thus, by altering the speed, it is possible to alter the shear rate; at each shear rate, the corresponding shear stress is measured. Non-Newtonian fluids are characterised by an apparent viscosity (μ_a), where $\mu_a = \dfrac{\text{shear stress}}{\text{shear rate}}$.

Non-Newtonian fluids fall into two major categories, these being time-independent and time-dependent. Time-independent fluids show viscosity characteristics that are independent of time, perhaps best illustrated by a pseudoplastic fluid which is the most common type of behaviour. When shear stress is plotted against shear rate, the rheogram of Fig. 4 is obtained. The apparent viscosity at any shear rate is determined from an equation, and can be plotted against the shear rate. Therefore, it can be seen that a pseudoplastic fluid shows a decrease in apparent viscosity as the shear rate increases; this is also called shear-thinning behaviour. If all the readings were repeated whilst reducing the shear rate, they would be identical to those obtained whilst increasing the speed; no hysteresis would be observed. This is another characteristic of a time-independent fluid. Most time-dependent fluids are thixotropic, showing a decrease in apparent viscosity under shear stress, followed by a gradual recovery when the stress is removed. Time-dependent fluids are usually measured at constant shear stress, and the

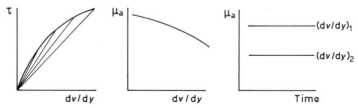

Fig. 4. Rheograms for a pseudoplastic fluid.

change in apparent viscosity is measured over a period of time. It is important to ensure that all the experimental conditions are recorded for time-dependent materials. The major types of non-Newtonian behaviour are summarised in Fig. 5, and plastic fluids are covered in the next section. Rotational viscometers are used to characterise non-Newtonian fluids.

Rotational viscometers rely on a spindle rotating in the test fluid, the force required to overcome the frictional forces being measured. The most common measuring heads are a single spindle, a concentric cylinder viscometer, or a cone and plate viscometer. The Brookfield viscometer is probably the most widely used. In its original design, it consisted of a series of interchangeable spindles, which were inserted into the test fluid; a choice of eight speeds were available. The equipment is fairly cheap and robust. Developments include a helipath stand to move the spindles up and down in the liquid, special T-piece adaptors for high viscosity fluids and gels, a low viscosity attachment and a special small sample adaptor. A cone and plate viscometer is also available from the

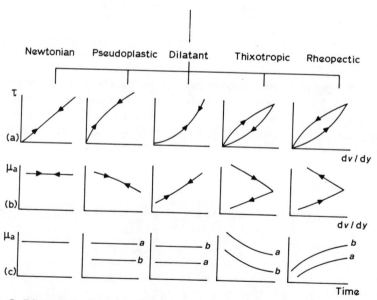

Fig. 5. Rheograms for Newtonian and non-Newtonian behaviour: (a) shear stress against shear rate; (b) apparent viscosity against shear rate; (c) apparent viscosity against time at two different shear rates, *a* and *b*, where $b > a$.

same manufacturers. The major disadvantage of using a single spindle is that it is not possible to predict the exact shear rate in the sample. This is overcome by use of the more expensive concentric cylinder, or cone and plate viscometers. Viscometers are described in more detail by Bourne (1982), and Brennan (1984). Many time-independent non-Newtonian fluids obey a power law:

$$\tau = k\,(\mathrm{d}v/\mathrm{d}r)^n$$

and a straight line relationship results when $\log \tau$ is plotted against $\log (\mathrm{d}v/\mathrm{d}r)$. The consistency index (k) and the power law index (n) are often used to characterise the behaviour of such fluids (Fig. 3(c)).

For all milk and milk products, there is considerable variation in composition and hence viscosity, so that it is recommended that the viscosity be measured wherever possible. The most convenient instrument for measuring low viscosity fluids is the capillary flow viscometer, which is sensitive enough to be able to detect the small changes which occur when milk is heated or homogenised.

Table IV shows some representative values for the viscosity of whole milk, skim-milk and cheese whey at different temperatures. Most of these fluids exhibit Newtonian behaviour over a moderate range of shear rates.

Homogenisation and heat treatment both tend to increase the viscosity slightly, with homogenisation giving the milk a creamier mouth feel. The effects of homogenisation become more pronounced as the fat content increases.

Bertsch and Cerf (1983) used a capillary flow viscometer to measure the viscosity of a variety of milk products in the range 70–135 °C; fat

TABLE IV
Viscosity of milk and whey at different temperatures
(average values) (mPa s)

	Temperature (°C)			
	10	*20*	*40*	*80*
Whole milk	2·79	2·12	1·24	0·68
Skim-milk	2·44	1·74	1·03	0·53
Whey	1·71	1·26	0·82	0·68

Interpreted from data in Kessler (1981).

contents ranged from 0·03 to 15·5 per cent, and some of the milks were homogenised.

A relationship of the following form was found between dynamic viscosity (μ) and temperature (θ):

$$\ln \mu = a\theta^2 + b\theta + c$$

The viscosity of homogenised milk and cream (up to 15 per cent fat) could be represented by:

$$\ln \mu = 3\cdot92 \times 10^{-5}\,\theta^2 - 1\cdot951 \times 10^{-2}\,\theta + 0\cdot666$$
$$+ F(-9\cdot53 \times 10^{-6}\,\theta^2 + 1\cdot674 \times 10^{-3}\,\theta - 4\cdot37 \times 10^{-2})$$
$$+ F^2(9\cdot75 \times 10^{-7}\,\theta^2 - 1\cdot739 \times 10^{-4}\,\theta + 9\cdot83 \times 10^{-3})$$

μ = dynamic viscosity (mPa s)
F = fat content (w/w)%
θ = temperature (°C)

Data are also presented for non-homogenised milk. High temperature data are useful for evaluating performance in UHT sterilisers, particularly pressure drops and residence time distribution in the high temperature sections.

An alternative semi-empirical approach for dispersions is outlined by Walstra and Jenness (1984). It assumes that the increase in viscosity caused by particles results from hydrodynamic interactions only:

$$\eta = \eta_0 \left(1 + \left(\frac{1\cdot25\,\phi}{1 - \phi/\phi_{max}}\right)\right)^2$$

where η = viscosity of suspension
η_0 = viscosity of solvent (water)
ϕ = total hydrodynamic volume
ϕ_{max} = hypothetical volume fraction giving close packing
($\phi = \phi_f + \phi_c + \phi_w + \phi_l$)

where ϕ_f = hydrodynamic volume of fat (1·11 ml g^{-1})
ϕ_c = hydrodynamic volume of casein (3·9)
ϕ_w = hydrodynamic volume of whey proteins (1·5)
ϕ_l = hydrodynamic volume of lactose (1·0)

This equation is useful for predicting the viscosity of dilute solutions; it breaks down when ϕ increases above 0·6, e.g. for liquids with high fat contents and for whey protein concentrates.

The viscosity of milk products increases as the concentration increases, and some data for concentrated skim-milk are given in

TABLE V
Dynamic viscosity of concentrated skim-milk at 25 °C

Total solids	Dynamic viscosity (mPa s)	Total solids	Dynamic viscosity (mPa s)
20	3·8	18·6	4·4
25	5·8	23·8	5·6
30	10·0	29·3	8·8
33	13·0	32·3	16·6
		39·8	59·3
		46·4	1280
Kessler (1981)		Allen (1980)	

Table V, which is compiled from two sources; there is a reasonable agreement between them. Notice the sharp increase in viscosity above 35 per cent total solids.

Kessler also records how the viscosity of skim-milk concentrate (25 per cent TS) changes with temperature.

The viscosity of full cream evaporated milk will depend upon the degree of forewarming, homogenisation conditions, the type of stabiliser used and the extent of the final in-container heat treatment.

Fernandez-Martin (1984) presents monographs for determining the viscosity of sheep's milk cheese whey over the concentration range 6·9–26·5 per cent total solids and temperature range 5–80 °C. The whey was concentrated by reverse osmosis, and most of the concentrates were found to be Newtonian over the shear rates measured. Problems were observed with protein denaturation at the highest temperatures.

$$\log \mu = 0{\cdot}2214 - 0{\cdot}0131\,\theta + 5{\cdot}5 \times 10^{-5}\,\theta^2 + (0{\cdot}02173$$
$$- 1{\cdot}04 \times 10^{-4}\,\theta + 1{\cdot}87 \times 10^{-6}\,\theta^2)s$$

μ = dynamic viscosity (mPa s)
θ = temperature (°C)
s = solids content (%).

Kjaergaard-Jensen and Neilsen (1982) discuss the factors that might influence the viscosity of recombined sweetened-condensed milks.

Viscosity is one of the main factors which limits the extent of concentration for ultrafiltration and reverse osmosis processes. The protein fraction makes the main contribution to the viscosity. Maubois (1980) examined the possibility of pre-concentrating whey by evaporation prior to ultrafiltration, and he showed how the viscosity changed during ultrafiltration, for four different concentrated whey samples of 29, 23, 21·4 and 16 per cent TS. In each case, viscosity was plotted against the percentage protein (dry weight basis) in the concentrate. If it is assumed that the concentration is not allowed to proceed once the viscosity has reached 20 mPa s, then the protein content attainable decreased as the total solids content of the feed increased. Evaporation appeared to offer no advantage for the production of high protein powders.

Cream

Freshly separated cream has a fairly low viscosity. The market cream is then standardised to the desired fat content, heat treated, homogenised, cooled and packaged. All these factors can significantly affect the viscosity of the final product, and each cream needs to be treated differently to obtain the best quality product.

For example, single cream is homogenised at fairly high pressures, usually after heat treatment; this increases the viscosity significantly. Filling into cartons using a piston filler will reduce the viscosity, probably due to shear breakdown, but the viscosity then increases during cold storage. Single cream has been shown to observe pseudo-plastic or dilatant behaviour, depending upon storage conditions. Cream cooled very quickly and stored at a uniform low temperature often shows a dilatant character. When the cream was warmed and then recooled, pseudoplastic behaviour was observed. Obviously the rheo-logical behaviour of cream is very complex, so it is important to record experimental data, shearing history, temperature, storage time and any other relevant information. Further details regarding other types of cream are given by Rothwell (1968) and Society of Dairy Technology (1975).

Kessler (1981) records data for the kinematic viscosity of creams of different fat contents at 20 °C (Table VI). The variation of kinematic viscosity with temperature is also shown.

TABLE VI
Kinematic viscosity of cream at 20 °C

Fat content (%)	Kinematic viscosity ($m^2 s^{-1}$)
20	$6·2 \times 10^{-6}$
35	$14·5 \times 10^{-6}$
45	35×10^{-6}

SOLID RHEOLOGY

Ideal solids are considered to be perfectly elastic, that is they revert back to their original length once the stress is removed, provided the elastic limit has not been exceeded. Two common properties used for elastic materials are the modulus of elasticity (E) and shear modulus (G), where

$$\text{modulus of elasticity} = \frac{\text{shear stress}}{\text{strain}} = \frac{\text{force/area}}{\text{extension/original length}}$$

$$\text{shear modulus} = \frac{\text{shear stress}}{\text{angular deformation}}$$

Further information on Young's modulus and shear modulus values for a variety of foods are given by Muller (1973) and Mohsenin (1970). Unfortunately most dairy products do not fall into this category, their behaviour being more complex; these can be described as plastic or viscoelastic.

Some materials will exhibit plastic behaviour. Below a certain limiting shear stress τ_0, they behave like solids; above τ_0 they behave like fluids and exert a plastic viscosity. The two most common forms are the Bingham plastic and the Casson plastic (see Fig. 6).

Products such as butter, spreads, set yoghurt and cheese are generally regarded as solids. However, most of these will breakdown and flow, but such breakdown may cause irreversible damage to the product.

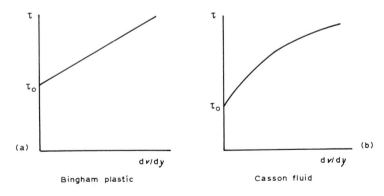

Fig. 6. Rheograms for a Bingham plastic (a) and a Casson fluid (b).

The Casson equation is:

$$\sqrt{\tau} - \sqrt{\tau_0} = \sqrt{\mu_p} \sqrt{dv/dr}$$

Flow curves for butter and margarine are given by Muller (1983).

VISCOELASTIC BEHAVIOUR

Viscoelastic materials exhibit both elastic and viscous behaviour simultaneously. For some materials, the viscous forces may predominate, whilst for others the elastic forces may do so. Figure 7 shows the strain–time relationship for a typical viscoelastic material, during the application of a constant shearing force and after the force is removed.

There are several methods for examining viscoelastic behviour, the main ones being:

1. To measure the change in strain at constant shear stress, whilst a sample is loaded, followed by removal of that load (Fig. 7). Creep function is the term given to describe the change of strain, at constant stress, and creep compliance is the ratio of the stress divided by the strain. A sample is said to exhibit linear visco-elasticity if there is a straight line relationship between the strain and the applied force, when the samples subjected to different stresses are measured after the same time interval.

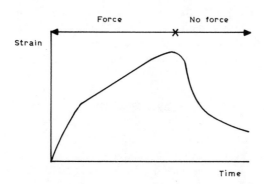

Fig. 7. Strain–time relationship for a viscoelastic fluid, during the application of a constant force, and after removal of the force.

2. To measure the change in stress at constant strain. The relaxation time (t) is defined as the time for the stress to fall to $1/e$ (36·8 per cent) of its original value. Muller (1973) states that:

$$t = \mu/G$$

High relaxation times are associated with materials where the viscous forces predominate, low relaxation times where the elastic forces predominate. If relaxation times are extremely long, some other time period may be selected, e.g. 50 or 70 per cent of the original value.

3. The use of dynamic tests, whereby the sample is subjected to a harmonic shear strain Y, amplitude Y_0, which gives rise to a harmonic shear stress, within the sample τ, amplitude τ_0, which is out of phase with the shear strain by an angle θ. Cone and plate or concentric cylinder instruments are useful for these tests, the test material being placed in the gap. The harmonic shear strain is set up in one of the elements and the corresponding shear stress is detected by the other element.

If the phase shift = θ

then θ = 90° for an ideal viscous liquid,

θ = 0° for an ideal elastic material,

$\tan \theta$ is known as the loss factor.

A storage modulus and loss modulus are defined as follows:

storage modulus = $\tau_0 \cos \theta / Y_0$: high for elastic materials

loss modulus = $\tau_0 \sin \theta / Y_0$: high for viscous materials

Rheological properties are useful, because as well as describing the flow characteristics of materials, they can be used to monitor certain processes, e.g. curd formation in cheese and yoghurt, as well as in the quality control and assessment of products.

FOOD TEXTURE

Texture is one of the main determinants in the quality of many dairy products. Texture is a psychological property, rather than a physical property. One widely accepted definition is that texture defines the attribute of a food material resulting from a combination of physical and chemical properties, perceived by the senses of touch, sight and hearing.

Strictly speaking texture should be evaluated by sensory methods using trained panellists. However, for routine work, sensory methods suffer many disadvantages, and a variety of other methods have been evaluated for a more rapid assessment of food texture. These include rheological methods, chemical tests and, more recently, observations from the microscope. It should be emphasised that results from such instruments are only valid if they can be shown to correlate with sensory evaluation data. Most of these instruments are designed to subject the food to some form of deformation. A few of them are designed to measure fundamental rheological properties, others imitate the forces that the food is subjected to during mastication, but the vast majority are empirical in nature, measuring properties that cannot be easily expressed in fundamental terms. Some of these instruments, which have been used for dairy products, are listed below:

Efflux viscometers, funnels	flow properties, yoghurt, creams
Torsion wire viscometers	custard
Extruders	butter and other spreads
Penetrometers	gels, yoghurts, butter

More sophisticated instruments which are widely used are the General Foods Texturometer and the Instron Universal Testing equipment. Texture measurement is comprehensively reviewed by Brennan (1984) and Bourne (1982); the latter gives a good account of the wide range of instruments available. The rheological properties of dairy products have been reviewed by Prentice (1979). Taneya (1979) describes the viscoelastic properties of Cheddar and Gouda cheeses.

Two instruments which have been found to give very useful estimations of the firmness of butter are the extruder (Prentice, 1954) and the cone penetrometer (International Dairy Federation, 1981). A relationship is given between the penetration depth and the yield shear stress (assuming plastic behaviour). Results from both these instruments were found to correlate very well with a sensory assessment of spreadability. On the basis of simplicity, the cone penetrometer has been recommended as a suitable instrument for assessing firmness. A further problem of oiling-off has been identified with some of the softer, more spreadable butters appearing on the market.

The rheology of gels is discussed by Mitchell (1980). Methods for measuring gelation properties, such as gel strength and gelling time are reviewed by Kinsella (1976). Rheological tests for dairy products involving structural failure have been discussed by Hamann (1983), whilst Prentice (1984) gives a history of some of the rheological work on cheese and butter.

SURFACE PROPERTIES

Milk is a good example of a complex colloidal system. Casein micelles are stabilised by colloidal calcium phosphate, and the dispersed fat phase by the fat globular membrane. The surface area to volume ratios of the dispersed phase components are very large, and Walstra and Jenness (1984) give the surface area to volume ratios and mean diameters of these dispersed phase components shown in Table VII. The same authors also show how homogenisation pressure will affect the size of the fat globules.

It is important to maintain colloidal stability during processing, so it is necessary to examine the forces acting at both the interface between milk and air, and that of the dispersed and continuous phase.

Surface tension is concerned with the forces acting within the fluid. Molecules at the surface of a fluid will be subject to an imbalance of molecular forces and will be attracted into the bulk of the fluid. Consequently, the surface is said to be under a state of tension.

The surface tension of a liquid can be regarded in two ways, either as the force per unit length acting on a length of the surface, or as the work done in increasing its surface area, under isothermal conditions. The SI units of surface tension are $N\,m^{-1}$ or $(J\,m^{-2})$. The conversion factor is $1\,dyne\,cm^{-1} = 10^{-3}\,N\,m^{-1}$ or $1\,mN\,m^{-1}$. It is the surface tension forces which cause most finely dispersed liquids to form spherical droplets; this is the shape having the minimum surface area to volume ratio.

TABLE VII
Surface area to volume ratio and size of dispersed phase components

Component	Surface area/volume ratio $(cm^2\,ml^{-1})$	Particle diameter
Fat globules	700	0·1–10 μm
Casein micelles	40 000	10–300 nm
Globular (whey proteins)	50 000	3–6 nm
Lipoprotein particles	100	~ 10 nm

Adapted from data in Walstra and Jenness (1984).

The methods for determining surface tension are discussed by Levitt (1973). Water has a surface tension value of $72 \cdot 6$ mN m^{-1} at $20\,°$C, whereas milk has an approximate value of 50 mN m^{-1} at $20\,°$C (Jenness *et al.*, 1974). The surface tension is lower than that of water due to the presence of casein, whey proteins and phospholipids, which are all surface active. Bertsch (1983) measured the surface tension of whole milk (4 per cent fat) and skim-milk, in the temperature range 18–135 $°$C, by a drop weight method. The surface tension was found to decrease in an almost linear fashion as temperature increased. There was very little difference between the results for whole and skim-milk, and the combined results could be represented by the equation:

$$Y = 1 \cdot 8 \times 10^{-4} \theta^2 - 0 \cdot 163\, \theta + 55 \cdot 6$$

where Y = surface tension (mN m^{-1})
θ = temperature ($°$C)

Jenness *et al.* (1974) concluded that the surface tension of milk decreased slightly as the fat content increased to 4 per cent, thereafter remaining constant. Sweet cheese whey was reported to have a similar value to skim-milk. Homogenisation was found to increase the surface tension slightly. Lipolysis and the liberation of free fatty acids decreased the surface tension, but heat treatment had little effect. The presence of detergent in milk would drastically reduce the surface tension.

There appears to be little published work on the effects of concentration on the surface tension of milk. This will be important in spray-drying operations, as it will affect the drop-size distribution of the spray.

An interfacial tension exists at the boundary of two immiscible liquids, again due to an imbalance of intermolecular forces. The interfacial tension between water and butter oil at 40 $°$C is $19 \cdot 2$ mN m^{-1}. Such a system is unstable and will quickly separate, as an interfacial tension below 10 mN m^{-1} is required to produce a stable emulsion. Surface active components in the milk serum will lower the interfacial tension (Powrie and Tung, 1976), the most effective being α-lactalbumin and interfacial protein. Other emulsifiers and stabilisers may be added and are important in aerated products, such as whipped cream, ice cream and milk concentrates. The interfacial tension of various emulsifiers, used in ice cream, against palm kernel oil–water, and palm kernel oil–casein solutions ($0 \cdot 1$ per cent) are given by Berger (1976). The emulsion stability in milk and milk products is discussed in more detail by Graf and Bauer (1976).

The work of adhesion (W_{AB}) is the work required to separate an interface and produce two distinct surfaces:

$$W_{AB} = Y_A + Y_B - Y_{AB}$$

A reduction in interfacial energy will increase the work of adhesion and help to stabilise the emulsion. Emulsions are further stabilised by a high viscosity continuous phase, and a similar, uniform charge on the surfaces of the particles.

Surface tension forces are important in size reduction and cleaning operations. The Weber number is a dimensionless group, which includes surface tension; its use in connection with high pressure homogenisers has been discussed by Loncin and Merson (1979). Masters (1972) presents correlations for mean particle diameters for pressure nozzles, two fluid nozzles and centrifugal atomisers. Hygienic design and cleaning of food processing equipment is considered in more detail by Troller (1983) and Jowitt (1980).

THERMAL PROPERTIES OF FOODS

Heat transfer plays an important role in many dairying operations, and in most situations, it is desirable to maximise the rate of heat transfer. This offers economic advantages and generally results in a better quality product. Therefore drying, heating, chilling and freezing operations are all influenced by the heat transfer processes taking place.

Some of the questions the dairy technologist may be required to answer are:

1. How much heat is required to process a particular material and what methods are available for conserving energy? In simple heating and cooling operations, this will involve only sensible heat changes, whereas in evaporation and freezing, latent heats will be involved.
2. What size heat exchangers are required for a particular heating duty? This applies to heaters, evaporators and refrigeration equipment. It involves the solution of steady-state heat transfer equations, and a knowledge of heat transfer mechanisms and resistances involved.
3. What are the heating times or freezing times? This involves the solution of unsteady-state heat transfer equations.

4. What are the requirements for steam, hot water, refrigerants, electricity and compressed air? This involves a knowledge of the thermodynamic properties of these fluids. The thermal properties of dairy products, relevant to these processes, will now be discussed in more detail.

Sensible and Latent Heats

Sensible heat changes are those which can be detected by a change in temperature, and involve no phase change. The amount of heat (Q) required to bring about a sensible heat change for a batch heating process is given by:

$$Q = \text{mass} \times \text{specific heat} \times \text{temperature change}$$
$$\text{(kJ)} \quad \text{(kg)} \quad \text{(kJ kg}^{-1}\text{ K}^{-1}) \qquad \text{(K)}$$

For a continuous process, mass is replaced by mass flow rate (kg s^{-1}) and the quantity of heat of duty is expressed as kJ s^{-1} (kW).

The specific heat of a substance is defined as the amount of heat required to raise a unit mass through a unit temperature rise. The value for water is 4·18 kJ kg^{-1} K^{-1} or 1 Btu lb^{-1} °F^{-1}. The specific heat of most substances is slightly temperature dependent; this can be overcome by using an average specific heat value for the temperature range being considered. If the variation of specific heat with temperature is known, the heat change can be determined by plotting the specific heat against temperature and evaluating the area under the curve ($\int C_p \, d\theta$).

The different components in foods have different specific heat values, so it should be possible to estimate the specific heat of a food from a knowledge of its composition. Water has the greatest influence on the specific heat.

Lamb (1976) gives a simple equation, considering the food to be a two component system — water (w) and solids (s), based on the mass fractions (m) of each

$$c = m_w c_w + m_s c_s$$

where c = specific heat.

Miles *et al.* (1983) make a further distinction, which is quite popular with dairy products, based on water (w), fat (f) and solids-not-fat (snf).

$$c = (0·5m_f + 0·3m_{snf} + m_w) \, 4·18 \text{ (kJ kg}^{-1}\text{ K}^{-1})$$

Kessler (1981) has recommended the equation:

$$c = 4 \cdot 18 m_w + 1 \cdot 4 m_c + 1 \cdot 6 m_p + 1 \cdot 7 m_f + 0 \cdot 8 m_A$$
$$\text{(water)} \quad \text{(carbohydrate)} \quad \text{(protein)} \quad \text{(fat)} \quad \text{(ash)}$$

Therefore, provided the chemical composition is known, specific heats can be estimated reasonably accurately. Values for frozen products can be obtained by substituting the specific heat of ice, in the respective equations. This, however, assumes that all the water is in the frozen form. Bertsch (1982) describes how the specific heat of milk changes over the temperature range 50–140 °C. Fernandez-Martin (1984) describes how the specific heat of sheep's milk whey, concentrated by reverse osmosis, is influenced by temperature and total solids.

$$c = 4 \cdot 216 + (-0 \cdot 0262 + 0 \cdot 092 \times 10^{-3} \theta)s$$

where θ is temperature (°C), s = total solids (per cent).

The specific heat of milk concentrates has been described by Fernandez-Martin (1972) as:

$$c = (m_w + (0 \cdot 328 + 0 \cdot 0027 \theta)m_s)4 \cdot 18$$

over the temperature range 40–80 °C and total solids range (8–30 per cent).

Latent Heat Effects

Latent heat changes are involved when phase changes occur, the major ones being the transition from solid (S) to liquid (L) and from liquid (L) to vapour (V). Under special conditions, the change from solid to vapour can also occur; this is known as sublimation and occurs when the water vapour pressure is maintained below 4·6 τ.

The major changes involved with dairy products are: the transition from water to ice (freezing): the removal of water during evaporation and concentration, and the phase changes involved in the fat fraction when products are cooled below 50 °C (crystallisation).

At atmospheric pressure water boils at 100 °C; the latent heat of vaporisation water is the amount of energy required to change 1 kg of water from liquid to vapour, without changing the temperature (2257 kJ kg^{-1}); this provides information on the energy requirements for evaporation. However as the pressure is reduced and the boiling point decreases, the latent heat value increases. At a pressure of 0·073 bar (absolute), the latent heat value is increased to 2407 kJ kg^{-1}.

Steam is used as the heating medium in evaporation, and the heat lost by the steam causes vaporisation of the liquid. In practice, the evaporation process can be regarded as an exchange of latent heat, and in a single-effect evaporator, it takes approximately 1 kg steam to remove 1 kg water vapour.

Many processes are performed under vacuum; this results in a reduction in the evaporation temperature, and an improved heat transfer rate due to the increased driving force between the steam and the evaporating liquid; the amount of energy required increases slightly. Steam economy is effected on evaporation plants by multiple effect evaporation, and the use of vapour recompression, either by steam ejectors or mechanical recompression.

Steam is a very efficient heat transfer fluid. It has both a very high latent heat value, and a high film coefficient. Mixing of air with steam will drastically reduce the heat film coefficient and, in thermal processing operations, may lead to under processing.

The thermodynamic properties of saturated steam are covered in the steam tables (ASHRAE, 1981). For saturated steam, with no air present, there is a fixed relationship between the temperature and pressure, so pressure gauges are often incorporated into steam lines and retorts as a back-up to temperature readings.

Higher processing temperatures require the use of higher pressure steam. For example, saturated steam for UHT processing (150°C) will be at a pressure of 4·76 bar(a), whereas steam for heating the inlet air to a spray drier (200°C) would be at a pressure of 15·5 bar(a). If steam is to be injected directly into a product, the steam should be saturated and free from any contaminating particles. Such addition results in a dilution of the milk, and this added water needs to be removed later in the process. Occasionally superheated steam may be required; this is where saturated steam is heated above its saturation temperature. One such use is in the sterilisation of cans, prior to aseptic filling in the Dole process, where the use of superheated steam avoids excessive condensation and wetting of the cans.

The thermodynamic properties of steam and other thermodynamic fluids, such as Freon 12, ammonia, carbon dioxide cryogenic fluids and immersion fluids are summarised by ASHRAE (1981).

Freezing of Foods

The latent heat of fusion for pure water is 335 kJ kg^{-1}; that is the amount of heat liberated when 1 kg of water is converted to ice at 0°C.

Unfortunately the situation for foods is more complex. The presence of solids depresses the freezing point, with most foods starting to freeze at about $-1\,°C$. This results in a concentration effect and a further depression of the freezing temperature. Therefore, the food does not freeze at a constant temperature; rather as freezing proceeds, so the temperature falls as more of the ice is converted to water; hence there is a concept of unfrozen water. Most of the water freezes over the temperature range $-1\,°C$ to $-10\,°C$, and by $-15\,°C$, more than 90 per cent of the water is frozen. The freezing point of milk is of considerable interest, because it is also used to detect any dilution of the milk. Some of the pitfalls of the method have recently been discussed by Kessler (1984). Freezing point calculations for ice cream and frozen desserts are covered by Arbuckle (1977). Smith and Bradley (1983) have described the effects of carbohydrates commonly used in frozen desserts on their freezing points.

Most foods contain substantial quantities of solid, whereas only the water contributes to the latent heat value. On this basis, Lamb (1976) suggested the following equation for determining the latent heat value of a food:

$$L = m_w \times 335 \,(\mathrm{kJ\ kg^{-1}})$$

One approach to operations involving sensible and latent heat change is to assume that the material has a melting point of $-1\,°C$, and that all the water is converted to ice at that temperature. The operation then comprises the sensible heat change (to the freezing point), a latent heat change, and a further sensible heat change to the storage temperature.

The fact that all the water does not freeze will give an over estimate, but this is balanced by the fact that all the food is considered frozen at $-1\,°C$, whereas in fact it is not, which will tend to under estimate the value.

Enthalpy (H) is a thermodynamic function, where $H = U + PV$, U = internal energy, P = pressure, V = volume.

It can be shown for operations taking place at constant pressure, that the enthalpy change is equal to the amount of heat (q) absorbed or evolved

$$\Delta H = q = \int C_p \, \mathrm{d}\theta$$

Enthalpy data for food are normally presented as the specific enthalpy $(\mathrm{kJ\ kg^{-1}})$ at different temperatures. Above $50\,°C$, virtually all fats are completely liquid, but as the temperature is reduced, crystallisation takes place and the fat solidifies. The extent of solidification will depend mainly upon the type of fat and the cooling conditions. The relationship between the percentage crystalline solids and temperature is known as the 'melting characteristic'. The melting characteristic and specific enthalpy for butter fat are shown in Table VIII.

Solid–liquid fat ratios can be measured by dilatometry, wide-line nuclear magnetic resonance (NMR) or differential scanning calorimetry (DSC). Ruegg *et al.* (1983) describe an improved DSC method for the estimation of the solid content of butter fat, which takes into account the temperature dependence of the heat of melting. Applying this correction gives data which agree more closely with those from NMR techniques. The fractionation of milk fat by cooling and crystallisation, and the composition and properties of the fractions are described by Badings *et al.* (1983a, b).

Figure 8 shows some thermodynamic data represented in the form of enthalpy against concentration for whole (full cream) milk. This chart is extremely useful, as it shows how the heat content changes with temperature and moisture content. In addition, when the temperature is taken below the freezing point, the amount of unfrozen water is also given. Full cream milk at 80 °C has an enthalpy value of 670 kJ kg^{-1}. If this is reduced in temperature to -10 °C, the new enthalpy value is 105 kJ kg^{-1}, and the fraction of water that is frozen is approximately 92 per cent. Therefore the amount of heat removed would equal 565 kJ kg^{-1}. Loncin and Merson (1979) reviewed other sources of data; some other charts are given in Rha (1975), most of which are based on the work of Reidel.

TABLE VIII

Enthalpy, percentage crystalline solids and apparent specific heat of butter fat

	Temperature (°C)								
	−40	*−20*	*−10*	*0*	*10*	*20*	*30*	*40*	*50*
Specific enthalpy (kcal kg^{-1})	3	11	17	24	32	45	54	60	65
Crystalline solids (per cent)	100	98	90	75	56	20	10	0	0
Apparent specific heat (kJ kg^{-1} K^{-1})	1·59	1·84	2·01	3·34	4·39	5·35	3·34	2·09	2·01

Adapted from data in Rha (1975).

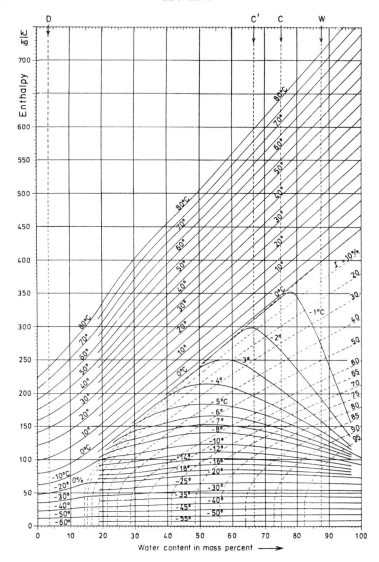

Fig. 8. Enthalpy–concentration diagram for the system dry whole milk/water. The dry solids contain 30 per cent fat. The value of the enthalpy is 0 at the reference temperature $-60\,^\circ$C. I is the percentage of the water which is frozen; D represents the composition of air dried, whole milk powder; C and C' are condensed milk with 7·5 and 10 per cent fat, respectively; W is whole milk. (From Riedel (1976), by permission of Verlag Hans Carl.)

TABLE IX
Specific heat and latent heat values for some dairy products

	Specific heat ($kJ\,kg^{-1}\,K^{-1}$)		Latent heat
	Above f. pt.	Below f. pt.	($kJ\,kg^{-1}$)
Cheese (37–38 per cent moisture)	2·09	1·30	125·6
Roquefort	2·72	1·34	183·8
Cheese-low fat	2·68	1·47	
Cream 15 per cent fat	3·85		
40 per cent fat	3·56	1·68	209·3
Ice cream (58–66 per cent moisture)	3·27	1·88	223·3
Milk	3·85		
Skim-milk	3·98	2·51	305
Butter	2·05		53·5

Compiled from data in Polley *et al.* (1980) and ASHRAE (1981).

An alternative approach is to use an apparent specific heat value, which accounts for sensible and latent heat changes, i.e. when considerable melting or crystallisation is taking place the apparent specific heat is high; when none is taking place, it is much lower.

Table VIII also shows how the apparent specific heat of butter fat changes with temperature. The apparent specific heat of whey, whole milk and cream containing 20, 40 and 60 per cent fat has been plotted against temperature over the range 0–50°C (Batty and Folkman, 1983).

Some specific heat values for milk products are given in Table IX.

Thermal Conductivity

This is a measure of the rate of heat transfer through a material when conduction is the controlling mechanism. It is applicable to solids, but it can also be measured for fluids when convection is eliminated.

Thermal conductivity is defined as the steady-state rate of heat transfer through an area 1 m^2 when a temperature driving force of 1 K is maintained over a distance of 1 m.

It can be measured for materials under steady-state or unsteady-state conditions. Methods for measuring thermal conductivity have been described by Mohsenin (1980) and Jowitt *et al.* (1983).

296   M. J. Lewis

Lamb (1976) gives the following equation for evaluating the thermal conductivity of a food from its moisture content:

$$k = 0.0801 + 0.568m_w$$

where m_w = fractional moisture content.

It was claimed that there were some large discrepancies below 50 per cent moisture content, and that there was no simple relationship between the thermal conductivity of frozen foods and moisture content. Water is the principal component affecting the thermal conductivity, since it has a higher value than the other chemical components. Ice has a value approximately four times higher than water. It is not so straightforward to predict the thermal conductivity from the composition of complex materials, as it depends on whether the components are considered to be in parallel or series (Miles *et al.*, 1983). Figure 9 shows how thermal conductivity changes with temperature, for a variety of products. In all cases, the thermal conductivity increases as the temperature increases; again it should be noted that as milk products become more concentrated, their thermal conductivity decreases.

MacCarthy (1984) has measured the effective thermal conductivity of skim-milk powder using a guarded hot plate technique. Values ranged from 0.036 to 0.109 W m^{-1} K^{-1} in the temperature range 11.8–49.7 °C for bulk densities between 292 and 724 kg m^{-3}. The effective thermal conductivity increased with temperature and with bulk density.

Thermal Diffusivity

Thermal diffusivity (a) is a composite property defined as (k/pc) with units of m^2 s^{-1}. It is an extremely useful property in unsteady-state heat transfer problems, because it is a measure of how quickly temperatures change with time, during heating and cooling processes. It is extensively used in unsteady-state heat transfer problems in a dimensionless form known as the Fourier number (Fo), where:

$$Fo = \frac{at}{r^2}$$

where t = heating time, r = characteristic dimension of food.

The importance of thermal diffusivity has been discussed by Singh (1982). Unsteady-state heat transfer methods for heating, chilling and freezing are discussed by Jackson and Lamb (1981), Lewis (1986) and

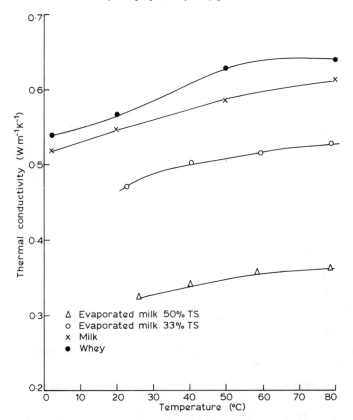

Fig. 9. The relationship between the thermal conductivity and temperature, for a variety of dairy products. (Adapted from data in ASHRAE (1981).)

Cleland and Earle (1982). Jowitt *et al.* (1983) have reviewed the COST 90 work on the thermal properties of foods. The thermal and physical properties of these foods play an important role in heat transfer processes, by affecting heat transfer by conduction and convection. Correlations for heat film coefficients are given by Kessler (1981) and ASHRAE (1981).

Sorption Isotherms

A sorption isotherm is a plot of the equilibrium moisture content of a food against relative humidity at a constant temperature. It can most easily be determined by equilibrating the food with atmospheres of different relative humidity and determining the moisture content, once the food reaches a constant weight. If the test material is initially dry, the isotherm is referred to as an adsorption isotherm; if it is wet, it is known as a desorption isotherm. In most cases the two isotherms are similar, but occasionally hysteresis is found. Usually moisture content is expressed as a percentage on a dry weight basis, i.e. (weight water/weight dry solids) × 100.

These isotherms are useful for determining the lowest moisture content attainable in dehydration processes, and for estimating the water activity of a food from its moisture content.

The water activity (a_w) of a food is a measure of the availability of water as a solvent for reactions in food. It is defined as

$$a_w = \frac{\text{water vapour pressure exerted by food}}{\text{saturated water vapour pressure at the same temperature}}$$

It can be seen that if a food is equilibrated in a sealed container (with a relatively small free volume) and the equilibrium relative humidity of the air equals RH, then $(a_w = \text{RH}/100)$.

Therefore, the sorption isotherm also gives the relationship between the water activity and moisture content. Two foods with the same moisture content will not necessarily have the same water activity, and it is the water activity which affects the chemical, microbial and enzymic reaction rates.

Milk has a water activity almost equal to 1·0, as do most other fresh foods; evaporated milk has a water activity of about 0·98, and ice cream mix (40 per cent solids) a value of 0·970. Sweetened condensed milk, containing 55–60 per cent sugar, has a value between 0·85 and 0·89. Most cheese varieties lie between 0·94 and 0·98, whereas dried milk (1·5–4·5 per cent moisture) lies between 0·02 and 0·2 (Troller and Christian, 1978; Walstra and Jenness, 1984). Humectants are compounds which, when added to foods, depress the water vapour pressure and hence the water activity; sugar and salt are two humectants in common use.

Intermediate moisture foods (IMF) are regarded as foods whose microbial spoilage is prevented by a low water activity, and which do not require rehydration before consumption. The water activity of such foods lies between 0·2 and 0·85.

There are many published equations relating water activity to moisture content, one of the most useful being the Brunauer–Emmet–Teller (BET) isotherm.

$$\frac{a}{m(1-a)} = \frac{1}{m_1 c} + \frac{c-1}{m_1 c} a$$

where a = water activity, m = water content (percent dry weight), c = constant, m_1 = monomolecular layer water content (as above). The factor m_1 is a measure of the monomolecular layer, i.e. the amount of water strongly bound to that material.

Iglesias and Chirife (1982) have compiled sorption isotherms for a wide variety of foods. For each isotherm, they used curve-fitting techniques to select the best two-parameter equation (from a choice of nine) to describe that isotherm. The results for some dairy products are summarised in Table X, together with calculated values of the monomolecular layer value. Some dairy products exhibit a broken isotherm, particularly during the adsorption stage. This is found with substances containing lactose, and is attributed to a transition from the amorphous to the crystalline form. No such change is noted for the desorption isotherm. Figure 10 shows examples for defatted milk and cheese whey.

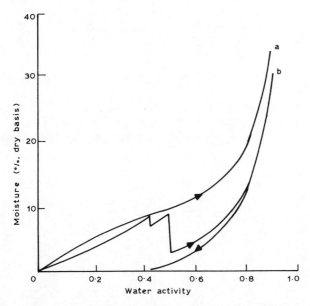

Fig. 10. Sorption isotherms for (a) defatted milk (30°C) and (b) cottage cheese whey (24°C). (Adapted from data in Iglesias and Chirife (1982).)

TABLE X
Summary of sorption isotherm data for dairy products

	Temperature (°C)	Type[E]	a_w range	Equation[D]	B_1	B_2	Monomolecular layer per cent (dry weight)
Cheese Emmental	25	A	0·1–0·8	1	1·1889	5·9967	3·3
	25	D	0·1–0·8	1	1·4435	11·9777	3·7
Cheese Edam	25	A	0·1–0·8	1	1·0668	4·9692	3·3
	25	D	0·1–0·8	1	1·2540	8·5716	3·5
Casein	30	—	0·1–0·8	2	2·1510	0·0044	7·6
β-Lactoglobulin[A]	25	A	0·1–0·8	2	1·5211	0·0166	6·6
Non-fat dry milk[AC]	30	D	0·1–0·8	1	1·9684	72·3080	6·5
	37·8	D	0·1–0·8	1	1·9927	67·8072	6·1
Skim-milk powder[AC]	20	A	0·1–0·8	4	−3·0113	1·3983	2·8
	34	D	0·1–0·8	1	2·0544	54·3870	4·7
	34	A	0·1–0·8	1	1·7764	23·8439	4·0
	14	D	0·1–0·8	1	2·6527	290·2579	

	Temperature (°C)	Type of isotherm[E]	a_w	Equation	$B(1)$	$B(2)$	
Whole milk[A][C]	24·5	D	0·1–0·8	4	−1·4503	3·1356	3·1
	24·5	A	0·1–0·8	1	2·1884	37·9004	3·5
Sweet whey[C]	24·5	D	0·1–0·8	3	3·1279	3·0619	—
Whey protein concentrate[A]	24·0	A	0·12–0·86	1	1·4806	17·6165	4·8
Yoghurt[A]	25·0	A	0·1–0·8	1	1·0529	6·4806	4·1
	45·0	A	0·1–0·8	4	−3·6752	0·2732	3·0

[A] Isotherms are also given at other temperatures, or alternatives are given.
[C] Broken isotherms.
[D] The following equations are relevant (where X = moisture content (% dry weight basis)):

(1) Halsey's equation $a_w = \exp(-B(2)/X^{B(1)})$

(2) Henderson's equation $1 - a_w = \exp(-(B(2)X^{B(1)})$

(3) Iglesias and Chirife's equation $X = B(1)[a_w/(1 - a_w)] + B(2)$

(4) Kuhn's equation $X = \dfrac{B(1)}{\ln a_w} + B(2)$

[E] Type of isotherm: A, adsorption; D, desorption.
Compiled from data in Iglesias and Chirife (1982).

Watt (1983) discusses the use of a three-parameter GAB equation for fitting sorption isotherm data. The results of collaborative trials on determining water sorption isotherms are reviewed by Jowitt *et al.* (1983).

ELECTRICAL AND DIELECTRIC PROPERTIES

Electrical Conductance

Electrical resistance and conductance provide a measure of the ability of material to transport an electric current, resistance usually being preferred for solids and conductance for liquids.

If the total electric resistance (R) is measured across a material, length (L) and cross-sectional area (A), then:

$$R = \frac{\rho_r L}{A}$$

where ρ_r = resistivity (ohm m).

The specific conductance (K) is the inverse of resistivity

$$K = \frac{1}{\rho_r} = \left(\frac{L}{RA} \right) : (\text{mho m}^{-1})$$

The reciprocal ohm is known as a mho or Siemen (S). The specific conductance is measured by resistance techniques, the cell usually being calibrated with a liquid of known specific conductance (Levitt, 1973).

The cgs unit mho cm^{-1} is still commonly encountered, where

$$1 \text{ mho m}^{-1} (\text{S m}^{-1}) = 10^{-2} \text{ mho cm}^{-1}$$

Most dairy products are poor conductors of electricity. Bovine milk has values ranging from 0·004 to 0·0055 mho cm^{-1} (Jenness *et al.* 1974). There is little difference between varieties, as the major contribution arises from potassium and chloride ions. It might be expected that increasing the concentration of milk solids would increase the specific conductance, but the relationship is not so straightforward. The conductance of concentrated skim-milk was shown to increase to a maximum value of about 0·0078 mho cm^{-1} at 28 per cent total solids, after which it decreased; this was explained by the extremely complex salt-balance between the colloidal and soluble phases.

The presence of fat tends to decrease the specific conductance, the specific conductance of milk fat being less than 10^{-16} mho cm^{-1}.

It has been suggested that conductivity measurements may be useful for monitoring processes where such changes occur. For example, in the decationisation of sweet whey using Amberlite IR 120 in its hydrogen form, the conductance rose from 5·2 to 8·15 mmho cm^{-1}. The demineralisation level was 75 per cent and the pH fell from 6·22 to 1·7. Increasing the pH to 3·0 using Amberlite IRA 47 in the OH$^{(-)}$ ion form, decreased the conductivity to 0·165 mmho cm^{-1}, and gave a demineralisation level of 94 per cent (Kanekanian, 1983).

The development of acidity occurring during many fermentations has also been observed to increase the conductivity. Mastitic milk also has an increased conductivity due to its raised content of sodium and chloride ions.

The conductivity of dairy products will be important in ohmic heating processes.

Dielectric Properties

The dielectric properties of foods are currently receiving more attention, due mainly to the advent of dielectric and microwave heating processes. The two properties of interest are the relative dielectric constant (ε') and the relative dielectric loss factor(ε'').

The relative dielectric constant (ε') is a measure of the amount of energy that can be stored when the material is subjected to an alternating electric field. It is the ratio of the capacitance of the material being studied, to that of a vacuum (or air) under the same conditions.

In an AC circuit containing a capacitor, the current leads the voltage by 90°. When a dielectric is introduced, this angle may be reduced. The loss angle (δ) is a measure of this reduction and is usually recorded as the loss tangent (tan δ). Energy dissipation within the dielectric increases as the loss tangent increases. A new factor, known as the dielectric loss factor (ε''), is introduced, which is a measure of the energy dissipated within the sample, where $\varepsilon'' = \varepsilon' \tan \delta$.

During microwave and dielectric heating, the power (P_0) dissipated within the sample is given by:

$$P_0 = 55·61 \times 10^{-14} f E^2 \varepsilon''$$

where P_0 = adsorbed power (W cm^{-3}), f = frequency (H_3), E = electric field strength (volt cm^{-1}). Materials which absorb microwave energy well are known as 'lossy' materials.

Mudgett *et al.* (1974) measured the dielectric properties of milk. At 1000 MHz, the dielectric loss factor increased from 22 to 32 over the temperature range 22–55 °C, whereas at 3000 MHz, the values were lower, decreasing linearly from 18 to 15 over the same temperature range. Loss factors at 300 MHz were much higher.

Values for the dielectric constant and the dielectric loss factors of a wide variety of foods are given by Mohsenin (1984) and Mudgett (1982). Both these properties are affected by the moisture content and temperature of the sample, and the frequency of the electric field.

REFERENCES

Allen, C. R. (1980). *Milk and Whey Powders*, Society of Dairy Technology, Wembley.
ASHRAE (1981). *ASHRAE Handbook, Fundamentals*, ASHRAE Inc., Atlanta.
Arbuckle, W. S. (1977). *Ice Cream*, 3rd Edn, AVI, Westport, Conn.
Badings, H. T., Schaap, J. E., Jong, C. De and Hagedoorn, H. G. (1983a). *Milchwissenschaft*, **38** (2), 95.
Badings, H. T., Schaap, J. E., Jong, C. De and Hagedoorn, H. G. (1983b). *Milchwissenschaft*, **38** (3), 150.
Batty, J. C. and Folkman, S. L. (1983). *Food Engineering Fundamentals*, John Wiley, New York.
Berger, K. G. (1976). *Food Emulsions* (Ed. S. Friberg), Marcel Dekker, New York.
Bertsch, A. J. (1982). *Lait*, **62**, 265.
Bertsch, A. J. (1983). *J. Dairy Res.*, **50**, 259.
Bertsch, A. J. and Cerf, O. (1983). *J. Dairy Res.*, **50**, 193.
Bertsch, A. J., Bimbenet, J. J. and Cerf, O. (1982). *Lait*, **62**, 250.
Bohlin, L., Hegg, P. O. and Ljusberg-Wahren, H. (1984). *J. Dairy Sci.*, **67**, 729.
Bourne, M. C. (1982). *Food Texture and Viscosity*, Academic Press, New York.
Brennan, J. G. (1984). *Sensory Analysis of Foods* (Ed. J. R. Piggott), Elsevier Applied Science Publishers, London.
British Standard Number 734 (1937).
British Standard Number 734 (1959).
Cleland, A. C. and Earle, R. L. (1982). *Int. J. Refrig.*, **5** (3), 134.
Egan, H., Kirk, R. S. and Sawyer, R. (1981). *Pearson's Chemical Analysis of Foods*, Churchill Livingstone, Edinburgh.
Fernandez-Martin, F. (1972). *J. Dairy Res.*, **39**, 65.
Fernandez-Martin, F. (1984). *J. Dairy Res.*, **51**, 445.
Graf, E. and Bauer, H. (1976). *Food Emulsions* (Ed. S. Friberg), Marcel Dekker, New York.
Hamann, D. D. (1983). *Physical Properties of Foods* (Ed. M. Peleg and E. B. Bagley), AVI, Westport, Conn.
Iglesias, H. A. and Chirife, J. (1982). *Handbook of Food Isotherms*, Academic Press, New York.
International Dairy Federation (1981). Document No. 135, Evaluation of the Firmness of Butter.

Jackson, A. T. and Lamb, J. (1981). *Calculations in Food and Chemical Engineering*, Macmillan Press, London.

Jenness, R., Shipe, W. F. Jr. and Sherbon, J. W. (1974). *Fundamentals of Dairy Chemistry* (Ed. B. H. Webb, A. H. Johnson and J. A. Alford), AVI, Westport, Conn.

Jowitt, R. (Ed.) (1980). *Hygienic Design and Operation of Food Plant*, Ellis Horwood, Chichester.

Jowitt, R., Escher, F., Hallstrom, B., Meffert, H. F. Th., Spiess, W. and Vos, G. (Eds) (1983). *Physical Properties of Foods*, Applied Science Publishers, London.

Kanekanian, A. D. A. (1983). PhD thesis, Reading University.

Kessler, H. G. (1981). *Food Engineering and Dairy Technology*, Verlag A, Kessler, Freising, Federal Republic of Germany.

Kessler, H. G. (1984). *Milchwissenschaft*, **39** (6), 339.

Kinsella, J. E. (1976). *Critical Reviews in Food Science and Nutrition*, **7**, 219.

Kjaergaard-Jensen, G. and Neilsen, P. (1982). *J. Dairy Res.*, **49**, 515.

Lamb, J. (1976). *Chem. and Ind.*, **24**, 1046.

Levitt, B. P. (1973). *Findlay's Practical Physical Chemistry*, Longmans, Harlow.

Lewis, M. J. (1986). *Physical Properties of Foods and Food Processing Systems*, Ellis Horwood, Chichester (in preparation).

Loncin, M. and Merson, R. L. (1979). *Food Engineering Principles and Selected Applications*, Academic Press, New York.

Lovell, H. R. (1980). *Milk and Whey Powders*, Society of Dairy Technology, Wembley.

MacCarthy, D. (1984). *Engineering and Food, Vol. 1* (Ed. B. M. McKenna), Applied Science Publishers, London.

Masters, K. (1972). *Spray Drying*, Leonard Hill, London.

Maubois, J. L. (1980). *J. Soc. Dairy Technol.*, **33**, 2, 55.

Mettler, A. E. (1980). *Milk and Whey Powders*, Society of Dairy Technology, Wembley.

Miles, C. A., Van Beek, G. and Veerkamp, C. H. (1983). *Physical Properties of Foods* (Ed. R. Jowitt *et al.*), Applied Science Publishers, London.

Mitchell, J. R. (1980). *J. Texture Studies*, **11**, 315.

Mohsenin, N. N. (1970). *Physical Properties of Plant and Animal Materials, Vol. 1, Structure, Physical Characteristics and Mechanical Properties*, Gordon and Breach, London.

Mohsenin, N. N. (1980). *Thermal Properties of Foods and Agricultural Materials*, Gordon and Breach, London.

Mohsenin, N. N. (1984). *Electromagnetic Radiation Properties of Foods and Agricultural Products*, Gordon and Breach, New York.

Mudgett, R. E. (1982). *Food Technol.*, **36**, 2, 109.

Mudgett, R. E., Smith, A. C., Wang, D. I. C. and Goldblith, S. A. (1974). *J. Fd Sci.*, **39**, 52.

Muller, H. G. (1973). *An Introduction to Food Rheology*, Heinemann, London.

Peleg, M. (1983). *Physical Properties of Foods* (Ed. M. Peleg and E. B. Bagley), AVI, Westport, Conn.

Polley, S. L., Snyder, O. P. and Kotnour, P. (1980). *Food Technol.*, **11**, 76.

Powrie, W. D. and Tung, M. A. (1976). *Principles of Food Science Part I, Food Chemistry* (Ed. O. R. Fennema), Marcel Dekker, New York.

Prentice, J. H. (1954). *Lab. Practice,* **3**, 186.

Prentice, J. H. (1979). *Food Texture and Rheology* (Ed. P. Sherman), Academic Press, London.

Prentice, J. H. (1984). *Measurements in The Rheology of Foodstuffs,* Elsevier Applied Science Publishers, London.

Rha, C. (1975). *Theory Determination and Control of Physical Properties of Food Materials* (Ed. C. Rha), D. Reidel, Dordrecht.

Riedel, L. (1976). *Chem. Mikrobiol. Tech. Lebensm.,* **4**, 177.

Rothwell, J. (1968). *Proc. Biochem.* (1), 19.

Ruegg, M., Moor, U. and Blanc, B. (1983). *Milchwissenschaft,* **38**, 10.

Singh, R. P. (1982). *Food Technol.,* **36** (2), 87.

Smith, K. E. and Bradley, R. L. Jr. (1983). *J. Dairy Sci.,* **66** (12), 2464.

Society of Dairy Technology (1975). *Cream Processing Manual,* Society of Dairy Technology, Wembley.

Taneya, S. (1979). *Food Texture and Rheology* (Ed. P. Sherman), Academic Press, London.

Troller, J. A. and Christian, J. H. B. (1978). *Water Activity and Food,* Academic Press, New York.

Troller, J. A. (1983). *Sanitation in Food Processing,* Academic Press, New York.

Walstra, P. and Jenness, R. (1984). *Dairy Chemistry and Physics,* John Wiley, New York.

Watt, I. C. (1983). *Physical Properties of Foods* (Ed. R. Jowitt, *et al.*), Applied Science Publishers, London.

Weast, R. C. (Ed.) (1982). *Handbook of Chemistry and Physics,* 63rd Edn, CRC Press, Cleveland.

Chapter 6

Modern Laboratory Practice — 1: Chemical Analyses

T. Andersen, N. Brems, M. M. Børglum, S. Kold-Christensen,
E. Hansen, J. H. Jørgensen, and L. Nygaard
A/S N. Foss Electric, Hillerød, Denmark

Since milk is used as human food, there is a natural interest in its composition. This interest may serve several purposes:

— Nutritive value of dairy products.
— Feeding of cows.
— Payment schemes.
— Processing, yield and product control.
— Breeding of cows.
— Control of hygiene and health status of herds.

A multitude of chemical analytical methods — gravimetric, volumetric, titrimetric — have been developed. A few have been established as reference methods, but others have been found suitable for routine analysis. A general feature of the chemical methods is, however, that analysis of large numbers of samples is inconvenient, instantaneous results are impossible, and cost per analysis is relatively high. Automation of these chemical methods has to some extent remedied the disadvantages, but the biggest leap forward has been made with the appearance of instruments based on the measurement of physico-chemical properties of the milk constituents, with no or very little sample preparation and handling. Typical examples are the infra-red (IR) and near-infra-red (NIR) instruments. These modern analytical methods perform quick, accurate and cheap analyses, making possible extensive control of cows, milk, processing and products.

308   *T. Andersen et al.*

CHEMICAL ANALYSIS

First a short review of traditional chemical methods will be presented. It should be noted that well-established procedures, like the Röse–Gottlieb method for fat determination and the Kjeldahl method for quantitative determination of total nitrogen and estimation of protein, are still the ultimate methods to be consulted when evaluating new instrumental methods. Calibration and performance checks of modern instruments also require the use of chemical reference methods. A selection of FIL-IDF and AOAC approved methods is listed in Table I.

TABLE I
A selection of (a) FIL-IDF standards and (b) AOAC methods

(a) FIL-IDF standards

Component	FIL-IDF	Principle	Status
Fat	1B:1983	Röse–Gottlieb	Reference
	22A:1983	Röse–Gottlieb	Reference
	16A:1971	Röse–Gottlieb	Reference
	9A:1969	Röse–Gottlieb	Reference
	13A:1969	Röse–Gottlieb	Reference
	116:1983	Röse–Gottlieb	Reference
	5A:1969	SBR	Reference
Protein	20:1962	Kjeldahl	Reference
	25:1964	Kjeldahl	Reference
	92:1978	Kjeldahl	Reference
	98:1980	Dye-binding	Routine
Lactose	28A:1974	Chloramine-T	
	79:1977	Enzymic	
	43:1967	Cuprous oxide	
Water	21A:1982	Drying oven	Reference
	4A:1982	Drying oven	Reference
	15A:1982	Drying oven	Reference
Somatic cells	Doc. 168:1984	Microscopic count	Reference
	Doc. 168:1984	Coulter Counter	Routine
	Doc. 168:1984	Fossomatic	Routine
	Doc. 168:1984	Auto-Analyzer	Routine

(b) AOAC methods

TABLE 1—*contd.*

Components	AOAC	Principle	Status
Fat, protein, lactose, water	16.078–.087	Infra-red absorption	IRMA Milko-Scan Multispec
Fat	16.063–.068 (+1st suppl.)	Light scatter	Milko-Tester Anritsu Milk Checker
Protein	cfr. 7.021–7.024 16.042–16.045	Kjeldahl Dye-binding	Kjel-Foss Pro-Milk
Somatic cells	46.086–46.104 46.105–46.109	Optical Fluorescence	Auto Analyzer Fossomatic

FIL-IDF: Federation Internationale de Laiterie — International Dairy Federation, Bruxelles, Belgium. Documents and standards, currently revised. AOAC: Official Methods of Analysis of the Association of Official Analytical Chemists, 13th edn, 1980, Washington, DC, USA.

Fat Analysis

Reference methods

The Röse–Gottlieb principle, a gravimetric method, is accepted as the reference method for fat, and all other methods should be checked against it. The procedure is, in short, as follows:

The sample is weighed precisely into a test tube. Ammonia solution and ethanol are added, and the milk fat is extracted two or three times with a mixture of diethylether and light petroleum. The combined extracts are then evaporated, and the weight of the oven-dried residue — which must be soluble in light petroleum — is determined.

The Mojonnier and Schmid–Bondzynski–Ratzlaff (SBR) methods are similar to Röse–Gottlieb. SBR is preferred when increased amounts of free fatty acids from decomposed milk fat are present, due to processing or to aging of the milk. In this method, hydrochloric acid replaces the ammonia solution.

The above methods are rather slow and time-consuming, and special attention must be paid to the highly flammable solvents used. Results are available after two days, although using a centrifuge to separate the water from the solvent reduces the delay.

Routine methods

The Gerber method, a volumetric method, is often used for routine analysis, using graduated and calibrated butyrometers.

Sulphuric acid and amyl alcohol are pipetted, along with the milk sample, into a butyrometer, which is then shaken and centrifuged. The volume of the separated upper layer of milk fat is read from the butyrometer scale at an elevated temperature.

The Babcock method, which does not involve amyl alcohol, is a similar, widely used method. These procedures provide quick results, and are reliable if carefully performed; deviations from the reference method are well documented. A disadvantage is the use of sulphuric acid, and the necessity of cleaning the butyrometers.

Protein Analysis

Reference method

The work of Johan Kjeldahl, described in 1883, is used as the basis for determining total nitrogen. One variant of this procedure, adopted by the IDF as a reference method, is as follows:

A precisely weighed quantity of milk is placed in a Kjeldahl flask containing mercury oxide (catalyst) and potassium sulphate (increasing boiling point). Concentrated sulphuric acid is added, and the mixture is heated and boiled until the sample is completely digested yielding NH_4^+. The flask is cooled, and the solidified contents are then redissolved in water. An excess of sodium hydroxide solution (including a sulphide to precipitate the mercury) is added, and the released ammonia is distilled via a condenser into a boric acid solution containing an acid–base indicator. The collected ammonia is titrated with standardised hydrochloric acid. The amount of ammonia present, and thus the amount of nitrogen, can then be calculated. For milk, the empirical factor of 6·38 is used to estimate the percentage of protein in the sample.

A manual Kjeldahl analysis takes several hours, and care must be taken to avoid accidents with the corrosive chemicals involved.

Routine method

The dye-binding method is suited for routine analysis, and the dyestuff used is Amido Black 10B, 'milk testing quality'. Raw, whole or homogenised milk or skim-milk with 2–5 per cent protein can be analysed directly. The principle is as follows:

Amido Black in excess forms an insoluble dye–protein complex by reaction with basic amino-acid residues (e.g. lysine with an ε-amino group) in the proteins at a buffered pH of 2·4. The light absorption of the buffered dye solution is recorded.

Milk and the dye solution are mixed, and the insoluble complex is filtered off (or, alternatively, centrifuged). The light absorption of the filtrate containing an excess of dye is measured. The percentage of protein is obtained from a calibration curve, which is prepared by determining protein content by the Kjeldahl method and corresponding read out of light absorption.

Although the Kjeldahl method determines total nitrogen and the dye-binding method quantitates basic amino-acid residues in protein (and not non-protein nitrogen), good correlation can be obtained between the two methods.

Lactose Determination

Many methods are available for the estimation of lactose. Most of them presuppose that no other interfering sugars are present, and some commonly used methods illustrate this.

Titrimetric method with chloramine-T
This method is based on the reducing property of lactose. Fat and milk proteins are precipitated with a tungstic acid reagent. To part of the lactose-containing filtrate, solutions of potassium iodide and a known quantity of chloramine-T are added, and the mixture is left to react. Hydrochloric acid is added to stop the reaction, and the unreacted chloramine-T is titrated with standardised sodium thiosulphate solution, using soluble starch to determine the end-point.

This method is applicable to normal fresh milk, and requires the minimum of laboratory equipment.

Copper reduction method (Munson–Walker method)
This method is also based on the reducing property of lactose. Ground cheese is dissolved in water, and the fat and protein are precipitated with Carrez reagents (zinc sulphate and potassium ferrocyanide solutions). Part of the lactose-containing filtrate is added to Fehlings solution

(cupric sulphate/alkaline tartaric acid solution). The mixture is then heated and boiled under strictly controlled conditions to precipitate cuprous oxide. The cuprous oxide is collected on a filter crucible and dried. By weighing the amount of cuprous oxide, the content of lactose can be calculated using an empirical factor from a table. This method is applicable to other milk products that contain no reducing sugars other than lactose.

Polarimetric method

This method is based on the ability of lactose to rotate polarised light. The fat and milk proteins are precipitated (acidic mercuric salt solution), and the optical rotation of the filtered solution, which is proportional to the lactose concentration, is measured. The lactose content can be calculated using the specific rotation. The method can be used for milk to which no other optically active components (sugars) have been added.

Enzymic method

The lactose in the sample is hydrolysed to glucose and β-galactose in the presence of a β-galactosidase (Gal) enzyme. The β-galactose which is formed is oxidised by nicotinamide adenine dinucleotide (NAD^+) in the presence of β-galactose dehydrogenase (Gal-DH). The amount of the reduced form of NAD^+, NADH — as measured by an increase in extinction at 340 nm — is proportional to the amount of β-galactose.

$$(1) \quad \text{lactose} + H_2O \xrightarrow{\text{(Gal), pH 6·6}} \text{glucose} + \beta\text{-galactose}$$

and

$$(2) \quad \beta\text{-galactose} + NAD^+ \xrightarrow{\text{(Gal-DH), pH 8·6}} \text{galactonic acid} + NADH + H^+$$

The method is applicable to products to which sugars other than β-galactose have been added. Added β-galactose can be determined by reaction (2), but L-arabinose interferes.

Products such as milk, condensed milk, cheese and ice cream can be analysed by dissolving the product, and precipitating the fat and protein with perchloric acid, trichloroacetic acid, or Carrez reagents.

Another enzymatic scheme can also be used:

$$(1) \quad \text{lactose} \longrightarrow \text{glucose} + \beta\text{-galactose} \qquad (\beta\text{-galactosidase})$$

$$(2) \quad \text{glucose} \longrightarrow \text{glucose-6-phosphate} \qquad (\text{hexokinase, ATP})$$

(3) glucose-6-phosphate + NADP⁺ ⟶ 6-phosphogluconate
  + NADPH   (glucose-6-phosphate dehydrogenase)

NADPH-extinction is then measured; added glucose can be determined by using reactions (2) and (3).

Total Solids and Water

The sample is weighed into a flat-bottomed dish and dried in an oven at 102 °C for several hours until constant weight. The content of water or of total solids is then calculated by simple subtractions.

Drying time is dependent on the product. Dilution with water, use of sand for even spreading, or preheating of the sample on a steam bath may be necessary (cheese and sweetened condensed milk).

Somatic Cell Count

A direct microscopic count of stained somatic cells (leucocytes and epithelial cells) is recommended by IDF as a reference method. The procedure is, in short:

10 μl of the milk sample is spread in two thin films (of 1 cm² each) on a microscopic slide and the films are dried. The fat globules are removed, and the cells are fixed and stained with methylene blue, all in one process. After being rinsed in water and dried, the prepared slide is placed under a microscope with high magnification. By counting the number of stained cells (nuclei) in a large number of microscopic fields, the number of somatic cells per ml milk sample can be calculated. A coefficient of variation of 5 per cent or less at a level of 500 000 cells per ml can be obtained by trained personnel for reference purposes.

MODERN ANALYTICAL METHODS

Today, most traditional, chemical procedures for analysis of the main milk components can be replaced by quick, easy and reliable instrumental methods. Attention will be focussed here on fat and protein analyses, and the count of somatic cells. The choice of instrumentation will be based on considerations such as: price of instrument, number of samples per hour, price per sample, auxiliary equipment required, method of presenting the sample to the instrument, output of data, staff

TABLE II

Some instruments available for the routine measurement of the components of milk and milk products

Instrument	Principle	Components	Manufacturer
Milko-Tester	Light scatter	Fat	1
Milk Checker			2
Kjel-Foss	Kjeldahl	Protein	1
Pro-Milk	Dye-binding	Protein	1
Milko-Scan	Infra-red	Fat, protein,	1
Multispec		lactose and	4
IRMA		water	3
Instalab	Near infra-red	Fat, protein,	6
Infra Analyzer	Reflectance	lactose and	5
Neotec		water	7
Fossomatic	Fluorescence	Somatic cells	1
Coulter Counter	Conductivity		8

1. A/S N. Foss Electric, Hillerød, Denmark.
2. Anritsu Electric Co., Ltd., Kanagawa, Japan.
3. Sir Howard Grubb Parsons & Co. Ltd, Newcastle upon Tyne, England.
4. Shields Instruments Ltd., York, England.
5. Tecnicon Industrial Systems, Tarrytown, NY, USA.
6. Dickey-john Corp., Auburn, IL, USA.
7. Neotec Corp., Silver Springs, MD, USA.
8. Coulter Electronics Ltd., Luton, England.
N.B. This list is not exhaustive.

required, and ease of repair and service. An additional factor is the extent to which the instrument must be regularly checked to ensure:

Accuracy: Agreement with a reference method must be acceptable. Recalibration may be necessary, for instance due to seasonal variations in the milk constituents.

Precision, as expressed by repeatability (repeated measurements on the same sample) and reproducibility (between laboratory agreement) must be acceptable (Grappin, 1984).

Some of the options are given in Table II, but before going into detail about the various instruments, a few basic concepts should be reviewed, namely the electromagnetic spectrum (Fig. 1), and the relationships expressed by the Lambert–Beer Law (Figs. 2 and 3):

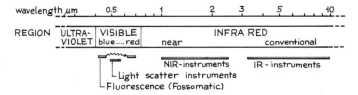

Fig. 1. Part of the electromagnetic spectrum: logarithmic scale.

$I = I_0 \times 10^{-\varepsilon \cdot l \cdot c}$ or $A = -\log_{10} I/I_0 = \varepsilon \cdot l \cdot c$, $T = I/I_0$

I_0 = intensity of incident light (radiation)

I = intensity of transmitted light (radiation)

$(I_s$ = scattered light, I_r = reflected light)

ε = (molar) extinction coefficient, absorption coefficient (varies with wavelength, solvent, temperature)

c = concentration of the solute

l = path (cell) length

A = extinction, absorbance, absorption

T = transmission, transmittance, transparency

Instruments for Fat Analysis

A widely used — and AOAC-approved — method, based on measuring the light scatter caused by milk fat globules in a homogenised milk sample, is represented by the Milko-Tester.

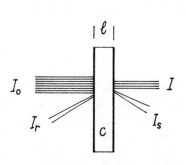

Fig. 2. Cuvette with milk, illustrating light reflection (I_r), scatter (I_s) and transmission (I).

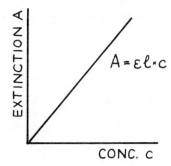

Fig. 3. Graphic representation of the Lambert–Beer Law.

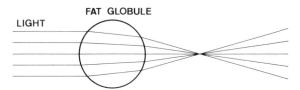

Fig. 4. Light refraction in a fat globule (idealised).

Light scatter photometry for fat determination

When visible light passes through a layer of milk, transmission is highly affected. This fact has motivated researchers to isolate the effect of fat content from the effects of other factors, in order to obtain a photometric test for fat in milk to replace the traditional Röse–Gottlieb and Gerber methods. Butterfat absorbs a negligible portion of energy from visible light, and the effect of fat, as measured by a visible-light photometer, is essentially due to refraction and reflection.

Refraction occurs when light passes from one medium into another with a different refractive index (RI). For butterfat the RI is about 1·454, while the surrounding medium with lactose, proteins and salt dissolved in water has an RI of about 1·33. This makes each fat globule behave like a condenser lens (Fig. 4).

When a thin layer of milk is placed in the light path in a photometer, the transmitted light forms a cone-shaped bundle of beams. The photocell receives a certain central part of these, and the amount is determined by the angle of acceptance of the photocell seen from the cuvette; the outer part of the bundle is lost. This produces a photometer response that reflects the presence of the fat globules (Fig. 5). Casein micelles also affect light scatter, but interference from casein micelles can be eliminated by solubilisation with a chelating agent (EDTA) at pH 9·5–10. This is uncomplicated, and a test on skim-milk confirms that this procedure works well. Even for a condensed skim-milk, casein scatter is completely suppressed.

The main problem in converting light scatter effects into a reliable photometric method for fat determination has been to understand and control the relationship between actual fat content, light scatter, fat globule size distribution, wavelength of the light, and the RI's for fat and milk serum. Within this complex problem, the effects of globule size distribution have been the subject of intensive research, theoretical as well as practical. Globule size distribution in raw milk varies significantly between breeds and between individual animals, with fat globule sizes of up to 10 μm. When light of, say 0·6 μm wavelength travels through globules of diameters of several wavelengths, it is a consequence of the RI difference between fat and serum that the light shows a phase shift depending on the distance it travels. Goulden (1958) has illustrated this by a graph of the scattering coefficient for monodisperse emulsions (Fig. 6).

Haugård and Pettinati (1959) used a bench homogeniser for raw milk to produce fat globules with diameters corresponding to the first linear part of this curve. They used one photometer with a small angle of acceptance and another one with a wide angle of acceptance; a nomogram was constructed to determine fat content and an homogenisation index.

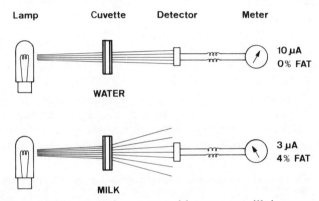

Fig. 5. Light scatter photometer with water or milk in cuvette.

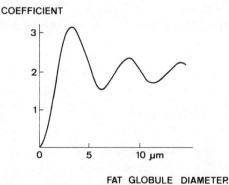

FAT GLOBULE DIAMETER

Fig. 6. Scattering coefficient at wavelength 0·5 μm and RI = 1·33/1·45 (not to scale). (Goulden, 1958.)

A photometer combining a wide angle of acceptance and a motor operated high-pressure homogeniser adjusted for optimal suppression of the globule size effect has been developed, and this formed a solid basis for the family of 'Milko-Testers', of which the first model was introduced in 1964.

The light scatter fat tester

A few years later improved models were marketed: the Milko-Tester Automatic for payment analysis of herd and cows' milk (180 samples per hour), the Milko-Tester Mk III for raw milk and dairy products, and the Milko-Tester Control for on-line standardisation in the dairy plant.

These instruments illustrate how basic photometric principles can be realised in practice.

(1) A 1·5 ml milk sample is diluted tenfold with a 5 per cent EDTA solution at pH 9·5, and this perfectly suppresses casein scatter, which otherwise would interfere with the fat determination. The dilution also makes it possible to use a wide light path (0·4 mm) in the cuvette, which can then be used for butterfat contents of up to 10 per cent without compromising the validity of the Lambert–Beer Law. The combination of a tenfold dilution by a basic solution and the 0·4 mm space of the cuvette eliminates the risk of the cuvette being blocked by acidified milk samples.

(2) High pressure homogenisation is carried out in four consecutive steps of 40 kp each at a stabilised temperature of 50 or 60 °C. The Milko-Tester Minor has a simplified one-stage, hand-operated homogeniser which gives slightly reduced performance.

The homogenisation changes the globule size distribution:

Parameter	Raw milk	Homogenised milk
Average size, d_{vs}, μm	3–4	0·3–0·4
Standard deviation, μm	1–2	0·2–0·25

This serves two purposes: variations in globule size distributions between individual animals and between breeds are suppressed, and all globules are reduced to a size where they contribute to light scatter with the same weight as they contribute to fat content, which is the meaning of the linear part of the curve shown in Fig. 6. This agreement holds for all kinds of raw milk samples, but for homogenised products, some Milko-Testers provide a calibration that is different from its raw milk calibration by about −0·3 per cent fat. This calibration is essentially unaffected when products from different homogenisers are tested, even if they have been homogenised at different pressures.

(3) Light scatter is determined by a photometer with a wide angle of acceptance, empirically designed for minimising the effect of globule size variations. Monochromatic light is not necessary, but a blue filter is used in some models to keep near-infra-red radiation from disturbing the stability of the photocell.

(4) Electronic data processing is used to produce the log function of the Lambert–Beer Law, to provide adjustments of linearity and sensitivity, and for analog-to-digital conversion.

Performance and limitations
A repeatability of 0·01–0·02 per cent fat is typical, and accuracy, compared to the Röse–Gottlieb reference method, is typically 0·06 per cent fat for single cow samples, and 0·02 per cent for bulk milk samples. For high-fat products, standard deviations of 1·0–1·5 per cent of the actual fat content can be expected. The Milko-Tester Minor has slightly higher standard deviations for precision and accuracy.

Sample preparation may be necessary: acidified products must be neutralised, condensed milk and cream must be diluted, powder must be dissolved, and cheese must be grated, dissolved and homogenised. In some cases, specific calibrations may be needed. This requirement applies first of all not only to homogenised products, but also to products containing additives that directly influence the optical density, such as chocolate, coffee and fruit flavours. Vegetable emulsions can be tested if a specific calibration is applied, but mixes of butterfat and vegetable fat can only be tested if the ratio between them is known and fixed. Air bubbles in the prepared samples must be avoided.

The accuracy and stability of the calibration are affected by the refractive index (RI) of the butterfat. The major part of the standard deviation of 0·06 per cent for the accuracy of single cow samples is due to RI variations, and this explains why accuracy is better for herd milk, and much better for bulk milk. The practical aspect of this is that seasonal RI variations are reflected as calibration problems for the fat tester. A typical winter RI can be as low as 1·4530, while the summer RI may be 1·4560. For a typical raw milk with 4 per cent fat, this means a difference of nearly 0·2 per cent fat between summer and winter calibrations, making it necessary to re-calibrate the instrument each spring and autumn. This is just as important for bulk milk tests as for herd and single cow milk tests.

A minor limitation of the light scatter fat tester is that it measures globular fat only, but for good milk samples, it is no problem to calibrate the fat tester using Gerber as a reference method. However, for samples that have been stored for a long time with inadequate preservation, the appreciable amounts of free fatty acids that can form are sufficient to explain the dramatic deviations that may be observed between fat tester and Gerber analyses. The light scatter fat tester is excellent for skimmilk tests in the dairy plant to check separator performance.

Instruments for Protein Analysis

One type of instrument is an automated or semi-automated version of the Kjeldahl procedure which measures total nitrogen.

Another type of instrument is based on the dye-binding procedure: Amido Black forms an insoluble complex with protein. Although an indisputable, versatile analysis method, the Kjeldahl analysis in dairy chemistry is mainly used to obtain reference values for protein. For routine analysis, quick and reliable alternative methods designed especially for milk analysis are available. Dye-binding is one alternative, and another, infra-red absorption, will be described later.

Automated Kjeldahl

Operation
In the Kjel-Foss Automatic (Fig. 7), six Kjeldahl flasks are placed in a carousel in one part of the instrument. A titration unit, containers with the necessary solutions of chemicals, and the electronics for operation occupy the other part. The Kjeldahl flasks rotate among the six positions with a holding time of three minutes at each position:
Position 1. Add three tablets containing catalyst and potassium sulphate to flask. Press dispenser arm for addition of approx. 10 ml hydrogen peroxide and 10–16 ml concentrated sulphuric acid (the volume depends on the fat content of the sample). Add precisely weighed sample: milk is added from a syringe; solid dairy products are wrapped in nitrogen-poor paper and dropped in. Close flask with lid.
Positions 2 and 3. Sample is digested at about 410°C. Heat is supplied by a gas burner in each position and after 2 × 3 min, the sample is fully

Fig. 7. Kjel-Foss Automatic for Kjeldahl analysis. (Courtesy of A/S N. Foss Electric, Denmark.)

digested. A fan at position 4 cools the flask and its contents. At the end of the 3 min period, the lid is opened and 140 ml of water is added for dilution.

Position 5. A solution of sodium hydroxide and sodium thiosulphate (50 ml) is dispensed into the flask. Steam is led to the flask and the ammonia is distilled through a condenser into a flask containing 50 ml of acid–base indicator solution. The indicator, which is originally red, gradually turns green as ammonia is absorbed. The original red colour is recorded by a photocell. When a colour change is registered, 1 per cent sulphuric acid will be added until the original red colour is regained. The amount of dispensed sulphuric acid is registered, and from this figure, the total nitrogen is calculated and displayed or printed.

A choice is possible between '0·5 g' or '1·0 g' as the sample size, and the results can be expressed in percentage nitrogen or percentage protein. Three built-in factors for calculating protein are available (one for dairy products, one for food products and grains, and one especially for wheat); a fourth may be defined by the user for special purposes.

The Kjeldahl flask is emptied in position 6, ready for another sample in position 1. The total time for analysis of one sample is 12 min. With the Kjel-Foss running continuously, a result will be displayed every 3 min corresponding to 100 samples every 5 h. Start-up and calibration last about 30 min.

The Kjel-Foss Automatic is suitable for the analysis of a great number of samples, but a semi-automated system may suffice if smaller numbers of samples are involved. Several units are available for facilitating the Kjeldahl analysis, including digestion units, distillation and titration units. A very long list of products ranging from earth and blood to food, grain and animal feeds can be analysed, but here we are mainly interested in milk and dairy products. Although protein, and thus nitrogen, is rather low in milk compared to other products, precision is good (CV less than 1 per cent), and accuracy and correlation to the manual, classical Kjeldahl method is completely satisfactory.

Most improvements of the Kjel-Foss Automatic have focussed on reducing safety risks, and on the growing concern about pollution by mercury. An alternative catalyst based on antimony (although not a completely harmless replacement) has been developed, and its properties are comparable to those of the mercury catalyst.

Semi-automated dye-binding method

Principle
This is the dye-binding method, using Amido Black, described in the 'Chemical Analysis' section.

Operation
The equipment, a Pro-Milk, is shown in Fig. 8. The procedure is, in short: turn the mixing assembly into the neutral position; place a piece of filter paper on the support and mount the transparent tube on top.

A milk sample (1 ml) is drawn-up into the syringe and dispensed into the tube. Dispense dye solution (20 g) into the tube, close the tube with a stopper, and tilt the mixing assembly two or three times for thorough mixing. The mixed content is filtered by applying air pressure to the tube with the rubber ball. The pressure is then released, and the protein content — as calculated by the reduction in blue colour — is read by a photometer in the measuring unit. The result, in percentage protein, is displayed on the read-out unit.

After removal (and cleaning of the tube and filter paper), the assembly is ready for a new sample.

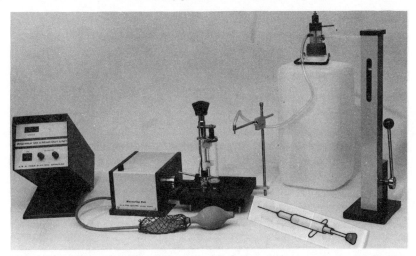

Fig. 8. Pro-Milk equipment for semi-automated determination of the protein by dye-binding. From left to right: read-out unit, percentage protein displayed; measuring unit with photometer; mixing and filtering assembly with tube and filter assembly and rubber ball; syringe for 1 ml milk sample; dispenser and container for dye solution. (Courtesy of A/S N. Foss Electric, Denmark.)

The Pro-Milk must be regularly checked against a standard dye solution and the actual working solution. Its measuring range is 2·0–5·5 per cent protein.

Products and performance
Directly applicable products are raw milk, homogenised whole milk and skim-milk. Other milk products like whey protein, ice cream mix and yoghurt can be examined following appropriate sample preparation such as dilution, precipitation or neutralisation (A/S N. Foss Electric). Investigations of the possibilities for applications have been carried out by the National Dairy Research Centre, Ireland (O'Connell and McGann, 1972; McGann and O'Connell, 1972; McGann *et al.*, 1972a; McGann *et al.*, 1972b) and the Institut für Chemie der Bundesanstalt für Milchforschung, Kiel, Germany (Thomasow *et al.*, 1971).

Repeatability, expressed as a coefficient of variation, is better than 0·5 per cent on milk, and the standard deviation for accuracy against Kjeldahl is 0·045 per cent. Fifty to sixty milk samples can be analysed per hour by trained personnel. Fully automated instruments based on the dye-binding principle were formerly produced by A/S N. Foss

Electric, but development of IR instruments for measuring fat and protein (and lactose) have made these superfluous.

Multi-component Analysis

Two types of instrument are available for multi-component analysis of dairy products.

One type, the infra-red (IR) instrument, makes use of the characteristic bands of absorption of the main milk components in the 3–10 µm wavelength range of the infra-red spectrum. Another type, the near-infra-red reflectance (NIR) instrument, is based upon diffuse reflectance in the near-infra-red range (more precisely 1–2·5 µm), of the spectrum. IR instruments are used for liquid or liquefied product applications, mainly raw milk, while NIR instruments may be the method of choice for solid materials, such as milk powders. The most versatile models of both types of instruments can determine both fat, protein, lactose and water/solids.

Infra-red instruments

The development of infra-red (IR) instruments for the determination of milk composition on a large scale started in 1961, when Goulden applied for a United Kingdom patent for the quantitative analysis of milk by infra-red absorption (Goulden, 1961). Goulden's work provided the basis for the first commercial instrument, the IRMA (Infra-red Milk Analyzer) developed by the firm Sir Howard Grubb Parsons & Co. Ltd. The method employed in the IRMA was officially adopted by the Association of Official Analytical Chemists (AOAC) in 1972 (Biggs, 1972), and since then, a number of IR instruments have been developed and marketed. Infra-red instruments are now used all over the world, both for centralised payment and dairy herd improvement tests, and for the analysis of milk and milk products at dairy plants.

Infra-red absorption

Vibrational energy changes in molecules occur in the infra-red region of the electromagnetic spectrum. The most useful vibrations, from the point of view of the organic chemist, occur in the narrow range of 2·5–16 µm (1 µm = 10^{-6} m). Functional groups have characteristic vibration frequencies within well-defined regions of this range. All infra-red milk analysers are basically spectrophotometers which measure the amount of infra-red energy transmitted through a sample. When the molecule is exposed to infra-red radiation at a frequency similar to that of the vibration within the atomic group, infra-red energy will be absorbed as a result of this oscillation.

The main interference effects of the IR method are due to the fact that water has a relatively high extinction coefficient in the IR region of the spectrum. Water is probably the strongest known infra-red absorbing compound, and even thin films show intense absorption throughout most of the 2·5–16 μm region.

Figure 9 shows three spectra recorded in a classical double beam infra-red spectrophotometer. Figure 9(A) is the spectrum obtained from water using a narrow path length (39 μm) cell in the sample beam and atmospheric air in the reference beam. Figure 9(B) shows the spectrum obtained from milk under the same conditions, and as can be seen, the spectrum of milk is almost identical to that of water. This is due to the fact that 85 per cent of milk is water. However, small differences can be seen around 3·5, 5·7, 6·5 and 9·6 μm. These differences are magnified in the last spectrum, Figure 9(C), which is a difference spectrum for milk versus water. This spectrum is obtained by placing a milk sample in one beam of the spectrophotometer and a water sample in a matching cell in the other beam. The instrument automatically subtracts the water absorption from the milk spectrum. The resulting spectrum shows the bands due to absorption by fat, protein and lactose at 3·5 and 5·7, 6·5 and 9·6 μm respectively.

The alternative wavelength at 3·5 μm for fat determination has been introduced recently (Nexø *et al.*, 1981).

Interfering components and water displacement

The main interfering effect of the IR method is explained by the high extinction coefficient of water in the whole IR region of the spectrum. Changes in the level of water concentration affect the readings obtained from all components, but since water concentration is dependent on the amount of other components present, equations which define the concentration of one component as a function of the readings obtained from all three components have been successful in correcting for these interference effects. For example, take the equation $F = bF_i + cP_i + dL_i$, where F is the fat content of the sample, and F_i, P_i and L_i are the instrument signals for fat, protein and lactose. As it reflects the absorption of energy by fat, b is large, whereas c and d are small, because

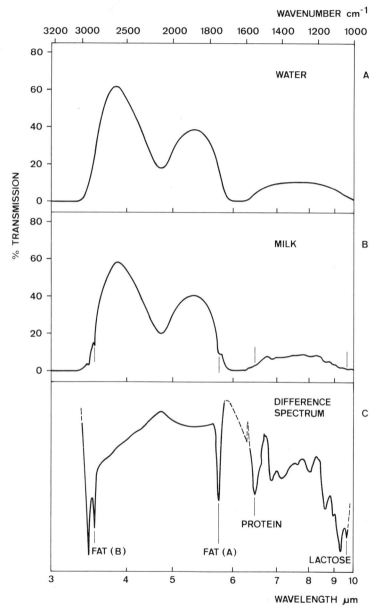

Fig. 9. IR spectra of water versus air, milk versus air, and milk versus water (see text).

they correct only for the interference effects of protein and lactose respectively.

Water absorption is very strong in the spectral region of interest. So the net result at any selected component wavelength is usually that the other two components, because they displace a more strongly absorbing medium, impart a negative displacement or absorptivity error to the result. There are two exceptions to this rule. One is protein absorption at the lactose wavelength, which is slightly stronger than the water it displaces (Shields, 1975). The other exception is at 3·5 μm, where the absorption of water is not so strong and the secondary components (protein and lactose) increase the total absorbancy additively (Sjaunja, 1982). Mineral components also interfere, partly because they displace water, but also because some of them can change the absorptivity of water. However, minerals are present in low and almost constant concentration in normal cow's milk, and can be accounted for when the instrument is calibrated.

Light scatter

In addition to attenuation by absorption, a beam of radiation passing through milk is attenuated by scattering from both fat globules and protein micelles (Goulden, 1964). Light scatter is proportional to $(d/\lambda)^4$, where d is the particle diameter and λ is the wavelength (Jøndrup, 1980). This means that the light scatter increases with increasing particle size and decreasing wavelength (Shields, 1975). To avoid light scatter, the particle diameter must as a rule be less than 1/3 of the wavelength, which for milk means a particle diameter less than 1·5 μm (Jøndrup, 1980). Small and homogeneous particle sizes are obtained by homogenising the milk samples before the absorption is measured. The most effective homogenisation is achieved when the milk fat is melted before the sample is introduced to the instrument. This is one of the reasons why samples must be pre-heated to 40°C before they are analysed. Another reason for pre-heating is that it can be difficult to make representative subsamples of cold raw milk where the milk fat often gathers at the top of the sample.

Instrumentation

There are different ways to compensate for the strong water absorption in the milk. Here three principles will be mentioned, and the types of instrument to which they are applied are briefly described. The different principles are shown schematically in Fig. 10.

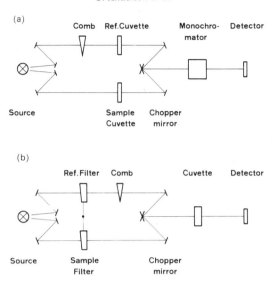

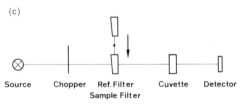

Fig. 10. Optics in various IR instruments. (a) Single-wavelength double-beam double-cell; (b) dual-wavelength double-beam single-cell; (c) dual-wavelength single-beam single-cell.

Single-wavelength double-beam double-cell principle

The first infra-red milk analyser, the IRMA, was based on a conventional double-beam spectrophotometer modified to automatically select, in turn, characteristic absorption wavelengths for the components in the milk to be analysed. In this way, the absorption of a milk sample at a specific wavelength was compared to that of water at the same wavelength.

In the IRMA, the energy from the infra-red source is divided optically into two beams: one, the sample beam, passes through a cell containing milk; the other, the reference beam, passes through a matching cell containing distilled water. The two beams are then recombined at the entrance slit of a monochromator by a reciprocating mirror which transmits the energy of the alternate beams at a frequency of 10 Hz. The monochromator, which includes a diffraction grating and a KBr prism, filters the energy into narrowly selected ranges of wavelengths, which are focussed onto a thermocouple that serves as a detector. Because of the difference in absorption by milk and water, there is an alternating energy level in the radiation reaching the detector. The resultant tiny voltage is amplified, rectified and used to drive a servo motor which moves an attenuating comb into or out of the reference beam until equal energy is detected in the two beams. The distance moved by the comb reflects the concentration of the milk component in the sample cell. A potentiometer fixed to the shaft of the attenuator translates the required beam attenuation into a corresponding DC voltage, which can be used to produce a read-out.

Dual-wavelength double-beam single-cell principle
In contrast to the principle mentioned above, the dual-wavelength system does not use a water cell for reference. Instead, a wavelength is chosen close to the sample wavelength where neither fat, protein nor lactose contributes to the signal.

In 1975, the first single-cell dual-wavelength infra-red milk analysers were introduced (Milko-Scan 203 and 300, Fig. 11). The same principle is utilised in the Multispec analysers (Shields, 1982). In these types of instruments, the radiation from the source is also divided into a sample beam and a reference beam. The beams are passed through interference filters — a sample filter and reference filter, respectively — which select narrow bands of wavelengths. The sample wavelength is characteristic for the actual component being measured, and the reference wavelength is, preferably, close to this, but not coincident with either fat, protein or lactose absorption.

The two beams are recombined using a rotating chopper mirror, and then passed through a narrow-path length cuvette containing the homogenised milk sample. The transmitted energy is collected by a mirror and focussed onto the detector. A servo system and an attenuating comb are used to balance the energy of the two beams, just as in the IRMA, and the signal generated is proportional to the movement of the comb. Following a logarithmic-to-linear conversion, the signals from each component are stored and used in pre-determined equations to electronically estimate component concentrations. There are several advantages of this system compared to the IRMA. One is that there is a

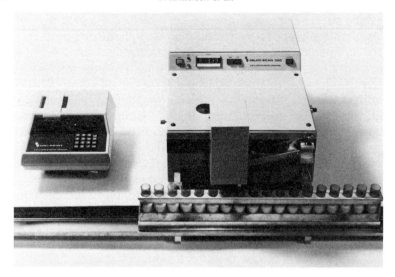

Fig. 11. 1975 version of IR instrument (Milko-Scan 300) for automatic fat and protein analysis. Measures 300 samples per hour. (Courtesy of A/S N. Foss Electric, Denmark.)

water displacement' effect in both beams, which to a great extent balances out variations in the water or solvent content of the sample. Another advantage of the dual-wavelength system is that the measurements are less sensitive to the degree of homogenisation, because light is scattered in both beams. However, the refractive index of butterfat varies in the region of an absorption band, and as light scatter depends on the refractive index, a difference arises between the sample and the reference beam, depending on the fat globule size. Finally, having only one cell means that any dirt in the cuvette windows will attenuate the sample and reference beam equally. A disadvantage of the dual-wavelength system is an incomplete compensation for water vapour absorption in the sample and reference beams. The absorption of infrared energy from water vapour varies with the wavelength and is, therefore, not the same at the reference wavelength as at the sample wavelength.

To eliminate fluctuations in humidity, the infra-red compartment of a milk analyser must be sealed, maintained at a constant temperature, and kept dry with silica gel.

Fig. 12. 1983 version of IR instrument (Milko-Scan 100) for three- or four-component analysis. Measures 90 samples per hour (fat, protein, lactose), manually presented to instrument. (Courtesy of A/S N. Foss Electric, Denmark.)

Dual-wavelength single-beam single-cell principle

In 1979, a principle was introduced that meant a dramatic simplification of the optical system (Milko-Scan 100 series, Fig. 12).

Where previous instruments compensated for the background 'noise' (water absorption) by attenuating the energy in the reference beam until null-balance with a comb was obtained, the read-outs of the single-beam instrument are calculated electronically from the ratio between the sample wavelength energy and the reference wavelength energy. The interference filters are arranged in a filter wheel which rotates to bring each filter in turn — two filters per component — into the path of the infra-red beam (Nexø *et al.,* 1980). In a single-beam instrument, the number of mirrors is reduced to two, which in turn reduces the travelling distance of the infra-red radiation to about 27 cm. This reduced light path, and the fact that the sample and reference beams follow exactly the same path, make this type of instrument less sensitive to water vapour than the earlier double-beam versions. The high stability and accuracy of the single-beam instruments is due to the reduced number of mirrors, the fact that the mechanically moving servo comb is no longer necessary, and to improved interference filters and electronics.

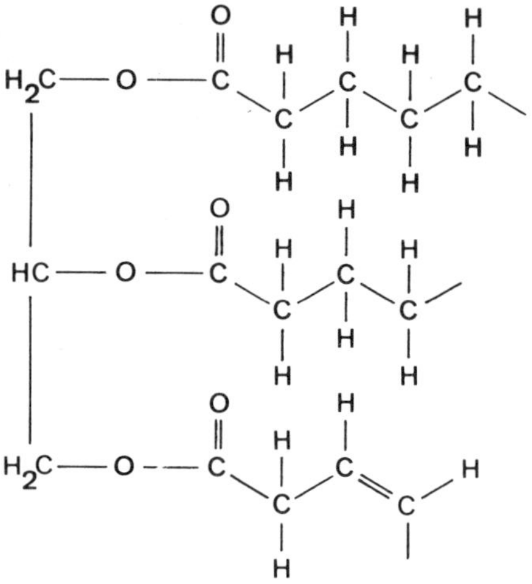

Fig. 13. Structure of butterfat molecule.

Reliability of data

Fat determination. As mentioned before, two different wavelengths, 5·7 μm and 3·5 μm, can be used to determine fat in milk. The absorption at 5·7 μm is due to stretching vibrations in the C=O bonds of the carbonyl group in fat, and this measurement 'counts' the number of fat molecules regardless of the length and weight of the individual fatty acids (Fig. 13).

If the average chain length (mean molecular weight) of the fatty acids is changed, the number of triglyceride molecules per unit weight will change too, and an error will occur in the results unless the change is compensated for by recalibrating the instrument. The composition of butterfat varies with season, region, breed, cow and stage of lactation (Grappin and Jeunet, 1972), and this means that an instrument using the 5·7 μm filter must be recalibrated when, for instance, going from winter to summer.

The absorption at 3·5 μm is due to stretching vibrations in the saturated C-H bonds of the fatty acid chains. This measurement is, therefore, related to both the size and the number of fat molecules in the sample, as the number of carbon–hydrogen bonds increases substantially in proportion to the molecular size. Both -CH₃ and -CH₂ groups absorb infra-red energy at 3·5 μm, but the C-H stretching is markedly reduced by the presence of double bonds adjacent to these groups. The absorption decreases as a function of the degree of unsaturation (Mills and Van de Voort, 1982), but even so, the 3·5 μm determination is less sensitive to variations in refractive index than the 5·7 μm determination is, because it reflects the variation in chain lengths (Fig. 13).

Another advantage of the 3·5 μm wavelength is that the measurement includes free fatty acids that may have formed during storage; these cannot be measured at 5·7 μm. Protein and lactose contribute to the absorption at 3·5 μm, but their interference is removed by means of suitable correction constants in the equation that calculates the fat content from the instrument signals. The difference in result between the two wavelengths is, in many cases, negligible, but analyses of individual cow's milk should be performed using the 3·5 μm filter for optimal performance.

Protein measurement. The wavelength for protein determination is 6·5 μm, and it is the nitrogen–hydrogen bonds within the peptide bonds that are responsible for the IR absorption (Fig. 14). Thus, the measurement represents the number of amino acids rather than their weight, but as the composition of protein in milk is fairly constant, this causes no problems. In contrast to the Kjeldahl analysis, which is the reference method, the infra-red measurement does not include non-protein-nitrogen.

Lactose measurement. Lactose is measured at 9·5 μm, and the absorption is mainly due to bending vibrations in the C-OH bonds (see Fig. 15) (Goulden, 1956).

Accuracy of results

Infra-red milk analysers have been tested and approved in several countries, and all of them have obtained AOAC approval (Biggs, 1972, 1978, 1979a, b; Van de Voort, 1980). Biggs has proposed a set of performance specifications based on manufacturers' claims (Biggs, 1979), and instruments with performances corresponding to these

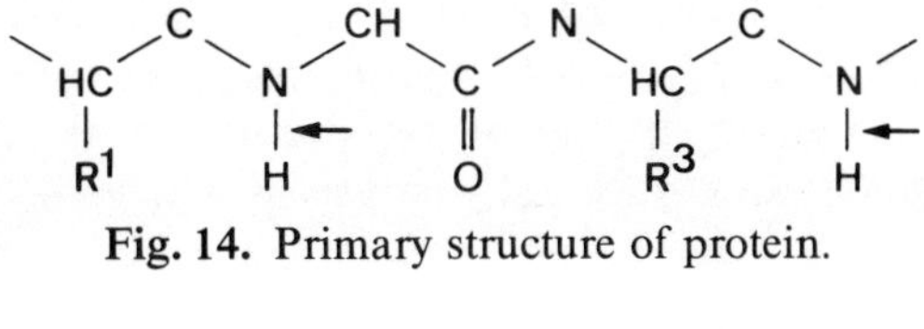

Fig. 14. Primary structure of protein.

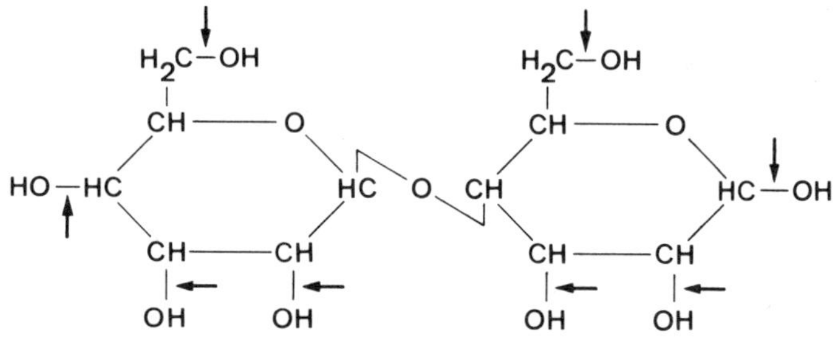

Fig. 15. Lactose.

specifications automatically comply with the requirements of the AOAC-approved method. The specifications are based upon analysis of eight different milk samples, and are as follows:

> Repeatability (standard deviation of duplicate instrument estimates): fat, protein and lactose: SD ≤ 0·02 per cent, conc. 2–6 per cent and total solids: SD ≤ 0·04 per cent.
> Accuracy (standard deviation of differences between instrument estimates and values found by reference methods): fat, protein and lactose: SD ≤ 0·06 per cent, conc. 2–6 per cent and total solids: SD ≤ 0·12 per cent.

The best accuracy is obtained with instruments that have been calibrated for the type of milk to be analysed (herd, individual cow, homogenised, unhomogenised, consumers' milk, etc.).

Operation of infra-red instruments

The operation of infra-red instruments is simple. In central laboratories where thousands of analyses are performed each day, fully automated milk analysers (Fig. 16) are used, whereas dairy plants with a limited number of samples prefer semi-automatic instruments. In both types of instruments, the zero-point and calibration must be checked regularly to ensure that no instrument drift takes place.

The zero-point is checked with demineralised water, and the calibration check is performed with pilot samples that have a known content of fat, protein and lactose.

Samples must be preheated to 40°C in a waterbath, and gently shaken or stirred to ensure an even distribution of the butter fat before they are presented to the instrument. Pushbuttons on the control panel of the instrument allow different analysis programs to be selected,

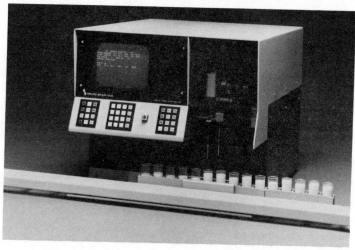

Fig. 16. 1984 version of programmable IR instrument (Milko-Scan 605) for automatic analysis of up to five components. Measures 600 samples (fat only) or 360 samples (fat, protein, lactose) per hour.

depending on which parameters the user wants to measure. Most infra-red instruments can electronically calculate total solids (TS) or solids-non-fat (SNF) from the measured contents of fat, protein and lactose, plus an added constant for mineral content. Certain instruments are also provided with a pair of filters for the direct determination of water.

Analysis of dairy products other than fluid milk

Any dairy product which is liquid, or can be converted into a liquid form, can be analysed in an infra-red instrument, provided that the TS content is low enough for the pump to work, and that the concentration of each single component does not exceed about 15 per cent. Soured milk products must be neutralised to redissolve the precipitated proteins; whipping cream must be diluted to bring the fat content down to within the instrument's measuring range; cheese must be grated, neutralised, diluted and homogenised; and milk powders must be dissolved in water. Because of the sample preparation (dilution, neutralisation) and/or the presence of lactic acid and various additives, the read-outs for most dairy products other than milk must be corrected.

Lactic acid, for instance, has absorption bands near the wavelengths for protein and fat (5·7 μm).

Infra-red milk analysers have gradually replaced turbidity tests for fat (e.g. Milko-Testers) due to the increased interest in protein content, and they have fostered new applications of fat, protein and lactose determinations, where tedious and time-consuming standard analyses previously limited the availability of these data. The newest micro-processor-controlled, infra-red instrument analyses 360 samples per hour for fat, protein, lactose and solids, and needs as little as 3·6 ml of milk per analysis. If the only results needed are for fat, the measuring speed of this instrument is 600 samples per hour (Fig. 16).

Somatic Cell Count

Mastitis

Mastitis is a complex disease of various degrees and consequences. There are many steps from the normal and healthy cow to the cow affected by serious mastitis. Two main diagnoses are:

1. Clinical mastitis: visible changes in the milk (acute or subacute).
2. Subclinical mastitis: no visible changes in the milk.

It is rather easy to detect the first group. The farmer will often have the cow treated with antibiotics, leading to a cure. Subclinical mastitis, on the other hand, is difficult to detect, and many investigations have shown that about 40 per cent of the average herd is subclinically infected.

Mastitis is due to the penetration of bacteria into the udder, and the host reacts to this by trying to combat the infection. This is done simultaneously in several ways. The main way is by secretion of white blood cells (leucocytes) from the udder into the milk, where an elimination of bacteria takes place. Consequently, by counting the number of these white blood cells, one can obtain a measure of how serious the infection is.

IDF classification of quarter milk samples is given in Table III.

Economic aspects

It is very well known that both clinical and subclinical mastitis induce a remarkable decrease in milk yield, and that this decrease is, in most

TABLE III

Classification	Somatic cell count	Pathogen present
Normal secretion	< 500 000 per ml	no
Non-specific mast.	> 500 000 per ml	no
Latent infection	< 500 000 per ml	yes
Mastitis	> 500 000 per ml	yes

cases, irreversible. It has also been shown that there is a very close connection between the number of somatic cells and yield. Not only is yield related to somatic cell count (SCC), but the milk constituents will also be influenced; the per-volume content of fat, casein and lactose decreases as the SCC increases (Fig. 17).

Fossomatic

Principle of electronic somatic cell counting
A/S N. Foss Electric, Denmark, has developed a fast and automated method to measure the number of somatic cells in milk. The method is

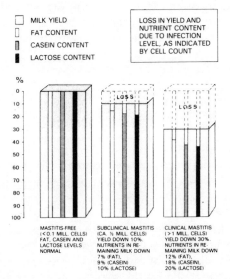

Fig. 17. (Courtesy of A/S N. Foss Electric, Denmark.)

called fluoro-opto electronic cell counting. A milk sample is warmed to 40 °C, stirred, and automatically pipetted, with a hot buffer solution, to a chamber in a revolving rack carousel. A dye solution (containing ethidium bromide) is added (Fig. 18). In the next carousel position, the mixture is stirred, and by means of air pressure, part of the solution is passed into a microsyringe and transferred to the edge of a vertical rotating disc. The tiny film of diluted milk, 0·5 mm wide and 10 μm thick, is then passed under a microscope. By means of a lens system and dichroic mirror, blue light (400–500 nm) is passed to the milk film. The dyed cells in the milk then fluoresce, emitting red light which is transmitted through the microscope to a photo-multiplier. The impulse from each cell is counted, and the total count is displayed, calculated as the number of somatic cells per ml milk. After each sample, the tubing is rinsed, and the mixing chamber in the carousel is washed and dried. The rotating disc is constantly rinsed and dried; the rinsing liquid is heated before use.

Two versions of the Fossomatic instrument are available: a fully automated version that can count cells in 180 or 215 samples per hour, and a semi-automated version (with a flow system different from that shown in Fig. 18) that can handle about 90 samples per hour; both are able to count up to 10 000 000 cells per ml. Repeatability is good, with the coefficient of variation as good as 4–5 per cent. The standard deviation, compared to direct microscopic somatic cell count (DMSCC), is less than 10 per cent; the correlation with DMSCC is 0·960. The precision and accuracy must be checked daily using pilot milk samples.

Coulter Milk Cell Counter

Principle
The Coulter principle is based on the difference in electrical conductivity between milk particles and added diluents, in that while the milk particles act as insulators, the diluent acts as a good conductor. By means of two immersed electrodes, an electrical current is established over a small aperture, and the milk particles suspended in an electrolyte are forced through the aperture. As each particle displaces the electrolyte in the aperture, a pulse proportional to the particle volume is produced. After initial fixation of each milk sample according to procedures recommended by the manufacturer, the Coulter Counter dilutes and incubates the sample, and counts the number of particles present in it.

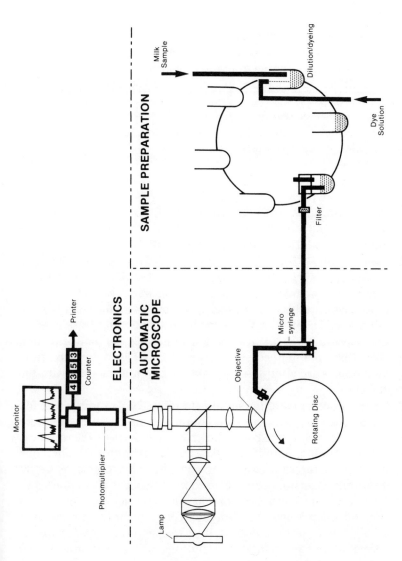

Fig. 18. Operation of a Fossomatic: schematic. (See text for further details.)

The Coulter Milk Cell Counter consists of the following sub-units:

1. *The pneumatic power supply*, which powers the circuits in the diluter section.
2. *The main unit*, composed of:
   (a) an electronic power supply that powers the electronic circuits in the Analyzer Section and Teleprinter;
   (b) a diluter section to accept milk samples which have been prefixed;
   (c) an analyser section;
   (d) a trolley holding containers for chemicals.
3. *A teleprinter* for printing results.

Operation

1. Samples are prepared by manually fixing the somatic cells, prior to incubation, in order to preserve their size: add Somafix® or formaldehyde solution to the milk sample. Heat sample for a short time, or leave at room temperature for 18–24 h before counting.
2. Sample vials (1–50) are placed in a carousel, and Somaton® blanks are added in two holes. After stirring, 2 ml of fixed milk is drawn through the sample valve, and approx. 45 μl of milk is transferred to a reaction tube with a precise volume of Somaton® to give a dilution of 1:100. (The sample valve is purged and rinsed before the next sample is taken.) The reaction tubes are incubated in a glycol bath maintained at 80°C for 10 min. During this reaction, fat globules are eliminated so that they do not interfere with the cell count.
3. On reaching the counting station, the diluted milk sample is gently stirred, and the cells are then counted using a 140 μm aperture tube. Suspended, fixed somatic cells pass through the aperture, and the resulting, amplified pulses are displayed on an oscilloscope monitor. A threshold value, linear with cell volume, is set to discriminate against pulses caused by fat, debris, air bubbles and contaminating substances, so that only particles the size of fixed somatic cells are counted. A digital register records the number of cells present in 0·3 ml of cell suspension, and the count is corrected and printed. After counting, the tube is rinsed. The capacity is 210 samples per hour.
4. Precision and accuracy should be checked daily.

FUTURE DEVELOPMENTS

A need for chemical analysis is indicated by:

- analysis for payment purposes and herd improvement;
- process control during manufacturing of dairy products;
- official demands and requirements.

Payment Purposes

Methods of settling accounts with the milk supplier have developed parallel to the appearance of analysis instruments with large capacities and direct output to EDP systems for processing of data and calculation of payments. The main component in milk payment was originally — and still is — butterfat, but protein content is gradually being introduced as a milk pricing parameter. There is a tendency towards greater control of raw milk quality, depending on the kind of product to be made.

In cheese production, not only as a whole but also with reference to the production of different specific types of cheese, there is a need to ensure that milk quality (especially protein content) is optimal for processing. Similar requirements as to the quality of raw materials can be made in the manufacturing of other dairy products, such as dried milk powder, condensed milk, etc.

Besides analysis of raw materials, there is an increasing demand for control of the hygienic quality of milk. Somatic cell counts and bacterial counts have been introduced in payment systems in a rapidly increasing number of countries. As more attention is paid to hygiene, accepted threshold limit values will be reduced, and this again increases the demand for instruments with better sensitivity.

Process Control

In the manufacture of dairy products, it is highly desirable to have ways of monitoring and controlling the production processes. Quick and reliable analytical instruments provide a basis for optimal processing.

Standardisation of milk according to fat content is controlled by analysing samples of the finished product from the pipeline in the dairy. Based upon the results, the flow of cream into the skim-milk can be increased or decreased to obtain the desired fat content. Increased instrument accuracy means that more narrow limits can be established for fat content in the finished product. A reduction of fat by 0·01 per cent may result in savings of more than $25 000 per year in some dairies.

In various processes, standardisation of protein content can be expected — for example, in the production of milk powder. Research and development in milk processing will add to our knowledge of the influence of individual milk components on the finished products, which may increase applications and even result in a need for measuring novel parameters.

Official Demands

A growing interest among consumers in the constituents, quality and age of food products has resulted in intensified control and legislation in many countries. Regulations that require declaration of contents and shelf-life are examples. To meet these demands, suitable analytical methods must be available. The introduction of new standards can, typically, come about in two ways:

1. In some cases, a producer of instruments develops and introduces to the market a new method of analysis which slowly gains greater use. The attention of the authorities is drawn to the method, and legislation makes it officially accepted.
2. In other cases, parameters are regulated by legislation which is to come into force within a specified time limit. This leaves time for potential producers of instruments to develop and market suitable apparatus to meet the demands made by the legislation.

There is, thus, a close relationship between arising needs for measurement and instruments offered on the market. Technological developments in physics, chemistry and electronics, including computer technology, offer many possibilities to the producers of analysis instruments. The technological possibilities within manufacturing of instruments are great. Not only single instruments, but also for total solutions, including sample handling, sampling, measurement, and recording and processing of data in integrated systems will, in the future, be offered. Great investments will be made in research and development to the benefit of milk suppliers, the dairy industry, and consumers.

REFERENCES

A/S N. Foss Electric. Pro-Milk Mk II Applications. DK-3400 Hillerød, Denmark.

Biggs, D. A. (1972). *J. Assoc. Off. Anal. Chem.,* 55, 488–97.

Biggs, D. A. (1978). *J. Assoc. Off. Anal. Chem.,* 61, 1015–34.

Biggs, D. A. (1979a). *J. Assoc. Off. Anal. Chem.,* 62, 1202–10.

Biggs, D. A. (1979b). *J. Assoc. Off. Anal. Chem.,* 62, 1211–14.

Goulden, J. D. S. (1956). *J. Sci. Food Agric.,* 7, 609–13.

Goulden, J. D. S. (1958). *Trans. Faraday Soc.,* 54, 941.

Goulden, J. D. S. (1961). *Nature,* 191, 905–6.

Goulden, J. D. S. (1964). *J. Dairy Res.,* 31, 273–84.

Grappin, R. and Jeunet, R. (1972). *Le Lait,* 52, 324–46.

Grappin, R. (1984). Challenges to contemporary dairy analytical techniques. Special publication no. 49, pp. 77–90, The Royal Society of Chemistry, London.

Haugård, G. and Pettinati, J. D. (1959). *J. Dairy Sci.,* 42, 1255–75.

Jøndrup, P. (1980). *Elektronik,* 8, 10–14.

McGann, T. C. A. and O'Connell, J. A. (1972). *Laboratory Practice,* 21, 489–90.

McGann, T. C. A. *et al.* (1972a). *Laboratory Practice,* 21, 628–31, 650.

McGann, T. C. A. *et al.* (1972b). *Laboratory Practice,* 21, 865–71.

Mills, B. L. and Van de Voort, F. R. (1982). *J. Assoc. Off. Anal. Chem.,* 65, 1357–61.

Nexø, S. A. *et al.* (1980). US Pat. 4,236,075.

Nexø, S. A. *et al.* (1981). US Pat. 4,247,773.

O'Connell, J. A. (1970). *Laboratory Practice,* 19, 1119–20.

O'Connell, J. A. and McGann, T. C. A. (1972). *Laboratory Practice,* 21, 552.

Shields, J. (1975). B.Ph.Thesis, University of York, England.

Shields, J. (1982). US Pat. 4,310,763.

Sjaunja, L. -0. (1982). Studies on milk analysis of individual cow milk samples. Report no. 56, Swedish University of Agricultural Science, S-75007 Uppsala, Sweden.

Thomasow, J. *et al.* (1971). *Milchwissenschaft,* 26, 474–81.

Tolle, A. (1971). *International Dairy Federation Annual Bulletin,* part 2, p. 3.

Van de Voort, F. R. (1980). *J. Assoc. Off. Anal. Chem.,* 63, 973–80.

Chapter 7

Modern Laboratory Practice — 2: Microbiological Analyses

Cadbury Schweppes plc, Lord Zuckerman Research Centre, Reading, UK

Milk and milk products are highly perishable foodstuffs which are of considerable economic and nutritional importance, especially in the Western hemisphere. In its original state, milk is an excellent foodstuff for man if it is obtained from healthy cows and produced under hygienic conditions. Although milk production methods have generally improved over the last few decades, the microbiological quality of some raw milk supplies still causes concern. Udder disease remains widespread, and consumers of raw milk still risk food poisoning. Refrigeration of farm milk can mask the effects of unhygienic practices, such as defectively cleaned milking equipment. Inadequate processing, poor packaging techniques and insufficient refrigeration can all substantially reduce the shelf-life of milk products and cause considerable economic loss.

Microbiological methods have an important role to play in the Dairy Industry. They are used to protect the Public Health, and can reduce economic losses by the early detection of inadequate processing, packaging or refrigeration. This can be achieved by monitoring the microbiological quality of raw milk supplies, bulk milk and finished milk products immediately after production and during storage.

THE NEED FOR MICROBIOLOGICAL TESTING

Public Health Aspects

Raw milk was once regarded as the most dangerous item in our diet, giving rise in England and Wales between 1912 and 1937 to about 65 000 deaths. The eradication of tuberculosis and brucellosis in cattle, the hygienic production and the heat treatment of milk have greatly reduced the incidence of food poisoning. However, there is recent evidence that the incidence of milk-borne disease has begun to increase (Galbraith *et al.*, 1982), mainly due to unpasteurised, defectively pasteurised or recontamination of the milk after pasteurisation. These outbreaks are mainly due to *Salmonella* or *Campylobacter* species. Liquid market milk and milk for further processing which has been correctly pasteurised will be quite safe provided that recontamination is prevented.

The major sources of bacteria in raw milk are the udder (interior and exterior) and the milking equipment (Cousins and Bramley, 1981). In 1979, a survey of some 500 British herds showed that about one third of dairy cattle were infected by subclinical mastitis. These infections may be caused by *Staphylococcus* sp., *Streptococcus* sp., *Escherichia coli* and *Campylobacter* sp., some of which are pathogenic for man. Between milkings the cows' teats may become soiled with dung, mud and bedding litter. If not removed before milking, this dirt, which may contain $> 1 \times 10^{10}$ bacteria per gram (of which $> 1 \times 10^9$ per gram may be psychrotrophic and $> 1 \times 10^6$ per gram may be spore-formers), contaminates the milk during milking. When milking equipment is solely responsible for high counts in the milk, then the cleaning and disinfection must be seriously defective, and the equipment would have a count of around 1×10^9 bacteria per square metre.

Clearly the microbiological monitoring of raw milk supplies is necessary to protect the Public Health, especially where the consumption of raw milk is concerned.

Economic Aspects

The numbers and types of micro-organisms in milk immediately after production directly reflects the microbial contamination during production. High numbers of certain types of bacteria in the milk can reduce the value of the raw supply to the processor by reducing the time for which the raw material can be stored, reducing the shelf-life of the product or, importantly, causing taints in the finished product.

There is a considerable variation in the incidence of thermoduric and psychrotrophic organisms in fresh raw milk, which probably reflects the different conditions on the individual farms. *Microbacterium lacticum* and bacterial spores show nearly 100 per cent survival after pasteurisation. Fortunately only a small proportion of the thermoduric organisms are also psychrotrophic (McKinnon and Pettipher, 1983), otherwise the shelf-life of refrigerated pasteurised products would be considerably reduced. Psychrotrophic organisms in raw milk supplies are of considerable importance to the Dairy Industry, as they reduce the time that the raw material can be stored before processing. Milk with an initial psychrotroph count of 1×10^4 per ml even at $5\,^\circ$C may contain $> 1 \times 10^6$ per ml after 3 days storage (Cousins *et al.*, 1977). Pasteurisation will destroy these bacteria, but heat-resistant enzymes may persist and could be detrimental to the quality of the milk products (Law, 1979).

Poor processing for example, inadequate pasteurisation, recontamination after pasteurisation or recontamination during packaging, may result in a finished product with a much reduced shelf-life. Clearly it is necessary to use microbiological methods to detect these mistakes as early as possible, in order that they can be speedily rectified and economic losses kept to a minimum.

THE MICROBIOLOGICAL TESTING OF MILK AND MILK PRODUCTS

On the basis of Public Health and economic considerations, it is necessary to test the microbiological quality of milk at a number of points along the chain from producer to consumer. These include the following:

(i) Raw milk supplies.
(ii) Bulk milk arriving at the processor's premises.
(iii) Monitoring bulk milk during storage prior to processing.
(iv) Monitoring the processing and quality control of finished products.

In many European countries in which the Dairy Industry is highly developed, the microbiological testing of farm supplies is carried out in central testing laboratories. It is the responsibility of the processor to monitor the microbiological quality of incoming bulk supplies, milk during storage and finished products.

Central Testing

In the United Kingdom there are six central testing laboratories, the largest is responsible for testing 12 000 supplies and the smallest 5000 supplies. Each laboratory is fitted with automated equipment and operates every day of the year. Test results are automatically recorded onto floppy discs and later transmitted to the Milk Marketing Board's Headquarters via a computer terminal (Fig. 1).

Milk samples are collected from each farm by the tanker driver once a week on unspecified days. After the bulk milk has been mixed, the samples are taken using a disposable sterile dipper and collected in disposable, sterile, plastic pots. Samples are identified with bar code labels which are unique for each producer. The bar code is read by an

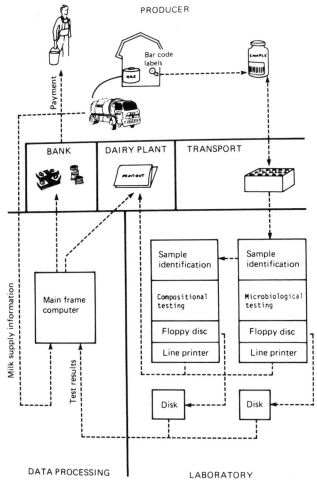

Fig. 1. Schematic diagram of the system of quality payment testing of milk.

infra-red light pen at the point of testing to avoid misidentification of the samples. The samples are transported in insulated containers fitted with ice packs to hold the temperature below 4°C for 48 h. Samples should reach the central laboratory within 36 h of collection.

Milk samples are tested for bacteriological quality by a plate colony count. A small volume of milk is removed from the sample and mixed with agar in a Petri dish. The Petri dish is then incubated for 3 days at 30°C, and the bacterial colonies formed counted automatically by a colony counter. Four bands of milk have been established for payment purposes, these are:

Band A — 2×10^4 bacteria per ml or fewer
Band B — $> 2 \times 10^4 - 1 \times 10^5$ bacteria per ml
Band C — $> 1 \times 10^5 - 2 \cdot 5 \times 10^5$ bacteria per ml
Band D — $> 2 \cdot 5 \times 10^5$ bacteria per ml

Band B supplies receive the standard payment and Band A supplies receive a premium. Supplies classified as Band C or D are subjected to penalties. Certain rules are applied to safeguard producers from the effects of a single high count in a month.

Antibiotic failures and unsatisfactory hygiene or compositional quality results are made known to the producers and buyers within 24 h of the test result being available. The producer's monthly payment from the Board is calculated from the volume of milk produced and its value, as determined from the results of the various tests.

Testing in Dairies and Creameries

Dairies and creameries still perform quality assurance tests on milk, but these are mainly on whole tanker loads and doubtful farm supplies. Incoming tanker milks are assessed for bacterial content either by the use of dye reduction methods or by a plate colony count. Both of these methods have major disadvantages; dye reduction methods are insensitive and unreliable, and cultural methods take 2–3 days to give a result. If the plate colony count method is used, the milk will probably have been processed before the bacterial content is known.

Increasing pressure is now being brought to bear on processors by their major customers regarding the bacteriological specifications laid down for both raw milk supplies used for processing and the finished products. Ideally, dairies and creameries need to use rapid methods for testing raw milk prior to acceptance, and methods which will successfully predict the eventual keeping quality of their finished products.

MICROBIOLOGICAL METHODS FOR THE TESTING OF RAW MILK

There are many different methods available to the microbiologist for testing the microbiological quality of milk. The choice of test will vary with the application. For example, it is preferable if tests used in a dairy or creamery are rapid so that any necessary remedial action can be taken quickly. If only small numbers of samples are involved, then increased labour requirements or running costs may be acceptable to achieve a rapid result. In central testing laboratories, rapidity is not as important as the tests on farm supplies are retrospective, the sample generally takes > 24 h to reach the laboratory. In this case, automation, low running costs and high accuracy and precision are given greater consideration than rapidity in selecting the best method to use.

Most of the microbiological methods available can be used to assess the microbiological quality of raw milk. Frequently the total bacterial count is measured, and if this is low, it is assumed that there is low risk to Public Health and that the quality of the milk is satisfactory. However, the initial total viable count is of little value in predicting the count after refrigerated storage. Many of the general microbiological methods can, with certain modifications, be used for specific applications for example, the detection of post-pasteurisation contamination, sterility testing and the detection of spoilage organisms such as yeasts. In order to simplify the discussion, the most widely used microbiological methods will be described for the application to raw milk, and then specific applications will be considered separately, as a number of microbiological methods can be used for a single specific application.

Methods for estimating bacterial numbers in raw milk can be divided into two groups, direct and indirect. Direct methods count cells directly, either by the ability of viable cells to grow and form colonies, or microscopically. Indirect methods measure either a chemical constituent, enzyme, metabolite or changes produced by bacteria during growth. This measurement is converted into bacterial numbers by reference to a calibration curve. The 'true' numbers of bacteria for the calibration curve are usually assessed by a direct method. Generally, direct methods are more sensitive and accurate than indirect methods, and cultural methods take longer to give a result than microscopic or indirect methods.

The adjective 'rapid' has been ambiguously used when describing microbiological methods. In terms of speed, 'rapid' has meant a few minutes or a number of hours. Miniaturised, modified and automated methods have also been described as 'rapid' since the processing time is shorter than the standard manual technique, even though the result may not be obtained for 1–3 days. The term 'rapid' should be reserved for tests which give a result in 1 h or less.

One of the first rapid methods in the food industry was probably the 'sniff and taste' test used at the creamery to determine the acceptability of raw can milk. Milks which smelled off and tasted sour, and therefore probably contained greater than 1×10^7 bacteria per ml, could be immediately rejected by this method. The test became inappropriate with the introduction of refrigeration as spoilage was no longer due to milk-souring organisms. Spoilage by psychrotrophic bacteria is less easily detectable by sensory assessment. Non-quantitative methods have no place in a modern Dairy Industry.

Direct

Viable

Plate colony count

The plate colony count isolates bacteria in a quantitative manner. A range of dilutions, usually 10-fold, of the sample is prepared in a sterile diluent, and 1 ml of each dilution is mixed with melted agar cooled to 45°C and then allowed to solidify (American Public Health Association, 1972). The organisms present in the sample are fixed within the agar gel. The pour plates are then incubated for 2–3 days at 30°C, during which time viable organisms replicate and form visible colonies. Individual colonies on a plate containing 30–300 colonies are counted, and then the count per ml calculated by multiplying the colony count by the appropriate dilution factor.

The plate colony count suffers from a number of disadvantages. Two or more adjacent cells may give rise to only a single colony. Some organisms may be unable to grow in the medium or at the incubation temperature selected, and some may be damaged by exposure to the warm agar and unable to replicate.

One of the major disadvantages of the plate colony count is that it takes 2–3 days to give a result. The plate colony count is very sensitive because, in principle, any viable cell when placed in an appropriate medium and incubated at a suitable temperature will give rise to a colony. No other microbiological method can detect with certainty the presence of a single, viable bacterium in a large volume of sample. The plate colony count is generally accepted as the reference method for the microbiological analysis of milk. Rightly or wrongly, new microbiological methods are compared against the plate colony count, and must show good agreement before they are accepted as suitable alternative methods. It is important to note that the plate colony count itself is also subject to errors, and provides only an estimate of the 'true' bacterial numbers. These errors, which are mainly due to sampling and decrease with increasing bacterial numbers, are often in the range ± 10–40 per cent.

Automation of the plate colony count enables an operator to perform analyses more rapidly than by the manual method. Using machines like the Colworth 2000, decimal dilutions of food suspensions can be made automatically and plated to a maximum of 2000 plates *per* day *per* technician with obvious savings in operator time (Sharpe *et al.*, 1972). However, for plate colony count methods, the lengthy time taken to give the result is mostly due to the incubation period required for colony formation. Automation of the preparation of the plates does little to decrease the overall time required for the technique.

Plate loop method
Savings in time and materials can be made by removing the need for the dilution series in the standard plate colony count. Various methods based on the Thompson plate loop method have been described for milk (Bradshaw *et al.*, 1973; Fleming and O'Connor, 1975). In the technique, two loops which retain 0·01 and 0·001 ml dip into the sample. A Petri dish is positioned under each loop and the loops are flushed with diluent, agar is then added and the contents of the Petri dish mixed. These machines can plate 10^{-2} and 10^{-3} dilutions from 300 samples per hour. The technique is suitable for enumerating bacteria in the range 3000–300 000 per ml. Of the various factors influencing the volume transferred by the loop, the speed of the loop as it emerges from the milk is the most important (King and Mabbitt, 1984). Increasing the speed two-fold results in a 23 per cent increase in the volume transferred. The depth of the dip and changes in the temperature of the sample also

cause a significant change in the volume transferred. These authors suggested that to obtain the transfer of 1 μl at the speeds commonly used in the plate loop method, the internal diameter of the loop should be reduced from the present standard 1·45 mm to 1·3 mm.

One of the most commercially successful automated plate loop machines is the Petrifoss (A/S N. Foss Electric, Denmark) which is widely used in the central testing laboratories of Europe. The instrument can 'plate' 300 samples per hour, and is, therefore, ideally suited to laboratories which analyse hundreds or thousands of samples each day. The manufacturer claims that the relationship between the Petrifoss result and that of a reference method has a correlation coefficient of 0·99.

Spiral Plate

The Spiral Plate method for enumerating bacteria also avoids the use of a dilution series. By using a varying rate of sample application, it needs only one plate to obtain counts over a range which would require 2 or 3 plates in the standard plate colony count. The Spiral Plater deposits a known volume of sample on a rotating plate in an ever decreasing amount in the form of an Archimedes spiral (Gilchrist *et al.*, 1973). After incubation different colony densities are apparent, closely packed or confluent in the centre to well isolated at the outside (Fig. 2). A counting grid which relates the area of the plate examined to the volume of the sample is used to convert the count in a given area of the plate to the number of bacteria per ml of sample. This can be done automatically in 1–4 s by using a laser colony counter. The count per ml is calculated from the area of the plate scanned to reach a pre-determined number of colonies. Using the Spiral Plater and laser colony counter, milks with bacteria in the range 500–300 000 per ml can be analysed in about one third the preparation time of the standard plate colony count. Contradictory results of comparison of the Spiral Plater with the plate colony count have been reported. Some workers found no difference between the two methods (Peeler *et al.*, 1977), whereas others found the overall geometric mean of the Spiral Plate count 33 per cent lower than the plate colony count (O'Connor and Fleming, 1979).

Droplet technique

Miniaturisation of the plate colony count also reduces processing times. In the Colworth Droplet technique, which can be used to enumerate bacteria in foods, the plates are 0·1 ml droplets of agar (Sharpe and Kilsby, 1971). The agar is used as diluent thereby saving materials. Five

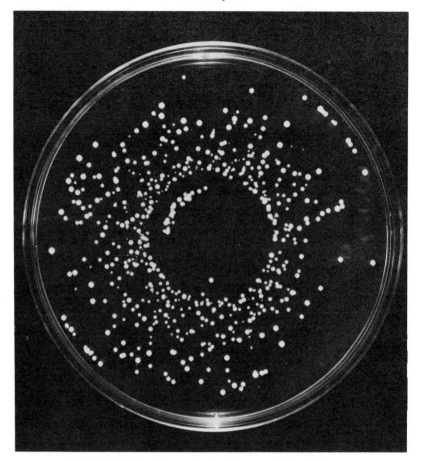

Fig. 2. *Escherichia coli* colonies on a Spiral Plate.

replicate droplets of each of three dilutions can be prepared in about 45 s, enabling operators to process about three times as many samples as by the standard plate colony count (Fig. 3). For milk, the droplet technique after 48 h incubation gives similar results to the standard plate colony count (Fondén and Strömberg, 1978).

Hydrophobic Grid Membrane Filter
One direct method for counting viable bacteria retained on membrane filters, following filtration of the sample, involves placing the filter on a

pad soaked in nutrient medium and counting the colonies formed after 1–3 days incubation. This method, which cannot be considered as rapid, has been used to enumerate coliform bacteria in raw and pasteurised milk (Fifield *et al.,* 1957; Claydon, 1975).

A variation on this method is the Hydrophobic Grid Membrane Filter (HGMF) described by Sharpe and Michaud (1974). The technique reduces the need for dilutions of the sample before enumeration, and hence reduces processing time, and gives better recovery than conventional filters, especially when high numbers of bacteria are involved (Sharpe and Michaud, 1975). The HGMF has a square grid pattern printed in hydrophobic material, such as wax, on a conventional

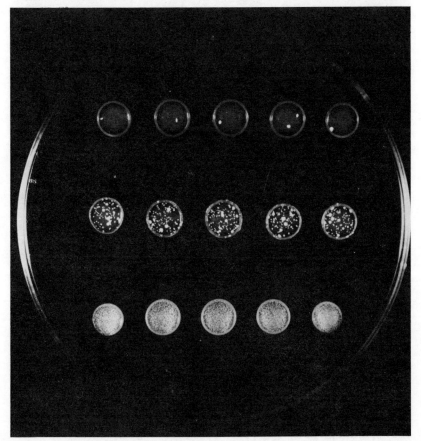

Fig. 3. *Escherichia coli* colonies in agar droplets.

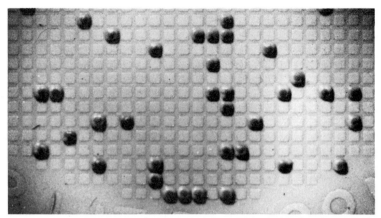

Fig. 4. *Escherichia coli* colonies on a Hydrophobic Grid Membrane Filter.

membrane filter. This divides the filter into a number of compartments, usually 2000–4000 depending on the size of the grids. The device functions as a most probable number technique, eliminating size variations in colonies and preventing lateral spread (Fig. 4). This greatly facilitates the automated counting of the colonies. Growth in a grid cell does not necessarily equal one colony in the plate colony count, since frequently a grid is inoculated with more than one bacterial cell. Coincident inoculation is allowed for in calculating the count per ml, e.g. growth in 3649 grid cells gives a most probable number of 30 000. Hydrophobic Grid Membrane Filters have been successfully used to enumerate coliforms in a variety of foods (Sharpe *et al.*, 1979). Pretreatment of milk and milk products with either a proteolytic enzyme, a detergent, or both, improves the filtration sufficiently to enable 3–5 g to be filtered (Peterkin and Sharpe, 1980). No bacteriological results for these products were reported. This method of assessing bacterial numbers in milk may have the advantage over the plate colony count in reducing operator time but, like all methods relying on bacterial growth, it requires a lengthy incubation period.

Microscopic

Breed smear
One of the first methods to use microscopy for counting bacteria in milk involved the preparation of milk films, staining with Methylene Blue, and then microscopic examination (Breed and Brew, 1916). The

method, which is still in use today, has the advantages of taking less than 1 h to complete, stained milk films can be stored as a relatively permanent record, and tentative identification of the types of bacteria can be made during examination. The disadvantages of the method include the use of a very small sample volume (0·01 ml) which leads to increased errors, a large microscope factor of about 500 000 which limits the sensitivity, failure of bacteria to stain, irregular distribution of bacteria in the films, and operator fatigue resulting from prolonged use of the microscope. This direct microscopic method is particularly popular in the United States of America, where it is used to differentiate raw milk supplies into three classes (American Public Health Association, 1972).

Direct Epifluorescent Filter Technique

The Direct Epifluorescent Filter Technique (DEFT) was originally developed as a rapid method for counting bacteria in raw milk (Pettipher *et al.,* 1980; Pettipher, 1983). The DEFT uses membrane filtration, fluorescent staining and epifluorescence microscopy. Pre-treatment of the milk with an enzyme and surfactant disperses somatic cells and fat sufficiently to enable at least 2 ml to filter through a 0·6 μm polycarbonate membrane filter; bacterial cells remain intact and are concentrated on the membrane. After staining with Acridine Orange, the bacteria fluoresce orange-red under blue light and can easily be counted using an epifluorescence microscope. The DEFT count is rapid, taking 25 min to complete, is inexpensive, correlates well with the plate colony count ($r > 0.9$), and is suitable for milks containing $6 \times 10^3 - 10^8$ bacteria per ml. The DEFT count showed >90 per cent agreement with the plate colony count in the classification of milk samples into groups of more or less than 10^4, 10^5 and 10^6 bacteria per ml. The method is both sufficiently rapid for monitoring tanker and silo milk, and sensitive enough for grading farm milk on the basis of bacteriological quality. No other method can give an accurate, sensitive count of bacteria in milk in as short a time as the DEFT.

One of the major disadvantages of the manual DEFT is operator fatigue associated with prolonged use of the microscope. This can be eliminated by the use of a semi-automated counting system based on an image analyser (Foss Electric UK, Ltd) which enables operators to count up to 50 DEFT slides per hour (Fig. 5). The semi-automated count of bacteria on DEFT slides agrees well with the corresponding visual DEFT count, $r = 0.94$ (Pettipher and Rodrigues, 1982).

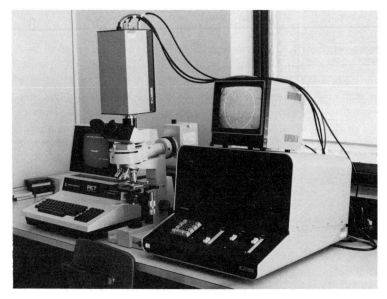

Fig. 5. Apparatus for the semi-automated counting of bacteria on DEFT slides.

The DEFT is now operating satisfactorily in more than 30 laboratories and in more than 9 countries. The Jersey Milk Marketing Board was quick to see the advantages of the method and has been using the semi-automated DEFT count to assess the bacteriological quality of farm milk since April 1982. The DEFT has been included in the revised version of British Standard 1984: 4285 'Methods of Microbiological Analysis for Dairy Purposes'.

Bactoscan

The Bactoscan instrument (A/S N. Foss Electric, Denmark) uses fluorescence microscopy for the automated counting of bacteria in milk. The Bactoscan works on the same principle as the Fossomatic, an instrument developed for automated somatic cell counts in milk. The basic difference between the two systems is that in the Fossomatic system, the somatic cells are stained directly in the diluted milk and subsequently counted, whereas in the Bactoscan, somatic cells, sediment and fat globules are first removed to prevent incorrect counting (Fig. 6). Somatic cells and casein micelles are dissolved

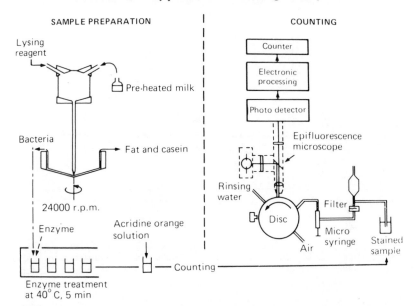

Fig. 6. Diagrammatic representation of sample preparation and counting by the Bactoscan instrument (from Kielwein (1982) by courtesy of the publisher, Verlag Th. Mann, Federal Republic of Germany).

chemically, and then the bacteria are separated by continuous centrifugation in gradients formed with solutions of dextran and sucrose. The bacteria recovered from the gradient are incubated with a protease to remove residual protein, and then stained with Acridine Orange and counted. For counting, the treated sample is applied in a thin film on the surface of a rotary disc and passed under a microscope objective. The fluorescent impulses in the microscope image are converted into electrical impulses and recorded. The instrument can analyse 68 samples per hour with an analysis time of 7 min. It would seem to be suitable for use in central testing laboratories.

For milk, the relationship between the impulses registered by the Bactoscan and the plate colony count correlates reasonably well, $r = 0.88$ (Fig. 7). The manufacturer claims a correlation between the Bactoscan count and the plate colony count of 0.8–0.9, and this was bettered during the evaluation reported by Saarinen (1984). The Bactoscan can be used to rapidly detect milks of poor bacteriological quality, and could be used to grade milks at the 100 000 bacteria per ml level, the first penalty band in

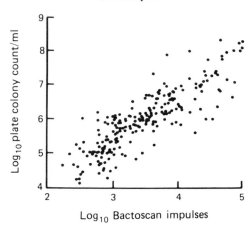

Fig. 7. Relationship between $\log_{10}$ plate colony count per ml and $\log_{10}$ impulses registered by the Bactoscan instrument for samples of refrigerated milk (from Kielwein (1982) by courtesy of the publisher, Verlag Th. Mann, Federal Republic of Germany).

a number of countries. In its current form, the instrument is not sufficiently sensitive to grade milks at the 20 000 bacteria per ml level, the upper limit of the premium grade in the UK.

Indirect

Dye reduction

Dye reduction tests are based on the ability of bacterial enzymes, such as dehydrogenases, to transfer hydrogen from a substrate to a redox dye which then undergoes a change in colour. The rate of reduction depends on the enzyme activity, and this has been used as an index of the number of bacteria present in the milk. In general, the reduction time is inversely related to the bacterial content of the sample when incubation with the dye commences.

Dye reduction tests were developed during the 1930s. The most commonly used redox dyes are Methylene Blue and Resazurin. Resazurin has the advantage over Methylene Blue in that reduction from purple through pink to colourless can be quantified using a comparator. A Resazurin test with a 10 min incubation time is used in England and Wales to grade milk on arrival at the creamery. The 10 min Resazurin test is rapid and requires small amounts of equipment, but only milks containing large numbers of actively growing bacteria ($> 10^6$ per ml) will fail. Methylene Blue reduction tests provide a reasonable estimate of the number of bacteria in non-refrigerated, bulked raw milk. The relationship between $\log_{10}$ plate colony count and Methylene Blue reduction time had a correlation coefficient (r) of 0·9 (Lück, 1982). Poorer agreement between the two methods was observed for refrigerated, bulked raw milk, $r = 0.62$ (Fig. 8).

There are a number of reasons why dye reduction times and numerical estimates of bacterial populations fail to correlate; the reducing activities of the different bacterial species vary, the reducing activity of bacteria is not lowered by clumping whereas the plate colony count and microscopic clump count are lowered, and substances such as antibiotics may inhibit bacterial growth. In addition, somatic cells at levels of about 1×10^6 per ml reduce Resazurin at a rate not dissimilar to that resulting from the same number of bacteria. Dye reduction tests are not suitable for classifying milk with low bacterial counts of $< 10^5$ per ml. Pre-incubation of the samples before testing improves the sensitivity of the method but prolongs the time needed, making it no longer a rapid method.

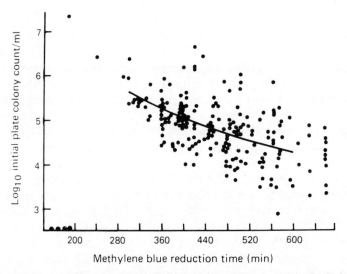

Fig. 8. Relationship between $\log_{10}$ initial plate colony count and Methylene Blue reduction time for samples of refrigerated raw milk. Line represents fitted regression curve (from Lück (1982) by courtesy of the publisher, Verlag Th. Mann, Federal Republic of Germany).

Electrical methods

The growth of micro-organisms results in changes in the composition of the culture medium as nutrients are converted into metabolic end products. Complex uncharged molecules, such as carbohydrates or lipids, are catabolised to smaller charged molecules such as lactic acid and acetic acid. Charged molecules, such as proteins and polypeptides, are converted via amino acids into ammonia and bicarbonate. As growth proceeds, these processes lead to a decrease in the overall impedance of the medium as conductance and capacitance increase. Electrical changes in microbial cultures provide a means of detecting micro-organisms and their metabolic effects.

Impedance is the resistance to flow of an alternating current through a conducting material. It is a complex quantity which is dependent upon the frequency of the alternating current employed, and can be described in terms of resistors, capacitors and inductors. The model of a resistor and capacitor in series seems to explain the electrical circuit formed when a pair of metal electrodes is placed into a microbiological medium (Hadley and Senyk, 1975). The relationship for this model is as follows:

$$Z^2 = R^2 + \frac{1}{(2\pi f C)^2}$$

where Z = impedance, R = resistance, C = capacitance and f = frequency of the alternating current. Impedance changes can be detected by passing a small alternating current through the medium and comparing the impedance of the inoculated medium with that of the uninoculated medium.

Electrical changes are a function of bacterial growth and replication. The threshold for detection depends upon the organism(s) and the media. In general, most media give detection thresholds at 10^6–10^7 organisms per ml. If the sample inoculated into the medium contains fewer organisms than this, there will be no detectable signal until the organism(s) have replicated sufficiently to reach the threshold number (Fig. 9). The detection time depends upon the initial number of viable organisms present in the sample and their specific growth kinetics. Variations in growth rates of different organisms may give rise to errors in the estimation of bacterial numbers. Similar numbers of a fast and slow growing organism will result in very different detection times. The incubation temperature used in impedance measurements is usually considerably higher than the storage temperature of the food. For

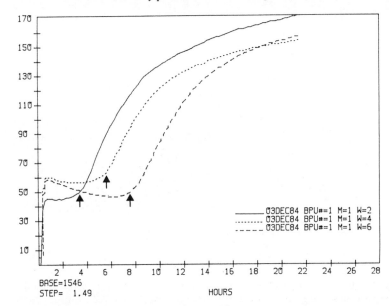

Fig. 9. Impedance curves for modules containing different numbers of *Escherichia coli*. Arrow indicates detection time. ————, 5×10^5; · · · ·, 5×10^4; — — — —, 5×10^3.

example, a temperature of 25 °C may be used for the assessment of a product normally stored under refrigeration. If the product has a mixed microbiological flora, then the types of organism(s) causing impedance changes may not necessarily be those causing spoilage of the product under storage conditions.

Generally, there is an inverse relationship between $\log_{10}$ plate colony count per ml or per gram and the impedimetric detection time. The method has an inherent delay whilst growth occurs, generally taking >6 h to detect 10^5 bacteria per ml. Impedance measurements cannot be regarded as rapid in the strict sense of the word, but they are particularly useful for the screening of large numbers of samples as little sample preparation is required.

Automated impedance measurements have been used to assess the microbial content of raw and heat-treated milk (Cady *et al.*, 1978; O'Connor, 1979). Using an 8·5 h detection time, milks could be classified into two groups containing either more or less than 10^5 bacteria per ml. There was an 81 per cent agreement between the

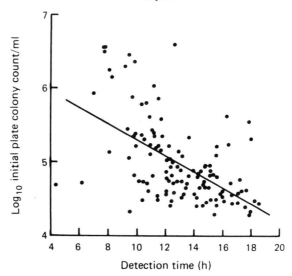

Fig. 10. Relationship between $\log_{10}$ initial plate colony count and impedance detection time for samples of refrigerated raw milk (from O'Connor (1979) by courtesy of the Agricultural Institute, Republic of Ireland).

classifications obtained by impedance and plate colony count methods, with the misclassifications equally distributed either side of the impedance detection limit (Fig. 10). Firstenberg-Eden and Tricarico (1983) have reported an improved correlation ($r = 0.95$) between impedance detection time and plate colony count. Samples containing $> 10^5$ bacteria per ml (total plate colony count) or $> 10^5$ bacteria per ml (mesophilic plate colony count) can be detected in *ca.* 16 and 4 h, respectively.

Bioluminescence

The functional significance of adenosine 5′-triphosphate (ATP) in the metabolism of living cells suggests that its assay should be an excellent monitor of biological activity in the sample. The firefly luciferase reaction, where light is produced by an enzymic reaction, is frequently used as an assay as it is specific for ATP. The reaction occurs as follows:

$$E + LH_2 + ATP \xrightarrow{\text{Mg}^{++}} E.LH_2\text{-}AMP + PP$$

$$E.LH_2\text{-}AMP + O_2 \longrightarrow E + L + AMP + CO_2 + h\nu$$

where E = firefly luciferase; LH_2 = reduced luciferin; ATP = adenosine 5'-triphosphate; $E.LH_2$-AMP = luciferyl adenylate complex; PP = pyrophosphate; L = oxidised luciferin and hv = light. The quantity of light recorded by a photometer is proportional to the concentration of ATP.

There are two factors which adversely affect the estimation of bacterial numbers by measurement of ATP: first, non-bacterial ATP and, second, quenching of the emitted light. Many biological materials, such as milk, contain non-bacterial ATP which must be destroyed before an accurate estimate of bacterial ATP can be made. This necessitates lysing the somatic cells and then incubating with the enzyme apyrase to destroy the ATP released. Following this treatment, the bacteria are chemically disrupted and the released ATP measured. Failure to destroy the non-bacterial ATP will give an elevated estimate of bacterial ATP, and hence bacterial numbers.

There is a relationship between the total ATP content and bacterial numbers in milk, but the correlation is poor (Britz *et al.,* 1980). This is presumably due to the varying levels of somatic cell ATP. Selective measurement of bacterial ATP in milk can be made by first destroying the somatic cell ATP (Bossuyt, 1981, 1982). For raw milk, the relationship between bacterial ATP and the plate colony count in the range 10^4–10^7 per ml had a correlation coefficient of 0·93 (Fig. 11). There is little increase in apparent bacterial ATP over the range 10^4–10^5 bacteria per ml, presumably because of residual somatic cell ATP influencing the

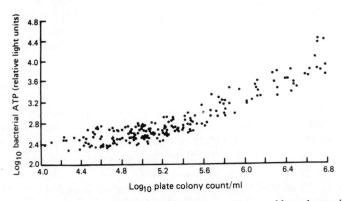

Fig. 11. Relationship between $\log_{10}$ plate colony count and $\log_{10}$ bacterial ATP content for samples of refrigerated raw milk (from Bossuyt (1982) by courtesy of the publisher, Verlag Th. Mann, Federal Republic of Germany).

result. Repeatability of the method is poor at levels of 10^5 bacteria per ml and below. Whilst the method may be useful for rapidly detecting raw milk of very poor bacteriological quality (> 10^6 per ml), it is unlikely to be sufficiently accurate for estimating bacterial numbers in good quality milk. Bioluminescence techniques may be suitable for detecting poor quality tanker milk supplied to the creamery, but they are unlikely to be useful for grading farm milks on bacterial content for payment purposes.

Limulus lysate test

The Limulus test can be used to rapidly and specifically determine the cumulative content of Gram-negative bacteria in foods. Gram-negative bacteria produce a lipopolysaccharide (endotoxin) which is a high molecular weight complex; it is not produced by Gram-positive bacteria. The lipopolysaccharide is generally released from the bacteria into the surrounding medium after death and lysis of cells, although some may be released by viable cells.

Present in the blue blood of the horseshoe crab, *Limulus polyphemus*, is a nucleated cell, called an amoebocyte, the cytoplasm of which is densely packed with granules. Limulus blood clots in the presence of bacterial lipopolysaccharide. All the necessary clotting factors are contained in an extract of the amoebocyte granules, called Limulus lysate.

The Limulus test is specific for lipopolysaccharide and very sensitive. As little as 10^{-12} g lipopolysaccharide per ml can be detected, occasionally even 10^{-15} g per ml. A single Gram-negative bacterium contains approximately 10^{-14} g lipopolysaccharide; because of the extreme sensitivity of the test, all utensils must be absolutely free from lipopolysaccharide.

For the Limulus Test, a 10-fold dilution series of the sample is prepared and equal volumes of Limulus lysate and diluted sample are mixed in a test tube. The tube is then incubated at 37 °C for 4 h, before being inverted and read. If the mixture remains unchanged and runs down the wall of the tube then that dilution of the sample does not contain lipopolysaccharide. If a firm opaque gel is formed which sticks to the bottom of the tube, then that dilution of the sample contains lipopolysaccharide. Generally, visual reading of 10-fold dilutions will give sufficient information about the level of lipopolysaccharide present in the sample. The accuracy of the method can be increased by using a two-fold dilution series.

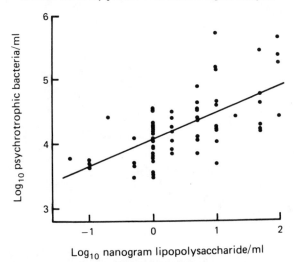

Fig. 12. Relationship between $\log_{10}$ psychrotrophic bacteria and $\log_{10}$ lipopolysaccharide for samples of refrigerated raw milk (from Hansen *et al.* (1982) by courtesy of the publisher, Cambridge University Press, UK).

There have been only a few reports of the use of the Limulus test for assessing bacterial numbers in raw milk (Terplan *et al.*, 1981; Hansen 1982; Hansen *et al.*, 1982). There is an indication that milks with higher numbers of psychrotrophic bacteria contain a higher level of lipopolysaccharide. However, for a given level of lipopolysaccharide the count of psychrotrophic organisms may vary by as much as 2 log cycles (Fig. 12).

Pyruvate

The determination of pyruvate, which is an intermediary metabolite in bacterial metabolism, has been suggested as a method of assessing the bacteriological quality of milk (Tolle *et al.*, 1972). The estimation of pyruvate is rapid, inexpensive, accurate, and can be carried out automatically (Suhren, 1982). Immediately after production milk contains 0·5–1·5 mg pyruvate per ml, but this value does not reflect the initial viable count. Both the initial pyruvate level and the increase in pyruvate after storage correlates with the Wisconsin mastitis score, which suggests that somatic cells contribute to the pyruvate content of milk. For individual farm and silo milks stored at refrigeration temperatures, there is not a close relationship between pyruvate values

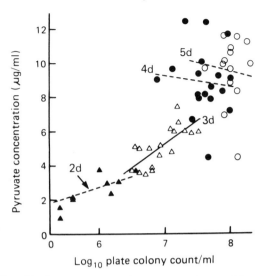

Fig. 13. Relationship between pyruvate concentration and $\log_{10}$ plate colony count for silo tank milks of 2 d (▲), 3 d (△), 4 d (●), and 5 d (○) of age stored at 5 °C. Line represents fitted regression line. ———, significant relationship ($P < 0.05$), - - - - -, non-significant relationship (from Cousins *et al.* (1981) by courtesy of the publisher, Cambridge University Press, UK).

and viable counts determined at intervals during storage (Cousins *et al.*, 1981). The relationship between the pyruvate content of milk and the bacterial count varies with both the temperature of storage and the age of the milk (Fig. 13). The pyruvate content of milk may be used to indicate milks with $> 10^6$ bacteria per ml, but it does not give an accurate estimation of bacterial numbers.

Radiometry

Most radiometric methods are based on the principle that microbial growth in media can be monitored by measuring $^{14}CO_2$ released during the metabolism of radio-labelled nutrients. These methods take from 2 to 24 h to complete depending upon the numbers of micro-organisms present in the sample. As with impedance measurements, the detection time (the time taken for the measurement of a specific concentration of $^{14}CO_2$) is generally inversely related to bacterial numbers. Radiometric methods are perhaps best suited to determining the sterility of products as they are particularly sensitive, given a sufficient period of incubation. They can also be used to detect samples containing large numbers of organisms within a few hours. The concentration or rate of production of $^{14}CO_2$ does not, however, always reflect the initial number of viable micro-organisms.

The usefulness of $^{14}CO_2$ production from several radiometric substrates has been assessed as an indicator of the bacteriological quality of raw milk (Cogan and O'Connor, 1977). The method took about 10 h to detect 1.9×10^5 bacteria per ml with very wide confidence limits of 2×10^4–1.6×10^6 bacteria per ml (Fig. 14). The method as described for milk seems to be of little practical use. The poor prediction of the initial viable count from the detection time may be due to a number of different factors; different species of bacteria in the milk have different growth rates, respiration of glucose yields $^{14}CO_2$ whereas fermentation may not, somatic cells may metabolise the radio-labelled substrates, and metabolism is related to the number of individual

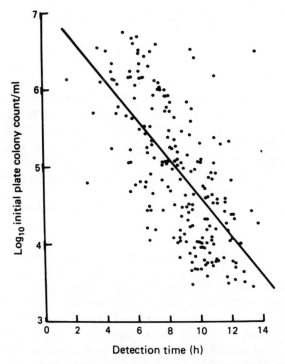

Fig. 14. Relationship between $\log_{10}$ plate colony count and radiometric detection time for samples of refrigerated raw milk (from Cogan and O'Connor (1977) by courtesy of the Agricultural Institute, Republic of Ireland).

bacteria, whereas the plate colony count records the number of viable clumps of bacteria.

SPECIFIC MICROBIOLOGICAL METHODS FOR MILK AND MILK PRODUCTS

Raw Milk

The microbiological flora of raw milk varies considerably with source, and the time and conditions of storage. High numbers of specific groups of organisms often indicate the likely source of contamination and, therefore, selective enumeration may aid trouble-shooting.

Streptococci are frequently the causative organsims responsible for bovine mastitis. These bacteria can be selectively enumerated by plating on Selective Streptococci Medium (Cousins, 1972). If the total count is high and these organisms form > 75 per cent of the bacterial flora of a bulk farm supply, then it is likely that there is mastitis in the herd. Generally, once the infected animal is detected and its milk excluded from the bulk milk, then the counts return to normal levels.

Coliform bacteria are also responsible for mastitic infections in cows. Coliforms can be selectively enumerated on Violet Red Bile Agar (American Public Health Association, 1972). This group of bacteria is less useful as an indicator of mastitis than streptococci, as coliforms are also introduced into the milk from faeces via dirty teats. Increased numbers of coliforms in the milk may, therefore, indicate mastitis or poor udder preparation prior to milking.

The numbers of spore-forming bacteria present in raw milk can be determined by heating a sample of the milk at 80°C for 10 min, and then plating on Milk Agar. Spore-forming bacteria generally originate in the faeces and gain entry to the milk via dirty teats. The numbers of spore-forming bacteria in milk are a possible index of faecal contamination. Generally the number of spores in the milk does not increase during storage, as these bacteria do not outgrow and sporulate under these conditions. However, the spore count may significantly increase as a result of contamination with milk and water residues, e.g. in tankers or equipment, where conditions may permit outgrowth and sporulation.

Psychrotrophic bacteria are mainly responsible for the deterioration in the microbiological quality of raw milk stored at refrigeration temperatures. Selective enumeration of this group indicates the time that the raw milk can be stored before processing. Various methods exist for enumerating psychrotrophs. The standard method is to plate on Milk Agar and enumerate colonies after incubation for 10 d at 7°C. As this method is time consuming, more rapid methods have been developed, e.g. enumeration of colonies after 24 h incubation at 25°C and the addition of crystal violet to inhibit the growth of Gram-positive bacteria. Impedance methods can be used to determine the level of psychrotrophic bacteria in raw milk. The detection times for 10^5 per ml, 10^4 per ml and 10^3 per ml are *ca.* 16, 19 and 23 h, respectively (Firstenberg-Eden and Tricarico, 1983).

Pasteurised Products

The major cause of spoilage of correctly refrigerated, pasteurised milk products are psychrotrophic bacteria which are post-pasteurisation contaminants. Assuming that the product is efficiently pasteurised and recontamination with the raw product is avoided, then these bacteria gain entry to the product from the equipment surfaces (notably the filler heads), the package or bottle, or from the air during filling. Very small numbers of psychrotrophic Gram-negative rods, mainly *Pseudomonas* species, can markedly reduce the shelf-life of these products. Commercial silo milks which were in-bottle pasteurised, thereby avoiding post-pasteurisation contamination, had a shelf-life of 14–32 days at 5°C, whereas the commercially bottled pasteurised milk had a shelf-life of 8–17 days (Schröder *et al.*, 1982).

The total viable count of pasteurised milk products is of little use in detecting post-pasteurisation contamination due to the higher numbers of heat-resistant bacteria (mainly spore-formers). Plate counts made using Violet Red Bile Agar can be used to detect gross contamination; 1 ml of product should be free from coliform bacteria. The technique can be made more sensitive by plating 5 ml of product on a large Petri dish. As the initial numbers of post-pasteurisation contaminants responsible for reducing the shelf-life of refrigerated milk products are below the detection limit of most microbiological methods, it is generally accepted that some form of pre-incubation is required to increase the numbers prior to enumeration.

One method which is widely used, particularly in the USA, is the Moseley Test in which the plate count is used to measure an increase in bacterial numbers in the product stored at 7°C (American Public

Health Association, 1972). This method is impractical in the context of a modern Dairy Industry as the pasteurised product in the distribution chain is spoiling at the same rate as that in the laboratory, and then it takes a further 2–3 days for the plate colony count to give a result. It is, therefore, likely that the consumer will detect any problems arising from poor production methods before the quality control laboratory. The modern Dairy Industry needs a rapid method for detecting post-pasteurisation contamination which, ideally, predicts the eventual shelf-life of the products.

Pasteurised dairy products contain relatively large numbers of Gram-positive bacteria, ca. 10 000 ml^{-1}, and small numbers of post-pasteurisation contaminants, often less than 1 ml^{-1}. Pre-incubation of samples of the products with chemicals inhibitory to Gram-positive bacteria, e.g. benzylkonium chloride and crystal violet, permits the selective enrichment of Gram-negative bacteria (Langaveld et al., 1976). Following selective enrichment, a number of microbiological methods have been used to detect an increase in bacterial numbers, electrical changes, or to predict the eventual shelf-life of the product. Biolumine-scence methods have been used to determine the numbers of post-pasteurisation contaminants in pasteurised milk using a Most Probable Number approach (Waes and Bossuyt, 1982), i.e. presence or absence in 1 litre, 100 ml, 10 ml and 1 ml portions.

Impedance methods have been used to determine the presence or absence of post-pasteurisation contamination in pasteurised milk (Bossuyt and Waes, 1983). Using crystal violet, nisin and penicillin as the selective inhibitors, Griffiths et al. (1984) showed that pre-incubation for 25 h at 21 °C followed by use of either the DEFT or bioluminescence method could indicate the extent of post-pasteurisation contamination of cream within 26 h of production. These methods could predict whether the product would keep for more or less than 7 days at 6 °C. A pre-incubated DEFT count has been used successfully to predict the eventual keeping quality of pasteurised milk within 24 h of production (Rodrigues and Pettipher, 1984). Samples could be divided into keeping quality groups of < 4·5, > 6, > 7·6, > 9·1 or > 10·6 days with 80–95 per cent correctly classified.

The use of selective enrichment techniques followed by rapid micro-biological methods offers the dairy industry the opportunity to predict the eventual quality of pasteurised milk products. The introduction of these methods should speed the detection of poor production techniques and reduce economic losses.

Products Treated at Ultra-high Temperature (UHT)

UHT products, such as milk and cream, should be free of viable micro-organisms if correctly processed and aseptically packaged. Occasionally poor processing or packaging techniques or contamination of the packaging material may lead to the presence of viable micro-organisms which can considerably reduce the shelf-life of these non-refrigerated products. Microbiological methods are insufficiently sensitive to detect very low levels of contamination without prior pre-incubation. In the Standard Method (Statutory Instrument, 1977) the package is pre-incubated at 30–37°C for 24 h and then 0·01 ml is plated, and if more than 10 colonies appear, the sample is considered to be contaminated and must be re-tested for confirmation.

The result of sterility testing of UHT products could be obtained faster if a rapid microbiological method were used in place of a viable count. The DEFT count has already proved useful for UHT milk (Pettipher and Rodrigues, 1981), and it is probable that bioluminescence and electrical methods could also be used. The use of a rapid method would permit the release of products 2 days earlier than if the standard method was used.

The Limulus lysate method for detecting the lipopolysaccharide of Gram-negative bacteria can be used on UHT products to give an indication as to the quality of the raw milk prior to processing. The lipopolysaccharide survives the heat treatment, and the concentration in the product is proportional to that in the raw milk. The lipopolysaccharide concentration is, therefore, a cumulative index of contamination with Gram-negative bacteria. The Limulus lysate test can be used to detect UHT products made from raw milk containing high numbers of Gram-negative bacteria (Südi, 1982).

Starter Cultures

Starter cultures of single or mixed strains of Streptococci and Lactobacilli species are widely used in the Dairy Industry, particularly for cheese and yoghurt manufacture. Although pre-prepared, freeze-dried cultures are commercially available, many manufacturers prefer to produce their own starter cultures on grounds of cost. It is advisable to check on the viability or activity of the starter culture before use, as a poor fermentation can result in considerable economic loss.

Generally mixed cultures are used on a rotational basis as inocula to reduce possible detrimental effects of phage build-up. Plate counts can be used to monitor the numbers of bacteria in starter cultures, but this is impractical as results are retrospective. Impedance detection times have been shown to correlate well with lactic acid production by starter cultures and, therefore, electrical methods may provide useful alternatives. The DEFT has been used to determine the numbers of bacteria in starter cultures, and it is probable that the bioluminescence method could also be used successfully.

Yeasts as Spoilage Organisms

Yeasts are the primary spoilage organisms of yoghurt, especially fruit yoghurt, due to contamination of, or growth in, the fruit mix before blending. It is, therefore, necessary to determine the numbers of yeasts in the fruit mix and final products. Conventional methods, such as plating on Rose Bengal Agar, take 3–5 days to give a result, often after the batch of fruit mix has been used. Electrical methods, based on either the impedance or capacitance signals, can be used to detect yeast cells in spiked fruit mixes. It takes *ca.* 14 h to detect 100 cells per ml (Fleischer *et al.*, 1984). Results suggest that pre-incubation to increase the initial numbers combined with an electrical method may be suitable for monitoring fruit mixes and finished products for yeast contamination.

POSSIBLE FUTURE DEVELOPMENTS

DEFT

It is likely that, over the next few years, the DEFT will become progressively automated and miniaturised. Results have shown that the sample volume and filter area can be reduced by at least ten times without affecting the count per ml. This could lead to a substantial saving in polycarbonate filter material and reagents. The development and marketing of an automated filtration and staining device would be of considerable benefit to DEFT users. Full automation of the sampling, pre-treatment and counting stages is also technically possible (Pettipher, 1983). Researchers are continuing to develop a method for discriminating between viable and non-viable cells in heat-treated products. If this were found it would substantially increase the applications of the DEFT.

Bactoscan

The Bactoscan, which is a fully automated instrument, would currently seem to be the method of choice for the routine testing of farm supplies in large central testing laboratories. The main disadvantage of the Bactoscan, at present, is its relative lack of sensitivity. The sensitivity of the instrument has been improved since its development, and it is now possible to grade milks at the 50 000 bacteria per ml level with reasonable accuracy. However, the current UK testing scheme has a premium grade for milks with less than 20 000 bacteria per ml, and the Bactoscan instrument is unable to accurately grade milks at this level. It is possible that future developments may further increase the sensitivity of the instrument, probably by improved sample pre-treatment, sufficiently to enable the Bactoscan to replace the automated plate loop method as the microbiological method used in central testing laboratories.

Bioluminescence

Bioluminescence methods are useful for many microbiological applications but they are less applicable to milk due to the high level of somatic cell ATP which is incompletely destroyed during pre-treatment of the sample.

It is possible that an improved technique for separating somatic and bacterial cells or that improved selective destruction of somatic cell ATP may improve discrimination and make these methods more useful to the Dairy Industry.

Electrical Methods

Improved methodologies have resulted in the use of electrical methods for the selective enumeration of groups of organisms in milk, e.g. total, mesophilic, psychrotrophic, Gram-negative bacteria and yeasts. It is likely that future developments, most likely in media engineering, will extend the number of applications of electrical methods to milk products. An increased understanding of what these methods actually measure would facilitate these developments.

Biosensors

There is a possibility that future developments may make biosensors the microbiologist's method of choice. Currently biosensors are mainly used for medical applications to detect low levels of enzymes, substrates or products (Lowe *et al.,* 1983). Solutions can be continuously monitored for chemicals as the response time of biosensors may be only a few seconds. It is probable that in the future biosensors which detect biomass will be available for industrial fermentations. It has been shown that a lactate sensor, which uses immobilised lactate oxidase, can measure the increase in numbers of lactate-producing bacteria (Matsunaga *et al.,* 1982). The complexity of milk may prevent the use of currently available biosensors, but it is possible that in the future biosensors which can continuously measure the numbers of micro-organisms in milk may be developed. This would permit the continuous monitoring of the microbiological quality of milk at all stages in the distribution network.

CONCLUSION

The Dairy Industry has been using conventional microbiological methods based on cultural techniques for many years. There has recently been a considerable improvement in the microbiological quality of farm supplies as a direct result of regularly applying quantitative microbiological analyses and, importantly, changing the payment schemes to reward hygienic producers.

Under pressure from their major buyers, the dairies and creameries are now changing from retrospective cultural techniques to rapid instrumental methods of microbiological analysis. Of the current methods, the DEFT, electrical and bioluminescence methods have applications in the processor's laboratory, whilst the fully automated plate loop method and Bactoscan instrument are better suited to the large central testing laboratories.

Rapid microbiological methods have considerable advantages over their conventional counterparts, providing results in as little as 30 min. This permits corrective action to be taken almost immediately, should this be required. These methods, perhaps with either modifications or sample pre-treatment, can also be applied to milk products. They can, therefore, be used by the processor to monitor incoming raw supplies, processing and the microbiological quality of the finished products. The increased use of rapid microbiological methods by the Dairy

Industry should lead to improved control of raw materials, processes and products, thereby reducing economic losses.

Looking to the future, the development of biosensors may revolutionise the microbiological analysis of milk. The creamery of the future may include the remote continuous monitoring of the microbiological quality of milk at all stages from receipt of the raw material to packaging of the finished product.

REFERENCES

American Public Health Association (1972). Standard Methods for the Examination of Dairy Products, Washington, D.C.

Bossuyt, R. G. (1981). *Milchwissenschaft,* **36**, 257.

Bossuyt, R. G. (1982). *Kieler Milchwirtschaftliche Forschungsberichte,* **34**, 129.

Bossuyt, R. G. (1983). *Journal of Food Protection,* **46**, 622.

Bossuyt, R. G. and Waes, G. M. (1983). *Journal of Food Protection,* **46**, 622.

Bradshaw, J. G., Francis, D. W., Peeler, J. T., Leslie, J. E., Twedt, R. M. and Read, R. B. (1973). *Journal of Dairy Science,* **56**, 1011.

Breed, R. S. and Brew, J. D. (1916). Technical Bulletin No. 49, New York Agricultural Experimental Station, Albany.

Britz, T. J., Bezuidenhout, J. J., Dreyer, J. M. and Steyn, P. L. (1980). *South African Journal of Dairy Technology,* **12**, 89.

Cady, P., Hardy, D., Martins, S., Dufour, S. W. and Kraeger, S. J. (1978). *Journal of Food Protection,* **41**, 277.

Claydon, T. J. (1975). *Journal of Milk and Food Technology,* **38**, 87.

Cogan, T. M. and O'Connor, F. (1977). *Irish Journal of Food Science and Technology,* **1**, 49.

Cousins, C. M. (1972). *Journal of the Society of Dairy Technology,* **25**, 200.

Cousins, C. M. and Bramley, A. J. (1981). *Dairy Microbiology,* **1**, 119.

Cousins, C. M., Sharpe, M. E. and Law, B. A. (1977). *Dairy Industries International,* **42**, 12.

Cousins, C. M., Rodrigues, U. M. and Fulford, R. J. (1981). *Journal of Dairy Research,* **48**, 45.

Fifield, C. W., Hoff, J. E. and Proctor, B. E. (1957). *Journal of Dairy Science,* **40**, 588.

Firstenberg-Eden, R. and Tricarico, M. K. (1983). *Journal of Food Science,* **48**, 1750.

Fleischer, M., Shapton, N. and Cooper, P. J. (1984). *Journal of the Society of Dairy Technology,* **37**, 63.

Fleming, M. and O'Connor, F. (1975). *Irish Journal of Agricultural Research,* **14**, 27.

Fondén, R. and Strömberg, A. (1978). *XX International Dairy Congress,* Paris, Publishers Congrilait, Paris, E 330.

Galbraith, N. S., Forbes, P. and Clifford, C. (1982). *British Medical Journal,* **284**, 1761.

Gilchrist, J. E., Campbell, J. E., Donnelly, C. B., Peeler, J. T. and Delaney, J. M. (1973). *Applied Microbiology, 25*, 244.

Griffiths, M. W., Phillips, J. D. and Muir, D. D. (1984). *Journal of the Society of Dairy Technology, 37*, 22.

Hadley, W. K. and Senyk, G. (1975). *Microbiology, 1975*, 12.

Hansen, K. (1982). *Kieler Milchwirtschaftliche Forschungsberichte, 34*, 138.

Hansen, K., Mikkelsen, T. and Moller-Madsen, A. (1982). *Journal of Dairy Research, 49*, 323.

Kielwein, G. (1982). *Kieler Milchwirtschaftliche Forschungsberichte, 34*, 74.

King, J. S. and Mabbitt, L. A. (1984). *Journal of Dairy Research, 51*, 317.

Law, B. (1979). *Journal of Dairy Research, 46*, 573.

Langaveld, L. P. M., Cuperus, F., Van Breemen, P. and Dijkers, J. (1976). *Netherlands Milk and Dairy Journal, 30*, 157.

Lowe, C. R., Goldfinch, M. J. and Lias, R. J. (1983). In: *Biotech. 83,* p. 633, Online Publications, Northwood, UK.

Lück, H. (1982). *Kieler Milchwirtschaftliche Forschungsberichte, 34*, 108.

Matsunaga, T., Karube, I., Teraoka, N. and Suzuki, S. (1982). *European Journal of Applied Microbiology and Biotechnology, 16*, 157.

McKinnon, C. H. and Pettipher, G. L. (1983). *Journal of Dairy Research, 50*, 163.

O'Connor, F. (1979). *Irish Journal of Food Science and Technology, 3*, 93.

O'Connor, F. and Fleming, M. G. (1979). *Irish Journal of Food Science and Technology, 3*, 11.

Peeler, J. T., Gilchrist, J. E., Donnelly, C. B. and Campbell, J. E. (1977). *Journal of Food Protection, 40*, 462.

Peterkin, P. I. and Sharpe, A. N. (1980). *Applied and Environmental Microbiology, 39*, 1138.

Pettipher, G. L. (1983). *The Direct Epifluorescent Filter Technique,* Research Studies Press, Letchworth, UK.

Pettipher, G. L. and Rodrigues, U. M. (1981). *Journal of Applied Bacteriology, 50*, 157.

Pettipher, G. L. and Rodrigues, U. M. (1982). *Journal of Applied Bacteriology, 53*, 323.

Pettipher, G. L., Mansell, R., McKinnon, C. H. and Cousins, C. M. (1980). *Applied and Environmental Microbiology, 39*, 423.

Rodrigues, U. M. and Pettipher, G. L. (1984). *Journal of Applied Bacteriology, 57*, 125.

Saarinen, K. (1984). *Meijeritietellinen Aikauskirja,* XLII (n:01) 33.

Schröder, M. J. A., Cousins, C. M. and McKinnon, C. H. (1982). *Journal of Dairy Research, 49*, 619.

Sharpe, A. N. and Kilsby, D. C. (1971). *Journal of Applied Bacteriology, 34*, 435.

Sharpe, A. N. and Michaud, G. L. (1974). *Applied Microbiology, 28*, 223.

Sharpe, A. N. and Michaud, G. L. (1975). *Applied Microbiology, 30*, 519.

Sharpe, A. N., Biggs, D. R. and Oliver, R. J. (1972). *Applied Microbiology, 24*, 70.

Sharpe, A. N., Peterkin, P. I. and Malik, N. (1979). *Applied and Environmental Microbiology, 38*, 431.

Statutory Instrument (1977). No. 1033, HMSO, London.

Südi, J. (1982). *Kieler Milchwirtschaftliche Forschungsberichte, 34*, 141.

Suhren, G. (1982). *Kieler Milchwirtschaftliche Forschungsberichte,* **34**, 117.
Terplan, G., Bierl, J., Von Grove, H. H. and Zaadhof, K. J. (1981). *Archiv fur Lebensmittelhygiene,* **32**, 15.
Tolle, A., Heeschen, W., Wernery, H., Reichmuth, J. and Suhren, G. (1972). *Milchwissenschaft,* **27**, 343.
Waes, G. and Bossuyt, R. (1981). *Milchwissenschaft,* **36**, 548.
Waes, G. M. and Bossuyt, R. G. (1982). *Journal of Food Protection,* **45**, 928.

Chapter 8

Technology for the Developing Countries

J. T. Homewood
Nestec Ltd, Vevey, Switzerland

The reader, especially if without experience in developing countries, may well ask why a separate chapter, under such a title should be considered necessary in a textbook on the dairy industry. Fresh milk has been a household article, often obtained from one or more animals kept by each family, in much of Asia, the Middle East and Europe since earliest historical times. European immigrants brought cattle to North and South America, and to Australasia, from the sixteenth century onwards. With their herds came the early dairy technologies of curd and cheesemaking, butter or yoghurt production, long established in Asia and Europe.

What has happened to justify an existing technology differential, as implied in the title, for those countries having centuries of dairying tradition, yet still categorised as developing? To answer this question, it has been necessary to recall some major steps in the evolution of the dairy industry, and related technology, in the developed countries. For those confronted with problems of a specific stage in their own country's development, this may serve to provide parallels with earlier experience. Errors and waste of resource may thus be avoided. The significance in this earlier experience of rapid growth of urban population and its needs, stimulating collection, transport, new processing methods and better distribution systems, is important for those looking to the future in the developing world. The benefits derived by the farming community from these systems should also be kept in mind.

The example of developments in the last 30 or 40 years in India, which by itself constitutes an important part of the developing world, is useful to illustrate how timely measures can at least mitigate problems arising from rapid urban expansion, to the benefit of urban health and rural productivity.

HISTORICAL EVOLUTION OF THE DAIRY INDUSTRY

A Common Starting Point

Dairy cattle are believed to have been first domesticated some 6000–10 000 years ago, and are recorded in early documents and artefacts in India, Babylonia, Egypt and Old Testament Palestine.

Milk was known as a food of high value — but was also recognised as being highly perishable. Therefore its transformation into products of somewhat greater keeping quality, such as cheese, butter and curds has been known and practised for many centuries. In India, many products such as Ghee (Samna), Khoa or Rabri (Anon., 1941) are still widely produced. These are suitable for domestic or small-scale, artisanal production. Marco Polo (Latham, 1958) records the use of sun dried skim-milk by Tartar cavalry in the mid-thirteenth century. Domestic or artisanal producers can dispose of whey or other residues arising in small quantities, for human or animal feeding, so that wastage is reduced. However, the products can be of uncertain keeping quality. Butter and cheesemaking techniques were certainly transmitted to North and South America with the arrival of colonists in the sixteenth century, and later to Australasia and New Zealand. It is of interest, especially when considering the recent questioning of the value of milk products, to recall that the 'Mayflower' settlers in North America in 1625 lost half their numbers in their terrible first winter — and all their children under 2 years of age. They brought no cows with them; but subsequent parties were better advised, and were indeed obliged to bring a certain proportion of cattle, sheep and goats for each family; they fared better. Although many of the earliest importations into North and South America were mainly for meat and as draft animals, some were milked for domestic needs. These multiplied fast; in the Valley of Toluca, Mexico, their numbers are recorded as doubling every 15 months — first introduced in 1538, they numbered 150 000 twenty years later. Improvement of dairy characteristics in the herds in Latin America followed in the late eighteenth and mid-nineteenth centuries; artisanal cheese manufacture started in 1875 in Brazil (region Serra da Mantiqueira Minas Gerais Province).

What, therefore, took place in Europe and North America to differentiate these regions from the developing world? Improved dairy cattle had been bred in Holland, in the United Kingdom and elsewhere. There was, however, no significant difference in the technology used in dairying in the first surge of expansion into North and South America and Australasia.

The Differentiation of Technology

The differentiation of technology in the developed countries followed the industrial revolution from which their new 'developed' status stemmed. Demand was stimulated by population growth and concentration in urban manufacturing centres. Rapid improvements in road and rail systems made it possible first to bring fodder in increased quantities to urban cowsheds, and to remove manure. In England, these cowsheds were judged, anyway until the 1860s, to provide the best milk in towns, as their cows were better fed and often better housed than those on farms surrounding the towns. In the 1840s and 1850s, milk was retailed round the streets of London by men and women carrying on a shoulder yoke a pair of wooden or metal tubs holding 8 or 10 gallons in all (Whetham, 1964).

The reduction in transport costs brought about by railways had a complex effect on the market. It was cheaper to move fodder and dung to and from urban cowsheds; but quantities of milk from distant farms (often of poor quality due to unsatisfactory milking conditions, transport and lack of cooling) competed wherever people required milk for cooking rather than drinking. At that time, a premium would have been paid for milk 'warm from the cow', and many consumers would otherwise boil any milk before consumption (as is still done in many developing countries). The increasing volume of sales in towns in the 1860s led to the use of hand or horse-drawn carts, carrying bulk containers from which milk was usually ladled by a measured dipper into the consumer's jug. Whetham (Whetham, 1964) says 'First consumers from a churn got all the cream and the last only skimmed milk. Further, all dairymen were plagued with the dishonesty of their roundsmen, who sold more milk than they were allotted, with the help of water'.

An impulse towards the elimination of town cowsheds came from a combination of the recognition of the often poor hygienic state of these sheds, and from diseases which inflicted heavy losses on the urban dairies' herds in 1865/66. Both factors led to a rapid increase in rail deliveries of 'fresh country milk', now held in higher esteem. Supplies to London by rail increased from 7 million gallons in 1866 to 20 million gallons in 1880 (Whetham, 1964).

A complex structure of contract purchase from farmers near to railway stations, and the purchase or sale of surplus quantities between retailers at the London railway terminal, led to the establishment of milk wholesalers, who provided churns, supervised transport and dealt with deficiencies or surpluses in supply. These wholesalers began (Whetham, 1964) in the late 1870s to set up depots at country railway stations, where milk delivered by farmers could be properly cooled. This cooling was carried out in the first place by well water, which in England probably permitted cooling milk to not more than 15 °C. Private milk retailing firms, growing in size, and wholesalers, provided a basis for the formation of limited companies and the provision of capital necessary for three technological advances which occurred at about this time.

First, in 1879, the centrifugal cream separator was exhibited at the Royal Agricultural Society's show, and this made butter manufacture a factory rather than a farm trade. Instead of standing milk for 24 h for cream to rise, milk could be run through the separator to give a continuous supply of fresh cream. Steam-powered churns and mechanical butter-workers could finish the task. The second new technology available in the 1870s to 1880s was mechanical refrigeration. At first, this facilitated imports of dairy products from America or Australasia, which depressed prices and caused farmers to turn from butter and cheesemaking to liquid milk sales; butter and cheesemaking co-operatives in the Midlands switched to cooling bulk liquid milk for the urban trade, and mechanical refrigeration, cooling milk to 4 °C, was the best guarantee of success wherever the capital and technical skills necessary could be assured. The third technological innovation was the milk condensing plant. In the 1860s, processes of vacuum concentration had been applied to milk in the USA, resulting in the development of so-called sweetened condensed milk and unsweetened (evaporated) condensed milk. These were at first distributed in cities in bulk exactly as was fresh milk. The American Civil War created a demand for army requirements, and here sweetened condensed milk, which was protected (like fruit jams) from bacterial development by the osmotic pressure exerted by its sugar content and packed in sealed, tin-plate containers, was extensively used.

This storable product, rapidly adopted for ships' stores, was also used to supplement fresh milk supplies in towns. Condenseries, built in country districts, could supply remote towns without insurmountable risks of spoilage, and many were constructed from the 1860s and 1870s onwards, in the USA, in Switzerland, England and France. These factories stimulated developments in tin-making on a factory scale, and the use of steam (as in the wholesalers' depots) for cleaning and sterilisation of farmers' cans. Trade in condensed milk soon crossed national frontiers, and a new range of dairy products, later expanded by milk drying techniques, first on steam-heated rollers and later by spray-drying techniques, was added to the existing international trade in butter and cheese.

These three technologies, separation, cooling and condensing/drying, enabled fuller use to be made of improved transport systems, and later of techniques such as pasteurisation (originating in the brewery industry). But their most important outcome was, perhaps, the increased requirement of capital involved by participation in the dairy business. Individual producers could (and still do) distribute their milk as producer-retailers. But the larger enterprises involved in wholesaling, transport and processing were obliged to develop tighter controls and suppress fraudulent practices, such as those described above by Whetham; increase in scale and capital involvement enhanced the importance of the assurance of a good reputation in order to retain the customer. The concentration of milk handling and retailing also facilitated the work of health, weights and measures, and other public authorities.

Whetham (1970) states: 'By 1914, the London trade was dominated by four or five wholesalers, equipped with country depots, a huge supply of churns, regular contracts with the railways, pasteurising plant and cold stores in their town dairies'. This should be contrasted with the situation described by Whetham as existing 15 years earlier: 'Bottled or specially treated milk provided a small fraction of the trade before 1900; only the medical profession and a few educated families were aware of the gross bacterial infection which contributed to such widespread diseases as infant diarrhoea and tuberculosis. Yet this growing interest in hygiene encouraged consumers to be suspicious of cheap milk, and to favour the large firms with cooling depots in country districts with steampowered plant for washing churns and delivery cans'.

The importance of the improvements achieved by 1915 and later can be assessed. Beaver (1973), commenting on the infant mortality rates for England and Wales (1840–1970) illustrated in Fig. 1, refers to the striking reduction recorded (from 150 to less than 20 per 1000) after 1900,

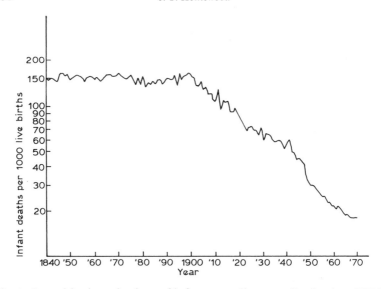

Fig. 1. Logarithmic scale chart of infant mortality rates, England and Wales, 1840–1970. (Reproduced with permission from Beaver (1973).)

and claims that this reduction can be attributed in large part to improved availability of safe milk (sometimes as fresh milk, but also in the form of suitable infant foods) and the diminishing presence of poor quality milk.

All in all, the result of concentration in the wholesaling, transport, retailing and processing of milk in England and Wales helped to bring about a marked — and justified — increase in public confidence in the end product. This result, and the change in the image of the milk vendor as at best a doubtful, and often a definitely untrustworthy, commercial partner, points the way for those wishing to build up dairy projects in countries where the conditions for such public confidence do not yet exist. Such confidence is as vital as the latest equipment, and can be harder to obtain.

The preceding paragraphs sketch only the broad outline of the main technological changes affecting the dairy industry in England and Wales. Something of the same sort occurred on a much larger scale in the United States and also, with local differences due to dietary habits and other causes, in most of Europe. A host of smaller changes have added to the total result, including detailed refinement of machinery and equipment, and in recent years, changes in marketing structures which are still not complete. The simplified version of events given in this section is intended to help readers wishing to define the state of development elsewhere. It may also, perhaps, encourage those facing apparently adverse conditions for a new venture, to concentrate on the essentials, and not to forget the importance of building, and sustaining, public confidence in milk and milk products. As has been found in the developed countries, this calls not only for an increase in public control functions, but also and above all for the building of a sense of responsibility amongst those working in the industry, at every stage from the farm to the consumer's doorstep.

The Return to a Common Technology

In the developing countries, the technology of the eighteenth century European and North American producers and processors has some-times persisted. This was, until 1947, broadly the state of affairs in India, which may be taken as an example of the rapid return to a common technology possible when circumstances are favourable, and Government policies also favour development.

Earlier conditions in India (Anon., 1941) can be judged from the following extracts from that report:

'Most of the village cattle are semi-starved and badly managed — about half the milk required in urban areas is produced on the spot under most uneconomical and insanitary conditions. ... duty-free import (of skimmed milk powder) is being used for the adulteration of whole milk — the measures at present used in the milk trade are both unsatisfactory and unsuitable — most milk is distributed under filthy conditions — vessels used have no lids and dirty straw, green grass and plugs made of old newspapers and rags are used instead ... pasteurised milk is at present available only at a few places but even at these the hospitals, etc., do not buy it always because of its high cost.'

Some pictures taken in 1957 (Figs. 2 and 3) show that poor handling conditions still existed. Some of the recommendations of the 1941 report were, however, already being applied. In particular, that calling for a tighter control of imports of skim-milk powder was put into force. This made it possible to use the powder under controlled conditions for adjusting the usually high fat content of Indian fresh milk (often above 4·5 per cent in cow's milk, and over 7 per cent in buffalo's milk). The skim-milk powder was used to produce 'toned' milk, which had a fat content corresponding to or below the European legal minimum compositions, but maintained the solids-non-fat content at about 8·7 per cent. The 1941 report was much concerned with improper use of the powder to cheat the customer, but the post-independence measures were sufficient to make this risk acceptable. Amongst important projects launched in the post-war years on the basis of toning were the Bombay Airey Dairy Colony and Bombay Milk Scheme which, based on legislation brought in in 1957, compulsorily re-housed animals formerly in urban stables in a large rural project, where some fodder could be grown, using manure (which would otherwise have been dried for use as fuel or wasted) as fertiliser. This project permitted enforcement of hygienic stabling and milking conditions. The concentration of production facilitated properly controlled processing by

Fig. 2. Milk collector, India, 1957. Foreign matter used to reduce spillage from milk containers carried on bicycle.

Fig. 3. Collection point for milk, India, 1957.

toning (to 2 per cent fat), pasteurisation and bottling (or later plastic pillow packs), with distribution through properly supervised sales points in the city. The scheme prospered and was complemented in the 1950s by the building, with New Zealand Government bilateral aid UNICEF support, of a modern processing plant (initially for milk powder and butter production) near Anand, some 200 km north of Bombay. Here the Kaira Co-operative had for many years supplied pasteurised fresh milk by rail to Bombay and continued to do so.

The concept of controlled toning of fresh milk was soon extended to other cities. However, the major innovation was the setting up of a *de facto* equalisation scheme for the quantities of skim-milk powder required. At this time, the United States, under Public Law 480, made available either on a bilateral basis, or through United Nations agencies such as the Food and Agriculture Organisation (FAO) under the so-called Operation Flood, large quantities of this powder which could be landed at low cost. The Indian National Dairy Corporation was set up and assumed exclusive responsibility for all imports of the low cost skim-milk powder and their subsequent use, charging users a higher price determined by demand. With the considerable sums derived from these sales, local milk production and the building of a chain of modern processing facilities was supervised and subsidised by the Corporation.

In addition, an element of private industry participation in setting up new plants, both from local and foreign firms, was encouraged. Through the various city milk schemes, the creation of new facilities using the latest available technology made possible a better urban supply of safe fresh milk, vitally important over a period when social and economic factors caused huge increases in urban population. Calcutta, Bombay, Madras and Delhi, each with about 2 million inhabitants in 1940, grew by the 1980s to 13 million, 11 million, 9 million and 8 million respectively. The chain of rural processing plants, some co-operatives or under State Dairy Corporations and some private firms, produced large quantities of milk powder and infant foods. In one year, one co-operative group alone converted more than 200 000 tons of milk into more than 45 000 tons of such products. This had a double effect. If fresh milk were lacking or scarce, the consumer, whether in towns or in the countryside, could buy a safe substitute, and many milk producers in villages could now find a guaranteed outlet for their production, surplus to domestic requirements. They could also often count on advice and assistance in increasing this production through the field services, veterinarians, fodder supply facilities and hygienic training often made available by the new processing plants (Anon., 1975, 1982; Khurody, 1977).

It is certain that in India, with about one-sixth of the world's population, many of whom have a very low standard of living, conditions are still far from ideal. It has also happened that projects using modern technologies were badly sited or badly managed. Nevertheless, centrally controlled use of available external aid in the form of low-cost supplies of skim-milk powder, and of modern technology in both the public and the private sectors, backed by adequate training at the national level, has made it possible to promote public health (Beaver, 1973) and to create the basis for transformation of an age-old industry. The example illustrates the principle of self-help, where improvement in agriculture can help to solve problems of urban nutrition which are likely to become of increasing concern (Fox, 1984).

Other examples of modernisation have been based not only on a supplementation of local fresh milk supplies by toning quantities of skim-milk powder, but by large-scale use of imported constituents wherever little or no local fresh milk can be obtained. This has allowed the setting up of factories planned to replace imports of condensed or powdered milks, using recombining techniques first fully developed for supplies to American forces in the Pacific area in 1941–45. Here, ice cream and fluid milk were the main requirements, but by the 1960s, trials had shown that sweetened condensed or evaporated milk, and even later milk powders and infant foods, could be manufactured by recombining butter oil and suitable skim-milk powders. There were powerful economic arguments in favour of this, whenever dairy surpluses were being made available, as under PL 480 from USA or later, under the EEC's surplus export scheme.

Such factories used existing modern technology with the few additional features necessary to ensure rapid and complete reconstitution of the dry powders or butter oil required. Some details of these devices are provided in this chapter. Plants of this type were set up in the 1960s in Thailand, Malaysia, Indonesia, The Philippines and Vietnam. They were also to be found from the 1970s in several West African countries. In the Far East, some increase in local milk production can be attributed to these plants, where governments have fixed progressively higher fresh milk prices. These can, perhaps, be borne for a time by local recombining industries if local fresh milk only constitutes a small part of their total raw material needs — in effect at the expense of the consumer of preserved milk products. Longer term, an equalisation scheme, allowing the burden of high local milk costs, if necessary to encourage an increase in production, to be shared by all users of skim-milk powder (which goes to ice cream and yoghurt manufacturers) will be needed for further progress (Rampini, 1978).

As stated earlier, the area where specific technology for developing countries is called for, is above all in fresh milk production and collection. The recombining plants themselves mostly use equipment and methods derived from their counterparts in the developed world.

THE ESTABLISHMENT OF A DAIRY INDUSTRY — PROBLEMS AND SOLUTIONS

In countries with some existing fresh milk production, the manufacture of local cheeses is generally the simplest way adopted for absorbing surplus milk at any distance from urban centres. This is usually a white soft cheese, farm produced, and with limited keeping quality, but requiring little investment for equipment or buildings. Such cheeses are consumed locally, and the whey can be fed to stock. The next development has often been that of a milk powder or condensed milk production, favouring the total milk production and its quality, but pushing the artisanal cheese production further away.

Urban development and the resulting demand for liquid milk and milk products can subsequently demand priority for pasteurised and UHT milk, soft cheese and yoghurts. The canned or dried milk operations then have to seek further afield for their milk supplies. Depending upon the scale of development in each particular instance, a check list which may be useful to those called upon to assess or plan a dairy project in a developing country, could be as follows. Only factors having a strong influence on ultimate success or failure of such a venture are considered.

Political and Economic Context

(1) Actual and estimated potential demand for end products of the venture; present and projected *per capita* consumption of such products and of alternative foods; present local production of potential raw or packing materials; possibility of procurement of individual ingredients.

(2) Existing trade in fresh milk; existing or potential price structures for fresh milk (see typical comparative figures illustrating differences in developed and developing countries, Table I), and for milk constituents, such as locally produced or imported butter, butter oil, and full cream or skim-milk powders; availability of ancillary constituents, such as sugar or lactose, and probable incidence of other cost components such as labour, energy, packing materials and taxes.

(3) Which competing products, locally produced or imported, already exist? What are their volume, value and price structures?

(4) Existing or potential price structures for finished products foreseen; do these permit economically self-sustaining operations or are subsidies or tariff protection necessary, and obtainable? In estimating cost of local production, the effect of variances of, for example, ± 10 per cent or even 20 per cent should be calculated for sales volumes, and for principal raw material and end product prices.

(5) Determine existing Government policies and practices affecting prices, particularly price controls or import constraints (quota systems) applied to finished products; do such constraints permit price or other

TABLE I
Farmgate prices (in US cents kg^{-1})

Jamaica	18	(1983)	Spain	24	(1982)
Holland	23	(1983)	Panama	24	(1982)
France	22	(1983)	Colombia	28	(1983)
Switzerland	43	(1983)	Brazil	20	(1982)
Australia	16	(1983)	Trinidad	59	(1983)
Ecuador	28	(1982)			

adjustments in face of variations in raw material costs, fluctuations in consumer demand, inflation in local or imported cost elements? What other support will be available to the project if constraints mentioned above threaten its economic survival?

(6) What are Government, trades union, or other bodies' policies and practices affecting, in the shorter or longer term, the effective management of the project?

Fresh Milk

If the outcome of a preliminary investigation, along the lines of the check list above, is broadly favourable, the next step may be to investigate the availability of fresh milk. This may in some developing countries present no problem other than that of assessing competitive demand (from other established industries) for an existing resource. However, some other such countries may not yet possess an organised fresh milk production; and for these instances, a further and more extensive check list may be useful; this is outlined below.

(1) What are the characteristics of existing herds of milking animals (e.g. cow, buffalo, yak, chowrie, sheep, etc.)? What, if any, schemes exist for improving these characteristics? What are the customary levels of animal nutrition?

(2) What are the characteristics of cattle owners — are human nutritional levels adequate to allow collection of milk surplus to domestic requirements? What is their level of literacy, and their openness to motivation and innovation?

(3) What are the preconceptions of potential suppliers concerning the sale of milk? Concerning the slaughter of surplus stock and sale of meat? What is the price ratio beef/kg live weight to milk/kg (should be around 4 or less)?

(4) Is there information concerning economics of milk production/ stock raising in the form of comparisons of yield/hectare with alternative crops? Take into account suitability of terrain/climate for such alternatives and possible under-employment of family labour resources; what would be the social impact of regular payments for milk versus annual payments for other crops?

(5) What, if any, system of credit (Co-operative, Agricultural Credit Bank, etc.) is available to producers? What are the interest rates charged? What chance to secure loans do small or even landless cow-owners have?

(6) What surfaces could be made available for fodder growing, or what possibilities exist for fodder collection by those lacking such surfaces? Resources such as crop residues or straw, cane tops, silage, molasses, trees and grasses from dividing strips in other cultivations such as rice paddies, should be assessed; what new varieties of grasses or legumes or other fodder crops could be introduced?

(7) Is there potential for increased production through improved animal health brought about by education of producers and by increased availability of veterinary care and supplies? What is their present fertility and calving interval?

(8) Is there a possibility of integration of increased animal production into existing mixed or other farming patterns? Is the value of animal manure for fertilisation of other crops understood, or is this used for fuel, or neglected?

(9) Is there some form of custom or planning defining arable and grazing areas? Are producers nomadic or semi-nomadic? If so, consider problems of overstocking likely to arise if revenues from fresh milk sales are directed to herd increase — any new venture for meat processing may be a desirable complement rather than a competitor in such instances.

(10) Are main and farm or village access roads adequate for all weather use? If not, what plans or possibilities exist at State or local level to remedy this?

(11) Can the milk collection area (milk district) be extended in future times? Can such extensions reach into areas of varying pluviometry or irrigation, or with varying soil types and traditional crop patterns, to facilitate protection against dominance by the fortunes of one or two major crops which may contend with animal fodders? Do soil analyses reveal major or trace element deficiencies? Will there be an improvement of soil structures through increased animal manure?

(12) What quantities of milk are likely to be available for collection initially in the district (litres $km^{-2} yr^{-1}$)? What competition is to be expected for supplies of fresh milk available in the district? What are the distances to major points of consumption of fresh milk from the district? Do the authorities concerned define a milk district for any project? Are there possibilities of agreements to limit uneconomic duplication of collection systems? What potential exists for long term expansion of intake per square kilometre of a district, to meet increased sales requirements, or possible future reductions in the district for the benefit of other milk users?

MEASURES TO STIMULATE MILK PRODUCTION

It will be seen from these check lists that to establish a dairy industry in a developing country calls for more than a good knowledge of modern dairy technology. A grasp of politics will be needed — since this is an industry which for a given capital input can involve relatively large numbers of producers and very large numbers of consumers, linked by a product which is more perishable than most, and in many countries is a basic foodstuff. Both producers and consumers are an element in politics. A sound commercial sense, some feeling for sociological questions, a good knowledge of animal husbandry and related animal health problems, and an appreciation of the evolution of infrastructural elements, will add to the chances of correct conclusions by a technologist assessing a potential milk district. Of course it will, in the end, be technology and economics which mainly determine the outcome of a project; but the technical man will do well to draw conclusions on the other aspects, from whatever resources are available. It is a great advantage if these resources are local ones, since people on the spot usually know best about their own region and its problems.

Analysis of the existing state of affairs may show adequate willingness by producers to sell surpluses of milk, promising potential for increased fodder production, and possibilities of increasing herd numbers by improved fertility (from better feeding) and by reducing calf mortality. There may be good future prospects for increased milk production per animal by improvements in veterinary care and through herd improvement (provision of bulls or artificial insemination with superior genetic characteristics). There may also be plans or, better, work in progress on infrastructures such as roads and irrigation systems.

Those preparing a project should not, however, assume that these necessary components of increased potential milk production will automatically be evoked by the creation of a new market for fresh milk, in the shape of a collecting, processing and product marketing venture. A well conducted campaign to inform potential suppliers, a sound collection system, fair dealing over quality and prompt payment for supplies are essentials for success; but it will still be necessary to make a cool assessment of factors which will help or hinder such a development. The comprehensive survey by McDowell (1981) may suggest other positive or negative factors in a particular instance. Where the State or Local Government bodies provide services, their effectiveness should be checked, and the cost of providing any necessary additional back-up included in the venture costing. It may, for instance, become evident that supplies of high quality seeds for fodder growing are not available (sometimes due to administrative, phytosanitary or import licensing obstacles). The possibility of overcoming such obstacles should then be assessed and costed.

After considering all likely obstacles to increased production, the additional cost of all necessary supports must be taken into account when estimating the total cost of fresh milk as a raw material in a dairy project. Typically Field Services and Inspection may add between 2 and 5 per cent to raw milk cost — and much more in the first years of a project.

Raw Milk Price Structures

In many developing countries, fat testing and payment per kg of milk fat is common. This helps to eliminate fraud (Rampini, 1978). However, where a rigid year-round price is paid, in a country where production drops seasonally in dry weather, suppliers are tempted by higher prices offered by itinerant milk vendors. This can best be met by a premium (which must be large enough to be perceptible — up to 15 per cent on basic price) to those suppliers who maintain deliveries at such times (see Fig. 4). Payment on milk solids-non-fat is rarely encountered.

Social Factors

It has been sometimes observed that in close-knit, village farming communities there are frequently affinities and antipathies which can influence milk collection. Experience has shown that by setting up more than one collecting centre, at suitable points in a village, quantities collected can be substantially greater. Not only is the collection system physically more accessible to suppliers; they also have a choice as to whom they sell their milk.

COLLECTION OF RAW MILK

The movement of available supplies of raw milk from the producer's shed, yard or the point in the field where milking takes place, to the processor's plant, often takes place under very different conditions in

Fig. 4. Seasonal shift of peak and increase of intake over one year, caused by 15% bonus paid for the first time in a low season, compared with pattern of intake collected in a larger area three years previously without premium: ----, original intake from 19 000 suppliers in 9000 km^2; ———, intake developed from 22 500 suppliers in 6000 km^2 with premium payments.

developed and developing countries. In the latter, road systems are liable to be less developed (see Figs. 5 and 6). The climate is often more arduous. Individual producers may supply only very small quantities, varying according to domestic needs. For all these reasons, methods

Fig. 5. Village road conditions in wet season, Madagascar, encountered when prospecting for a milk district.

Fig. 6. Deterioration of road surfaces may prevent mechanisation of collection.

adopted must be appropriate, as must the evaluation of possible technological aids. This evaluation may differ from that common in developed countries.

For those preparing a milk collection system, it must first be emphasised that this activity can cost more than 10 per cent of the price paid for raw milk at the processing plant. This figure can be substantially higher if intermediate receiving and cooling stations, where quality is assessed and suppliers' deliveries are weighed, are to be provided and maintained. Even in those products with the highest added value, milk powders and concentrated canned milk, raw milk delivered to the processing plant often accounts for two-thirds of the total product cost ex factory. For pasteurised and sterilised whole milk in carton or bottle, the proportion is even higher. The situation is therefore clear; over-investment in, or inefficient organisation of raw milk collection will constitute a burden ultimately to be borne by the individual producer or consumer, and which may be too heavy to bear.

The milk collection system must correspond to the scale of the processing plant. This will be dictated in the first place by the market for its products; but also by the need to use sophisticated equipment, anyway for sterilised milks in carton or bottle, or for condensed or dried products. Here a certain minimum throughput must be maintained or else overheads and capital charges will be too heavy. Thus, the size of milk district will depend upon a calculation of available or potential litres per day per km, of collecting route, the road system, and the economic and marketable throughput necessary for the plant to prosper.

Very often, such simple calculations will be complicated by factors such as existing and potential competition for raw milk supplies, delimitation of collection areas by the responsible authorities, and legal prescriptions concerning milk collection and cooling, often based on those established in developed countries, but sometimes inappropriate to local conditions. All factors having been taken into account, a detailed survey of each route or sub-district, recording the location of each potential supplier and the quantities anticipated, must be prepared. Where suppliers are far from an all-weather road, some means of intermediate transport, either to a village collecting station or roadside collecting point (see Figs. 7 and 8) is necessary. This may involve hand-carried or head loads, bicycle or (in hilly districts) mule or horse transport suitable for the tracks available.

In many developing countries, it will very rarely be possible to collect large quantities of cooled, fresh milk from farmgate, as in Europe, where

Fig. 7. All-weather transport in Latin America — without all-weather roads.

Fig. 8. Village transport, transfer to truck from village roads, Chiapas, Mexico.

power, water supplies and supplier's size and means combine to make this possible. In a few exceptional instances, such as corral farms installed in certain Gulf states, the complete processing and carton or bottle packing of pasteurised or sterilised milk is done at the farm, using technologies identical with those common in the developed countries. For the other developing countries, an intricate system, often starting from village collecting stations, will be needed. But this may be more a matter of organisation than of direct investment. Indeed, before considering the glittering array of technologically advanced equipment available to the would-be investor at any dairy equipment exhibition, the need for technology to remain appropriate must be firmly kept in mind.

For a new venture in a developing country, where even the market for the end-product, and often the potential for raw milk production, is not yet certain, modesty in capital investment both in the milk collection activity and in the processing plant itself, is often imperative. The first indulgence should, perhaps, be in securing a site for processing large enough for extensions in the future.

As an example of the risks of over-investment, an impressive milk receiving, cooling and pasteurising venture seen in Western India in 1957, in the first stages of that country's adoption of modern dairy technology, may be instructive (see Figs. 9 and 10). The plant, excellent in itself, had a staff of about 25, including a General Manager, Sales Manager, Production Chief and plant engineer, and a throughput of about 600 litres per day. It is to be hoped that this venture subsequently developed the raw milk supply and sales it needed. However, the point to be taken is that in the dairy industry in developing countries, the least costly solutions must be sought to the problems of collection of raw milk, delivered to the processing plant in good condition.

As recorded earlier, in the same way that European and North American dairy processing was revolutionised by improved rail systems and the creation of an associated network of collecting and chilling stations, improvements in road systems and road haulage organisation will determine possibilities in any developing country (other than the few, such as Southern Mexico, Venezuela or Thailand, where water transport is sometimes also feasible). The road or waterway network will determine how far the collection system can extend into a given area, since with an all-weather road, supplies can be moved, say, 50 or even 80 miles to the plant in the time that it takes to bring these same supplies from milking to a village collecting point and thence by bicycle, bullock cart or horse- or tractor-drawn vehicle over rough tracks from the village to the all-weather road. Generally a maximum of 5–6 h from time of milking to cooling should be allowed, and this only if hygiene at the

Fig. 9. Milk receiving, cooling and pasteurising plant, Western India, 1957.

Fig. 10. Modern equipment in plant with inadequate intake, Western India, 1957.

farm and in transport is good. Efforts to teach this hygiene will take time and money; Figs. 11 and 12 show how improvement in roads and motorisation can speed collection.

At this point, it should be mentioned that practical experience seems to show that milk from relatively low-yielding animals in warmer climates, where there is little feeding of concentrates and extensive grazing, can be surprisingly resistant to souring when compared with that produced by typically high-yielding animals in developed dairying areas. Those dealing with projects where large numbers of improved

Fig. 11. Small scale motorisation and better roads speed collection in Thailand, 1984.

Fig. 12. Village transport transfers to trucks, India, 1970s.

animals are to be brought in to a hitherto little developed region should, therefore, not only plan for the necessary (and too often inadequate) feeding, housing and veterinary care, but also watch for an adverse change in milk quality at the plant, if timely steps are not taken to speed-up collection. Conversely, where no major genetic uplift is anticipated, measures based upon European or North American experience may prove to have been unnecessary in planning collection systems. Experience in Northern India in 1962 with a special tanker using flake-ice as a coolant were soon terminated as the vehicle could not stand up to the prevailing road conditions; but a conventional churn collection proved adequate for the particularly resistant buffalo milk being collected.

The village collecting station or roadside pick-up point (see Fig. 13) should also be the focus of quality control and quantity measurement, since it is vitally important that the supplier sees evidence that the quality of his milk is fairly judged. Therefore, after weighing, often in the presence of the supplier, the first quality checks should take place here. For these purposes, the simplest equipment in the form of bottles with preservative for fat content samples (on the assumption that there are no great fluctuations in supply or quality, aliquot samples are not necessary) is sufficient. For keeping quality in cow's milk, after simple organoleptic checking, the alcohol test (Bodex tester) can be applied to churn quantities or suspect deliveries at this point. The village centre,

Fig. 13. Roadside collection of village supplies, with testing, Madagascar 1976.

operated by a reliable agent, who is himself usually also a producer, can be the source of advice to suppliers, of payment and as a link to the veterinary and field services (Anon., 1975).

Transport by contractors' vehicles (bicycles, carts, tractor/trailer and for hard roads, motor vehicles) can improve flexibility for seasonal fluctuations, as well as for extension into promising areas and curtailment of those which do not develop. This may also help to reduce initial investment in transport by the project. It has to be remembered that a once or twice daily employment of, say, 2–5 h is hardly sufficient to amortise the cost of a vehicle. Own investment in special vehicles (e.g. tankers) should be based on proven possibilities of employment and economics compared with contractors's offers.

It is assumed that on arrival at the processing plant all supplies, either in churn or bulk, will be tested first for keeping quality, for example by alcohol, clot-on-boiling or reductase tests. Supplies intended for products having legally defined compositions will also be checked for compositional quality. But it is the keeping quality tests which will give the first warning as to the need for measures to improve the handling of raw milk. These warning signs may come at times when the milk is liable to be unstable, that is, when changing from dry to green fodder (grass milk) and at periods of calving. Faced with these warnings, those responsible will have a range of options from which to choose. First should come reorganisation or improved discipline in the existing collection structure, more rigorous checks on suppliers' milking practices and utensils, and on any intermediate handling. Any shortening of the time taken to bring milk to the plant will help. A review of road systems, collection routes, utensils and vehicles may suggest how to achieve this.

When no more, or not much more, can be expected from these measures, some means of checking bacterial development during inevitable transport delays must be sought.

For some years, United Nations agencies (principally the Food & Agriculture Organization and the World Health Organization) have considered the use of hydrogen peroxide as an additive which slows the development of acidity in cow's milk (Anon., 1980). Despite strong reservations in this report as to quantity, quality and method of use, the observation of which it might be difficult to ensure in practice, hydrogen peroxide has been used by suppliers and collection systems in some developing countries, to allow collection in difficult terrain from small suppliers without capital investment in receiving and cooling stations.

Amongst milk processors there is, however, a residual suspicion (so far unsupported by complete scientific evidence) that the action of hydrogen peroxide in checking development of acidity does not arrest other changes, enzymatic or chemical, which can take place progressively over time in raw milk at high ambient temperatures. It is felt that the adverse effects of such changes may develop in products with a long storage life. Therefore, for such products, it is felt better to avoid the use of hydrogen peroxide.

A better-founded resistance to the use of this product is that resting on the feeling that the use of sterilising agents, before raw milk keeping quality is assessed, masks the effects of careless handling which should, reflected in acidity development, provoke follow-up and correction at farm or collecting centre, by the inspection and field service. In effect, hydrogen peroxide attacks the symptoms, not the disease.

When everything other than investment in cooling facilities, intermediate between the supplier and the processor, has been tried, the type of such investment likely to be most cost-effective must be considered. If electricity and water supply systems, refrigerant supplies and maintenance, are available and reliable at the village collecting centres, then the possibilities will be determined by the scale of collection. If, as was often the case in the past, such systems are absent or unreliable, the simplest cooling arrangements can be those using ice-blocks to produce chilled water for circulation through a surface or plate cooler, or even, in the most primitive circumstances where no electricity is available, for cooling of milk in an ice water tank, in churns with periodic agitation by hand. Ice blocks are often available in small towns in the developing countries, especially where cold stores are operated for perishable agricultural products such as poultry, dairy products and seed potatoes, and where itinerant soft drink vendors use ice for their delivery tricycles. Thus, additional investment for ice-making can often be avoided. Ice blocks can be distributed to village collecting centres by the collection vehicles.

Such simple cooling arrangements are the least costly. When, however, efficient cooling to 4°C is sought, and power is available or can be provided by diesel power, the farm cooling tank provides the first item of modern dairy technology in the collection system. These are available in capacities from 200 litres up to more than 5000 litres capacity, with either integral or free-standing compressor units. Generally, up to 5 HP, these units are air-cooled, reducing complexity for maintenance. This complexity is further reduced if the direct expansion type is used. The alternative ice-builder type, with better cooling characteristics and requiring less horse-power (which is important if a diesel-powered generator has to be provided) have the disadvantage of the additional water circulating system which also needs maintenance.

Cooling characteristics typical of direct expansion and ice-builder tanks are shown, under different conditions of usage, in Figs. 14–17. It will be seen that these tanks usually require more than 2 h to bring the milk down to temperatures below which bacterial multiplication is substantially reduced. This is a drawback of this type of lightweight cooling set-up, when compared with the more expensive separate ice-building refrigeration systems using a plate cooler for heat exchange.

Approximate costs of wholly owned receiving and cooling stations, equipped with direct expansion farm tanks, are shown in Fig. 18 (expressed in Swiss Francs at the then exchange rate of 3.50/£1 Sterling, in mid-1982). It will be seen that site and buildings can form a substantial part of the investment. Where the cost of such stations has seriously limited expansion of a milk district, it has sometimes been found possible to interest groups of would-be suppliers in providing a suitable room or building, in which the processor can install equipment.

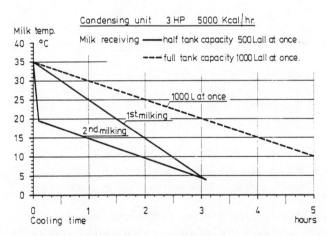

Fig. 14. Typical performance characteristics farm cooling tank, 1000 litres, with direct expansion unit, under different loading conditions. (European and US Standards specify: two milkings in 1 day, 35–4 °C, 3 h; (pick-up every 2 days) four milkings in 2 days, 35–4 °C, 2 h. Direct expansion tanks cannot therefore be used for a single loading of warm milk to tank capacity.)

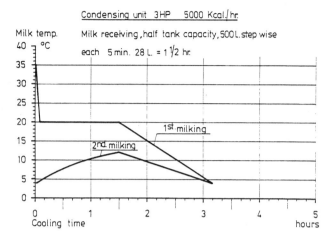

Fig. 15. Typical performance characteristics farm cooling tank, 1000 litres, with direct expansion unit, with stepwise loading.

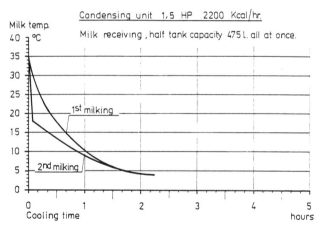

Fig. 16. Typical performance characteristics farm cooling tank, 950 litres, ice bank type, loaded in two stages, each half tank volume.

In some instances, rural co-operative societies have preferred to set up such facilities on their own account, perhaps with credits and technical assistance from the processing plant. In this way, a group of suppliers will feel freer to negotiate sale of their milk to their best advantage. At the same time the processor who provides credit or technical assistance has every incentive to provide adequate continuing field service and veterinary support, as a means of retaining his milk supply.

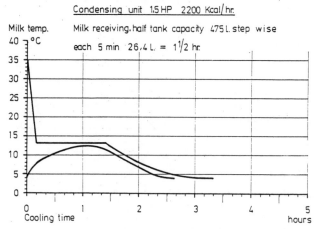

Fig. 17. Typical performance characteristics farm cooling tank, 950 litres, ice bank type, loaded stepwise.

More substantial receiving stations of the 'instant cooling' type are equipped with mechanical refrigeration (from 12 up to about 20 000 litres day^{-1}, with NH_3 systems for larger stations where the cost of refrigerant gas becomes more important) in air-cooled units up to 5 HP (water cooling above this power), serving ice-builders with a water circuit to plate cooler all allowing immediate reduction of milk temperature from 35 °C to 4 °C. As a rough guide, and assuming that such stations are equipped with stand-by diesel generators, and in the larger sizes above 20 000 litres day^{-1} with reject sour milk segregation and separation facilities, as well as rotary, steam churn washers and related services (which accounts for their relatively higher cost per unit of throughput), costs of such instant cooling stations were estimated by one processor in mid-1982 as given in Table II.

TABLE II
Capital cost of milk cooling stations designed to handle the
volumes indicated

Litres day^{-1}	Litres h^{-1}	Total capital cost	Capital cost per litre per day
5 000	2 000	£57 000	£11.40
10 000	4 000	£75 000	£7.50
20 000	8 000	£149 000	£7.45
40 000	14 000	£207 000	£5.20

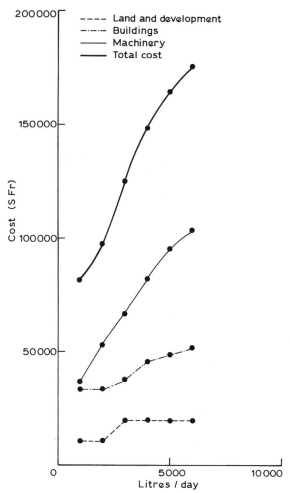

Fig. 18. Typical 1982 land, building and machinery costs for milk cooling stations using farm tanks, expressed in Swiss Francs at Sw.Frs.3.50/£1 Sterling.

Generally, advantages of scale are equally great in terms of operating cost, since the expense of reliable operating supervision will not greatly differ within the range of quantities handled, indicated above. These operating costs can vary in developing countries from less than 1 per cent to as much as 7 per cent of the raw milk price paid to the producer.

The figures quoted in Table II, for different types and dimensions of receiving and cooling facilities, may, it is hoped, help those who have to confront cost calculations involving economies of scale in processing facilities versus increase in collection costs for increasing quantities of raw milk. For certain very large-scale operations, for instance for large milk powder specialities plants, it may be found necessary to resort to pre-condensing (where feasible and permitted for the end-product in question) to reduce milk haulage costs. For rough calculations of the cost of providing pre-condensing facilities, in units handling between 100 000 litres and 200 000 litres per day, a capital cost, excluding site, of about £1 million plus £5 per litre of daily throughput (1982 values) can be used. This includes normal provision for steam, water and power supplies. No provision is included in this figure for effluent treatment, which should conform to specific local requirements.

In concluding this section, it should again be emphasised that once economic problems are solved, the most important factor in building up a raw milk supply in a developing country is not a matter of technology but of psychology. The confidence of the supplier, that he will receive fair treatment, prompt payment, and support from field and veterinary services, that the collection system will be reliable and that his supplies will be accepted within whatever quota or other limitations he has agreed to observe, takes time to build. Once achieved, this confidence must be maintained by the efficient organisation and adequate equipment of the field, veterinary, collection and accounting services; without it, such investments will count for little until confidence is established. Suppliers have long memories; every possible step should be taken to avoid betrayal of their confidence.

ROLE OF RECONSTITUTED MILK CONSTITUENTS

Economic and Political Background

In this chapter, much emphasis has been placed on means of developing a local raw material supply for a dairy industry. This has become increasingly a preoccupation of governments, after a period in which import substitution was the main aim, even at the continuing expense in foreign currency of milk constituent imports. In past years, the fact that some dairy products have constituents such as sugar, cereals or fruit fillings which can be locally purchased, and are packed in glass, tinplate, paper complexes or cartons which can also be of local origin, has given a sufficient local content even when the main dairy constituents had still to be imported. If local energy, labour and services were taken into account, such products could show a substantial saving in foreign exchange if compared with the cost of imports. If such imports were of fluid products, such as sterilised or condensed milks, the high cost of transporting the water content would also favour local manufacture.

Historical Development

The basic dairy constituents used in recombining are skim-milk powder and butter or, more recently, butter oil. This 'oil' is a more than 99·6 per cent pure product obtained from cream or butter by a process of emulsion-breaking, centrifugal separation and vacuum drying. Packed in airtight drums, it will keep for months at ambient temperatures. As stated before, recombining resulted mainly from the needs of the United States armed forces in the Pacific Area in 1940–45.

In addition to pre-war processing plants a number of large-capacity spray drying plants were erected in the United States during the war years, originally for full-cream milk powder. The report on Milk Utilisation published by the British Productivity Council in 1953 after a team visit to the US in 1952, records the rapid expansion of skim-milk powder production using these facilities. Already at that time, strict quality grading served to enhance the usefulness and acceptance of skim-milk powder for many purposes in different sectors of the food industry. The liquid skim-milk, previously returned to suppliers for stock feeding, was available from the butter manufacturers who handled a quarter of the more than 50 million tons of fresh milk then produced in the United States.

Skim-milk powder was exported first as food aid to devastated areas of Europe and Asia after 1945, and later under Public Law 480 in subsequent food aid programmes.

The competitive situation resulting first from the United States' exports, and subsequently from increased production in the European Community, in Australia and New Zealand, led to lower prices for the constituents, easily stored and exportable in bulk, than for the conventional finished dairy products. This, coupled with the stimulus given by governments in importing countries for the establishment of local industries, often by tariff barriers against conventional imports of dairy products, quickly led to the setting up of a series of recombining plants first for evaporated and sweetened condensed milk. These were built in Malaysia, Philippines, Indonesia, Thailand, the Caribbean and Sri Lanka in the 1960s and 1970s, and were later complemented by recombining operations for sterilised milk in bottles or cartons, milk powders and infant foods. Yoghurts and ice cream also developed in these countries on the basis of recombining.

Present and Future Economic Situation of Recombining

Figure 19 shows the fluctuations experienced in the world market prices of butter oil and skim-milk powders in the period 1971–84. It will be obvious that this fluctuation must cause problems for industries dependent on these raw materials.

Supplies are now predominantly from Europe and are influenced by the levels of restitutions on exports (or, when prices on the world market are high, export levies) determined by the EEC (see Fig. 20). There is some pressure in the trade associations concerned to prevent undue favouring of export of bulk constituents, but in fact over the period 1978–84, the restitution level fixed has more often been slightly favourable to bulk constituents than otherwise (see Fig. 20).

Another economic influence, frequently exercised by governments on recombining industries in developing countries, involves an obligation to accept any quantities of fresh milk offered by suppliers, often at much higher cost than the equivalent imported constituents. This can be a considerable stimulus to fresh milk production. A problem often experienced in such instances is the uneven response of individual processors to such obligations. Bakeries and ice-cream producers, using substantial quantities of imported milk constituents, find it difficult to absorb fresh milk. Ultimately those industries properly complying with such obligations may find the financial burden intolerable. Therefore an equalisation scheme (as instanced for India earlier in this chapter) whereby *all* imports of milk constituents are taxed for the benefit of local fresh milk production for *all* purposes, is highly desirable. However, some guarantees that such taxes will indeed benefit fresh milk production, and not disappear into the revenue, are imperative.

Application of Recombining

Within the limits imposed by such factors, recombining can be invaluable in developing a modern dairy industry in developing

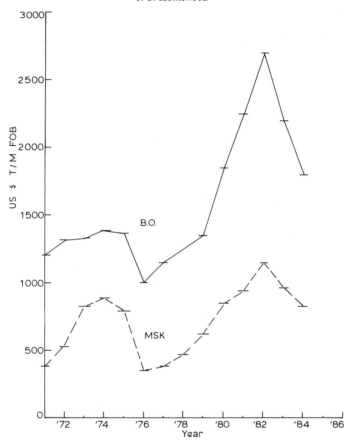

Fig. 19. Fluctuations in prices of raw materials for recombining 1971-84.

countries. There are two principal ways in which this can come about. The operation at economic capacity of a new plant can be ensured, up to the limit set by marketing opportunities, throughout the long period needed to build-up a sufficient fresh milk supply. If large seasonal fluctuations in fresh milk availability persist, as is often the case in developing countries, recombining can maintain supplies of products with relatively limited keeping quality. As a second possibility, those plants which cannot be based on local milk constituents, but which find sufficient justification from secondary raw and packing materials, energy, labour, or other locally available constituents, can produce a

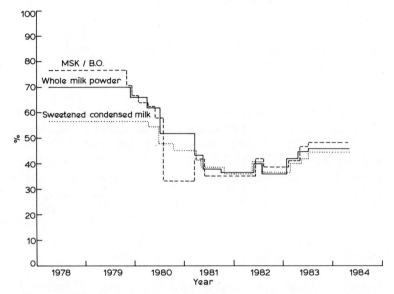

Fig. 20. Relative stability achieved in EEC export aid for milk constituents compared with finished milk products, over the period 1978–84.

range of dairy products to reduce import costs whilst improving availability of milk products.

Specific Technology for Recombining

The first technological problem involved in recombining was that of dissolving large quantities of skim or full cream milk powder in water or milk. Dumping funnels and fluid recirculating systems were already recorded in use in the USA in 1952 (Anon., 1953). The International Dairy Federation has in recent years published a series of monographs detailing appropriate technology for specific products (Anon., 1979).

Air entrainment was one difficulty encountered in recombining operations. This not only affects efficiency of subsequent homogenisation, but can also provoke coagulation under heat treatment of high-concentration premixes, and adversely affect keeping quality of long-life products. To eliminate air, some form of deaeration is normally required if not already available in a concentration step under vacuum. However, most modern recombining plants for sweetened condensed milk, evaporated milk or milk powders no longer incorporate a concentration step, unless dilution to allow filtration, to remove impurities in local sugar supplies, for instance, dictates this.

Although full cream milk powders are often used in recombining, on account of price advantage or simplicity of use, the shorter storage life of these powders, due to the relatively high exposure to oxidation of their fat content, leads many processors to prefer the separate butter oil and skim-milk powder as constituents. For butter oil or other fats, simple heating devices — usually hot rooms — ensure a fluid state for pumping; this allows accurate dosage. The elimination of any dwell-time of warm fats with fresh milk, or milk constituents containing lipase is essential for good keeping quality of the end-product. Typical specifications for butter oil are available from the International Dairy Federation (Anon., 1983). For skim-milk powders, product requirements will dictate a suitable level of heat treatment, as defined in the American Dried Milk Institute (ADMI) standards. These are based on differentiation in the whey protein nitrogen index (WPNI).

In recent years, the merits of Sweet Buttermilk Powder, which contains 8–10 per cent of butter fat and about 2 per cent of lecithin, have been appreciated, since this powder has often been available at about the same price as skim-milk powder with only about 1 per cent butter fat. It is not only a price advantage which recommends buttermilk powder, as from a technological point of view, buttermilk is an emulsifying aid for the fats in a recombined milk product. Buttermilk can also, in appropriate proportions, favourably influence organoleptic quality. It also restores an element otherwise missing when recombining, in order to achieve a composition corresponding to the original whole milk used for butter and skim-milk manufacture, and, in certain instances, buttermilk powder can assist in stabilising premixes under heat treatment.

One other product range, in addition to those already mentioned, which can now be based on recombining, is that of cream and natural cheeses. Cheese base powders, produced by the removal of milk serum by ultrafiltration, can eliminate costly whey disposal problems in natural cheese production. This may also make it possible to avoid the need for imported natural cheese in the manufacture of processed cheese, usually based on a percentage of ripened natural cheese, with milk powders, caseinate and butter oil.

As mentioned earlier, economics are vital, and it should be repeated that, within the EEC, considerable pressure has been exercised by dairy products manufacturers to prevent undue advantages in export restitutions for milk constituents as compared with finished milk products. Figure 20 shows that despite a decline in overall levels of restitutions, these have maintained some advantage for milk constituents compared with more sophisticated products. This may be taken, by those planning recombining operations, as some grounds for confidence in the future of such operations. A recent publication of a US Co-operative concerning marketing of constituents in the Caribbean area (Anon., 1984) seems to confirm similar grounds for operations based on US exports.

SELECTION OF PRODUCTS FOR MANUFACTURE

The choice must, in most instances, depend upon a market survey, designed to illustrate present product availability (whether fresh milk, dairy products already manufactured locally, or imports) and to investigate attitudes to new products, including perceived value. If the new project forms part of a state financed, planned expansion of production for social reasons — as was the case in most of the major Indian dairy projects — the choice of products, mainly butter or ghee and skim-milk powder, is determined by the overall state plan. Where no strong market or official constraints exist, choice of products will often depend upon distribution systems potentially or actually available.

Normal temperature distribution (as for dry groceries) can handle condensed and dried milk in countrywide sale. To a lesser degree, and if stock rotation is strictly controlled, UHT milk and processed cheese can be handled in normal temperatures. Chilled (+4°C) and frozen (−20°C to −30°C) distribution can only be envisaged for urban sales, where volume per kilometre of delivery route justifies the expensive vehicles, point of sale cabinets and supervision involved.

Despite the expense, the prospect (Fox, 1984) that by the year 2025, 80 out of 93 cities with more than 5 million inhabitants will be in the developing world, warrants examination of the possibility of chilled or frozen dairy products distribution in such centres of population.

Therefore the comparative incidence of distribution in capital and operating costs for chilled and frozen products has been established in tabular form (Fig. 21). Where alternative distribution systems can be considered (e.g. for UHT milk and processed cheese), this has been shown.

The estimates of costs and capital investment are, as elsewhere in this chapter, based on industrial-scale operations of the dimensions shown, and for production units exclude site costs, power or steam supply,

PRODUCT RANGE	OPERATING CONDITIONS	MIN. SALES VOLUMES	CAPITAL INVESTMENT PRODUCTION £Stg/kg/y	DISTRIBUTION £Stg/kg/y	COST OF DISTRIBUTION % OF TURNOVER
BLOCKS JARS	1 shift 2000 h/y $+\eta = 100$	$\sim$ 500 To/y	0.85	0.13	5 – 10
PLASTIC CUPS	1 shift 2000 h/y $+\eta = 0.7$	$\sim$ 3 Mio cups or 300 To/y	0.70 – 0.85	0.10 – 0.25	10 – 20
CUPS	1 shift 2000 h/y $+\eta = 0.85$	$\sim$ 500 To/y	0.65 – 0.70	0.10 – 0.25	12 – 15
STICKS BULK	Seasonal Fluctuation See Fig 22	$\sim$ 1 Mio l/y	1.0 *	0.25 – 0.40*	15 – 25

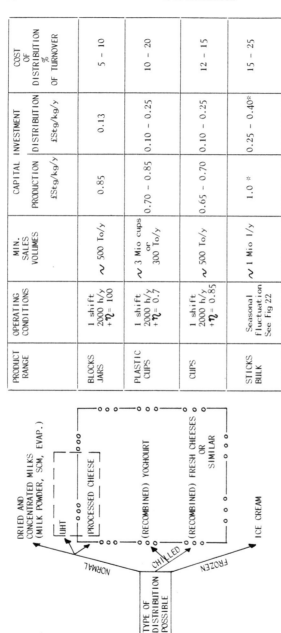

DRIED AND CONCENTRATED MILKS (MILK POWDER, SCM, EVAP.)

UHT

PROCESSED CHEESE

(RECOMBINED) YOGHOURT

(RECOMBINED) FRESH CHEESES OR SIMILAR

ICE CREAM

TYPE OF DISTRIBUTION POSSIBLE

NORMAL

CHILLED

FROZEN

— — — } combined distribution
— o o o — } feasible

$+\eta$ Effective use of theoretical capacity due to weekly or seasonal demand fluctuation

* £Stg/l/y

Fig. 21. Comparison of capital investment and operating costs for production and distribution of normal temperature, chilled and frozen products (1984 values).

administration or social buildings and effluent treatment since these are specific to each project. Customs duties and transport charges for imported equipment are also excluded as being specific. Due caution should be excercised in extrapolating from the figures given — the effects of scale are considerable.

As a reminder of project parameters which could be overlooked in a choice of products, the effect of seasonal demand fluctuations cushioned by seasonal two-shift working and medium term storage, a pattern of sales, production and stocks, applicable in a Northern Hemisphere temperate climate ice cream market, is shown in Fig. 22. This illustrates the need to work out an operating pattern and to allow for less than the theoretical maximum output of production facilities when dealing with products of limited keeping quality. This type of limitation must also be applied to outputs of condensed and dried milk plants, where raw material supplies are restricted for economic or political reasons to the yield of a milk district, which may be subject to fluctuations of 1:5 in low and high seasons. As a result, two-thirds of theoretical capacity is a good level of utilisation for such plants.

It is an advantage of recombining operations that a much higher utilisation of installed capacity can be attained than would be possible

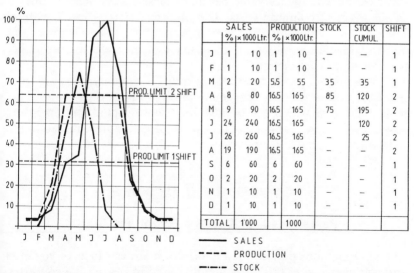

	SALES %	SALES ×1000 Ltr.	PRODUCTION %	PRODUCTION ×1000 Ltr.	STOCK	STOCK CUMUL.	SHIFT
J	1	10	1	10	—	—	1
F	1	10	1	10	–	–	1
M	2	20	5,5	55	35	35	1
A	8	80	16,5	165	85	120	2
M	9	90	16,5	165	75	195	2
J	24	240	16,5	165	–	120	2
J	26	260	16,5	165	–	25	2
A	19	190	16,5	165	–	–	2
S	6	60	6	60	–	–	1
O	2	20	2	20	–	–	1
N	1	10	1	10	–	–	1
D	1	10	1	10	–	–	1
TOTAL		1000		1000			

———— SALES
– – – – PRODUCTION
—·—·— STOCK

Fig. 22. Influence of seasonal demand on production programmes and storage requirements in a typical temperate climate, Northern Hemisphere, ice cream operation.

in a plant tied to the natural fluctuations, due to feed availability, calving periods and climate, common in fresh milk production. Partial recombining can be envisaged, to maintain a stable rate of production, wherever funds and licences are obtainable, and economic or political considerations permit. Thus gaps in supply, for instance to the urban fresh milk consumers, can be avoided in the period necessary to build-up a surplus — used for manufacturing longer life products — of local milk production, so that in the low season, all demands for urban consumers can be met. Some comments on individual chilled and frozen products listed in Fig. 21 may be useful. Processed cheese production requires skill and experienced staff and a supply of ripened natural cheese. As mentioned elsewhere, this can now be produced without the problem of whey disposal, which would otherwise be expensive wherever pig-keeping is excluded on religious grounds, by the use of cheese base powders; but small projects will be assisted by any possibility of imports of natural cheese. These have to be purchased when the market price is favourable and stored at controlled low temperature, which is expensive. For yoghurts and desserts, it may be possible to ease the burden of distribution costs by common delivery systems for these products and soft cheeses. The higher value per kg of these allows them to carry a useful share of common costs. The limiting factor is usually filling capacity, and with short shelf-life products, seasonal and shorter-term demand fluctuations often call for a capacity 40 per cent or more higher than average requirements.

For traditional products with longer keeping quality, the need for a relatively high level of management ability implies only fairly large-scale operations which can be justified from a capital and operating cost point of view. Capital costs for these products are outlined in the following section on effects of scale.

CONSIDERATIONS OF SCALE

Pasteurised, Chilled and Frozen Products

Artisanal methods permit home production, for instance, of yoghurts and soft cheese, and this is frequently found in developing countries, either using local fresh milk or milk powder. Cultures are carried over from day to day.

Small industrial operations can start up with a few hundred litres per day. However, the increasing difficulties of a widening distribution system, and the crippling cost of returned, spoiled goods, limit such ventures, which require skilled commercial and operational management. For larger-scale but still limited ventures, success may depend upon the possibility of adding to an existing similar activity in distribution, say of butter and pasteurised milk, where the knowledge and management ability required is already available. Some rough estimates of the effect of scale in the capital cost of manufacturing plants for certain products, made in 1982, may serve as an example. These estimates exclude, unless otherwise stated, costs of acquisition of plant sites (which may vary greatly according to the country and location), of effluent treatment plant, and of any staff housing or special social facilities. Import duties and transport charges for equipment are also excluded. The figures can only be seen as a guide to the effect of scale in relatively large operations — their extrapolation to much smaller units would be misleading.

Thus, a typical, minimum size project might be a yoghurt and dessert production unit which is to be added to an existing, pasteurised milk operation. A start-up capacity of about 300 tons yr^{-1} is assumed (effective, taking into account seasonality) to be extended to 1000 tons later. As stated in Fig. 21, the initial capital investment for manufacture only will be in the order of £700 per ton per year. It is probable that, in the beginning, only natural and flavoured yoghurts will be produced. These, if heat treated (a slightly more complicated procedure, with gelifying additives, if permitted) will tolerate some breaks in the distribution cold chain — up to 18–20 °C for up to 12 h — such as are likely to occur in a developing country with a difficult climate. They will also have a much longer shelf-life, but less health appeal than yoghurts with living cultures.

For larger-scale yoghurt and dessert production, in about 75–25 per cent proportions, full industrial-scale can be reached at the approximate investment figures quoted in Table III (costs of distribution vehicles and chilled shop cabinets are excluded). These figures illustrate the effect of scale on capital investment. The largest units may also offer operating economies through the use of 'form and fill' machines, which give lower packaging material costs. There may be other scale advantages, especially in distribution and publicity, as the size of the operation increases.

For fresh cheese production on an industrial scale, the rough

TABLE III
Capital costs for yoghurt and dessert production

Capacity (tons yr^{-1})	3 000	10 000	20 000
Investment £Stg ton^{-1} yr^{-1} (1982 values)	£560	£312	£250

TABLE IV
Capital costs for fresh cheese production

Capacity (tons yr^{-1})	2 000	4 000
Investment £Stg ton^{-1} yr^{-1} (1982 values)	£900	£700

estimates given in Table IV apply (under the same reservations), and show that here there is rather less advantage in capital cost differentials.

Ice cream production is also possible on an artisanal scale, and a wide range of small-scale industrial equipment is available, mainly from continental suppliers. The product is, however, demanding in terms of cold chain stability. This costs money and, as said before, requires strict organisation down to the retail outlet or vendor (for street articles). Small ventures rely on street impulse sales — industrial ventures also, but to an increasing extent on so-called 'take-home' sales of larger packs, so far as domestric refrigeration permits. For relatively large-scale industrial operations, capital investments for production, but excluding long term storage space, distribution vehicles and freezer cabinets in retail outlets (allowing for 50 per cent impulse items) will be of the order of those given in Table V. It will be seen that the advantages of scale are somewhat greater in yoghurts than in ice cream, at least as far as capital investment for production is concerned.

TABLE V
Capital costs for ice-cream production

Capacity (litres yr^{-1})	12 mio	21 mio	50 mio
Capital investment £Stg litre^{-1} yr^{-1} (1982 values)	£0.75	£0.56	£0.41

Products with Longer Keeping Quality

Fluid sterilised milk

Processing lines for the UHT treatment of whole fresh or recombined milk do not differ greatly in cost for larger or smaller hourly throughputs. If, however, varying sizes of pack have to be filled, involving varying hourly throughputs on a given line, this may increase costs (use of variable speed drives, split heat treatment apparatus or sterile intermediate storage). There will also be increased charges for additional filling and packing equipment, which is generally not adjustable for different sizes of pack. Filling units are commonly only available on a rental basis plus per unit costs on the packing material, supplied or licensed by the machinery leasor. In order to make a rough estimate of scale effect in such plants, the base rental (payable on installation) has to be included as part of the capital cost.

Recombined sweetened condensed milk

The capital cost of such plants is often strongly influenced by the need to provide infrastructures available in developed countries. Since some of these, such as container manufacture, are available in certain developing countries, a rather low cost per ton (for a product containing a high percentage of sugar and at least 25 per cent water) can be reached if container manufacture is excluded (Table VI).

Full cream milk powder (see Table VII)

There is relatively less benefit from an increase in scale of such milk powder plants, due to the need for increased concentration, drying and container manufacturing equipment.

TABLE VI
Capital costs for recombined SCM production

Capacity (tons yr^{-1})		
Finished product	11 500 tons	34 000 tons
Capital cost £Stg ton^{-1} yr^{-1} (1982 values)	£500 ton^{-1}	£250 ton^{-1}

TABLE VII
Capital costs for full cream milk powder production
(produced from whole liquid milk with own container
manufacture)

Capacity (*tons yr*$^{-1}$) Finished product	7 200 tons	11 500 tons
Capital cost £Stg ton^{-1} yr^{-1} (1982 values)	£1 400 ton^{-1}	£1 100 ton^{-1}

CHOICE OF MANUAL AND AUTOMATED SYSTEMS

Today, almost any fluid circuit, water, steam, or refrigeration, in common use in dairy plants, will incorporate some simple, automatic controls. Thus, a degree of automation is already inherent in the service equipment available to those planning a dairy industry. In the milk circuits, circulation cleaning (CIP) is now assumed in the design of equipment. The tendency is to construct systems with a minimum of joints, always a difficulty in cleaning, and to limit these to inlets and outlets of valves, pumps and other pieces of equipment such as tanks, homogenisers or heat exchangers. For circulation cleaning, and for process control, simple sequential controllers are commonly employed. The maintenance of such limited automation has to be ensured by adequate stocks of spare parts, where no manufacturer's representation can be ensured.

Thus any new project will therefore almost certainly comprise some automation. The choice, for all but artisanal production, is between limited or advanced automation, where more elaborate sensing devices and feed-back systems are used to correct divergences from standard operating conditions. In general, a project should be designed to take advantage of the degree of sophistication of the infrastructure in the country concerned, especially those providing maintenance, for electrical and electronics systems. Often this will mean keeping to the simplest systems available.

The second factor encountered is manpower available. There may be an extreme shortage of skilled or unskilled industrial workers. Where this is the state of affairs, manual systems (particularly for handling of raw materials) would be too costly, and investment in bulk handling systems may be justified, as well as a higher degree of automation. Elsewhere, governments are often inclined to favour labour-intensive investments. There are, however, strict limits to the increase in staffing possible inside a food factory, where every extra worker is a potential source of contamination and of errors in processing. Such operations as milk collection and transport are not so much under this handicap.

Within a factory, individual steps, such as vacuum drying, may be carried out by hand-loaded batch ovens in place of continuous band driers, in order to increase employment. But such choices in process are relatively limited. Thus, the choice of manual or automated systems of greater or lesser complexity will be conditioned by availability and cost of unskilled and skilled labour, by the type of infrastructure for maintenance, and by the process involved and the capital cost of installations in relation to their effective output.

PACKAGING, STORAGE AND DISTRIBUTION

The distinction between such activities carried on in a developed or a developing country is likely to arise from differences in ambient temperature (often higher in developing countries), in infrastructures such as transport, mechanical and electrical maintenance (often weaker in developing countries), and in density of sales outlets (less kg per km of van route in developing than in developed countries). Thus, it is evident that sterile or dried or preserved products, relatively unaffected by ambient temperature and requiring less frequent delivery to retailers, will not call for special precautions, at least so far as hard packs (tinplate or adequate plastic/paper/foil complexes) are concerned. Rodent attack and damage from boring insects will set limits to the use of paper or complexes in replacement of tinplate. Many products will suffer if exposed to direct sunlight or very high temperature, and this will set limits to the use of lighter packing materials and the type of storage used. Condensation caused by exposure to air of high humidity and high temperatures, such as occurs when cargoes are shipped in winter from Europe (temperatures from 4°C upwards) to West Africa (80 per cent humidity and 30°C air temperature) will cause rust spots on tinplate unless ventilation and heating are arranged during transport to bring cargo temperatures up.

This problem will not affect local distribution, but even for this there are similar problems when tropical seaboard plants ship to high plateaux. Increased altitude can also cause apparent swelling of the sealing membranes used for tins of powder products, often a source of complaints or rejection, although there may be no defect in the contents.

For milk preserves, suitable for distribution at ambient temperatures, a comparison of the cost of packaging materials per litre of fresh milk equivalent has been derived from experience in certain developing countries in the 1980s (Table VIII). These figures are for materials only, and exclude capital charges or rentals, operating and maintenance costs. They have been converted at rates current in 1984.

As stated earlier, the relative fragility of packs should be considered in relation to the type of distribution system foreseen. Thus, a lighter pack with shorter life and higher risk of product loss through leakage or souring may be acceptable in a well organised distribution, where the final vendor can eliminate defective units and can account for such losses.

There is constant pressure to diminish packaging costs, despite risks, and this is understandable in the context of low consumer purchasing power in a developing country. Thus, for liquid pasteurised and even sterilised milk, pouch packs in plastic have been employed, despite some difficulties in handling both in distribution and in the home. Plastic pouches have also been used extensively for milk powders, despite a relatively rapid taste deterioration and ultimate rancidity due to inadequate barrier properties against light and oxygen.

TABLE VIII
Packing materials cost per litre of fresh milk equivalent

Pack	Cost per litre (p)
Sweetened condensed milk 400 g tin	3·9–4·6
Full cream milk powder 800 g–1 kg tin	2·8–4·2
Sterilised whole milk in cartons 1 litre	5–6
Sterilised whole milk in cartons 1/4 litre	10–16

For shorter life products, such as yoghurts and other chilled products, packaging will probably depend upon availability of preformed cups. In the larger projects, substantial economies can be achieved by the use of 'form and fill' machines, thermo-forming pots from continuously fed reels of appropriate complexes. The use of thermo-adhesive labels allows a reduction of film substance, and can lead to packaging material economies of up to 10 per cent or more compared with the cost of preformed cups.

In ice-cream distribution, the limited availability of home freezers in most developing countries may make it necessary to concentrate on street, impulse-buy articles, bulk sales being limited to hotels and institutional outlets. Capital investment in distribution will often, at least in the early years, be almost equal to those in production facilities, with about 60 per cent going to cabinets and 40 per cent to vehicles. The annual amortisation and maintenance charges for cabinets and vehicles may amount to nearly half their original cost. Generally, for storage and distribution, conditions required depend upon the product and packaging chosen.

FUTURE OF MILK PRODUCTS IN DEVELOPING COUNTRIES

The last statistics available from the Food and Agriculture Organisation of the United Nations, Rome (Anon., 1978) show, in contrast to relative stagnation in milk consumption per capita in the developed world, a marked increase in consumption in the developing countries. New statistics and projections are, it is understood, likely to be available in 1985. The pattern of consumption in the developed countries shows a shift towards yoghurts, chilled desserts, UHT sterile milk and various cheeses. There is a decline in butter consumption and long-keeping preserved milk.

In developing countries, where consumption was in 1978 somewhere between 0 and 80 kg per capita per year fresh milk equivalent (contrasting with an average of about 200 kg per capita per year in developed countries) growth in this consumption was expected to continue. In Latin America, with an average of 80 kg per capita per year, the largest and fastest growing segments were pasteurised and UHT sterile whole milk. Milk preserves (condensed and dried) continued to expand slowly. There was also slow growth, from a low level, of cheese consumption. In other areas of the developing world, especially where local fresh milk production is inadequate or non-existent, condensed and dried milk consumption continued to increase, with some development of UHT sterile fluid milk where transport and distribution possibilities favour this bulkier and less resistant form of packing. Cheese consumption was very little developed in these areas.

Thus, the total picture is one where considerable new investment to meet increasing demands, either through development and processing of local fresh milk or through import replacement by recombining, may be expected to take place in the developing world. It may also be expected that rapid growth in urban populations in the developing world, and the need for better structures in agriculture, may stimulate fresh milk production where this is possible, and the development of new marketing systems, such as those for chilled and frozen products, where these offer economic and social advantages.

Whatever the means chosen, the examples quoted in this chapter of increased benefits in agricultural development, improved planning of processing structures and distribution, may, it is hoped, assist those who will be engaged in new ventures using modern, and appropriate, dairy technology in the developing world.

REFERENCES

Anon. (1941). *Agricultural Marketing in India,* Government of India.

Anon. (1953). *Milk Utilisation,* British Productivity Council, London.

Anon. (1975). *Nestlé in the Developing Countries,* Nestlé SA, Vevey.

Anon. (1978). Projections of Product Development to 1985; Milk and Milk Products Supply, Demand and Trade, FAO Rome ESC., PROJ/78/3rd.

Anon. (1979). Monograph on recombination of milk and milk products, IDF Bulletin No. 116, International Dairy Federation.

Anon. (1980). Evaluation of certain food additives, 24th Report of the joint FAO/WHO Expert Committee on Food Additives, WHO, Geneva.

Anon. (1982). *The Etah Development Programme,* Unilever, London.

Anon. (1983). Milk fat products, Standard A2, IDF Publication D-DOC-188.

Anon. (1984). Improving Jamaica's dairy industry, *Land O'Lakes Mirror,* February.

Beaver, M. W. (1973). Population, infant mortality and milk, *Population Studies,* **27** (2), 243–54.

Fox, R. W. (1984). The world's urban explosion, *Nat. Geographical Magazine,* (August) 179–85.

Khurody, D. N. (1977). Increased production and rationalised consumption of milk in Asia, *Indian Dairyman,* **29** (3), 149.

Latham, R. E. (1958). *Travels of Marco Polo,* Penguin Books, Harmondsworth, UK.

McDowell, R. E. (1981). Limitations for dairy production in developing countries, *Journal of Dairy Science,* **64,** 2463.

Rampini, F. (1978). Nestlé in Indonesia, *Politica ed Economica,* Anno IX No. 3, Rome (trans. Nestlé Vevey).

Whetham, E. H. (1964). *The London Milk Trade 1860–1890, Economic History Review,* **17,** 369–80.

Whetham, E. H. (1970). *The London Milk Trade 1900–1930,* Reading.

Index